Scanning Probe Microscopy
of Polymers

ACS SYMPOSIUM SERIES **694**

Scanning Probe Microscopy of Polymers

Buddy D. Ratner, EDITOR

University of Washington

Vladimir V. Tsukruk, EDITOR

Western Michigan University

Developed from a symposium sponsored by the Division
of Polymer Chemistry, at the 212th National Meeting
of the American Chemical Society,
Orlando, FL,
August 25–29, 1996

American Chemical Society, Washington, DC

Library of Congress Cataloging-in-Publication Data

Scanning probe microscopy of polymers / Buddy D. Ratner, editor, VladimirV.
    Tsukruk, editor.

    p.   cm.—(ACS symposium series, ISSN 0097–6156; 694)

    "Developed from a symposium sponsored by the Division of Polymer
Chemistry at the 212th National Meeting of the American Chemical Society,
Orlando, FL, August 25–29, 1996."

    Includes bibliographical references and indexes.

    ISBN 0–8412–3562–7

    1. Polymers—Surfaces—Congresses. 2. Polymers—Microscopy—
Congresses. 3. Scanning probe microscopy—Congresses.

    I. Ratner, B. D. (Buddy D.), 1947– . II. Tsukruk, V. V. (Vladimir
Vasil'evich). III. American Chemical Society. Division of Polymer Chemistry.
IV. American Chemical Society. Meeting (212th : 1996 : Orlando, Fla.) V.
Series.

QD381.9.S97S23 1998
547.7—dc21   98–15684
                                                  CIP

The paper used in this publication meets the minimum requirements of American National Standard for
Information Sciences—Permanence of Paper for Printed Library Materials, ANSI Z39.48–1984..

(MH)

Foreword

THE ACS SYMPOSIUM SERIES was first published in 1974 to provide a mechanism for publishing symposia quickly in book form. The purpose of the series is to publish timely, comprehensive books developed from ACS sponsored symposia based on current scientific research. Occasionally, books are developed from symposia sponsored by other organizations when the topic is of keen interest to the chemistry audience.

Before agreeing to publish a book, the proposed table of contents is reviewed for appropriate and comprehensive coverage and for interest to the audience. Some papers may be excluded in order to better focus the book; others may be added to provide comprehensiveness. When appropriate, overview or introductory chapters are added. Drafts of chapters are peer-reviewed prior to final acceptance or rejection, and manuscripts are prepared in camera-ready format.

As a rule, only original research papers and original review papers are included in the volumes. Verbatim reproductions of previously published papers are not accepted.

ACS BOOKS DEPARTMENT

Contents

POLYMER MORPHOLOGY AND STRUCTURE:
MOLECULAR FILMS AND INTERFACES

PROBING OF LOCAL SURFACE PROPERTIES OF POLYMERS

SPM TECHNIQUES: CURRENT TRENDS

INDEXES

Preface

Scanning probe microscopy (SPM) is a new experimental technique for the qualitative and quantitative characterization of surface microstructure and properties with atomic resolution. Invented in 1982 (as scanning tunneling microscopy) and 1986 (as atomic force microscopy (AFM)), this technique now is used in virtually all branches of natural science. New modes (friction, force modulation, electrostatic, thermal, and others) were added making a family of SPMs with wide versatility. The history of SPM investigation of polymeric materials began in 1988–1990 with the revolutionary works by Hansma et al., Reneker et al., and Cantow et al. Since then, exciting results have been obtained in virtually all fields of polymer science. Several representative applications are folding in polymer single crystals, local mechanical properties of polymer films, nanodomain morphology of multicomponent polymers, viscoelastic behavior, nanoprobing of chemical composition, and intermolecular interactions of biopolymers.

Taking into account the wildfire spread of SPM applications in polymer science within the past several years, we felt that it was the right time to summarize results and discuss trends. The goal of the symposium organized in Orlando, Florida, August 25-30, 1996 was to stimulate discussion of SPM applications among a wide community of polymer scientists. A total of 65 contributions were presented by participants from 12 countries (United States, Germany, Holland, Japan, France, Switzerland, United Kingdom, Canada, Belgium, Ukraine, and Italy). We believe that the 21 chapters selected for this volume represent the spirit of this forum. A variety of SPM studies of different polymer surfaces ranging from perfectly ordered single crystals to polymer composites with heterogeneous microstructure are collected in this volume.

An introductory chapter (Overney and Tsukruk) serves as a comprehensive guide for the reader and introduces basic principles, explains specific terminology, discusses current trends and limitations, and provides an extensive bibliography in the field of SPM studies of polymeric materials.

The first three sections of this volume titled Polymer Morphology and Structure are devoted surface morphology and microstructure of various polymer surfaces as studied by different SPM modes. The first section includes a review on morphological studies of a classical representative of polymer single crystals, polyethylene written by Reneker and Chu. Polypropylene crystallization is discussed in the Stocker et al. chapter. Two chapters written by Vansco et al. and Goh et al., respectively, consider microstructure of polymer and biopolymer fibers. The second section deals with polymer composites and compliant materials such as natural rubbers (Hild et al.), carbon filled polymer blends (LeClere et al.), thermoplastics (Everson), anisotropic polymer films (Biscarini), and biomolecules (Domke et al.) and degradable polymers (Bourban et al.). SPM studies on organic molecular films are presented in the third section which includes polymerizable Langmuir–Blodgett monolayers (Takahara et al. and Peltonen et al.) and LB films studied at the air–water interface (Eng). The fourth section, Probing of Local Surface Properties of Polymers, discusses the nanoprobing of polymer surface properties. Bliznyuk et al. discuss current approaches to quantitative characterization of surface properties. Aime et al. and Hammerschmidt et al. present their results on mechanical probing of the deformational and viscoelastic behavior of polymer networks and films. The final section, SPM Technique: Current Trends, includes several chapters which discuss current trends and modern developments in SPM applications. Resolution limits of the SPM technique are considered in the Luthi et al. chapter with demonstration true-atomic resolution of crystal lattices. Vezenov et al. review CFM applications to various chemically modified surfaces and Tsukruk et al. discuss surface interactions between various chemical groups in different environments (pH and ionic strength). Boland et al. demonstrate examples of nanolithographic fabrication with SPM tips, coupled with recognition-specific microscopy.

Finally, we acknowledge numerous sponsors who made this symposium happen: the Division of Polymer Chemistry/ACS, The Petroleum Research Fund, Ford Motor Company, Digital Instruments, Inc., Burleigh Instruments, Science Application International Corporation, and Park Scientific Instruments.

We hope that this volume, representing the first such collection in polymer science, will be a valuable guide for the expanding community of polymer SPM users.

BUDDY D. RATNER
Department of Chemical Engineering and
    Center for Bioengineering
University of Washington
Seattle, WA 98195

VLADIMIR V. TSUKRUK
Department of Materials Science
Western Michigan University
Kalamazoo, MI 49008

INTRODUCTION

Chapter 1

Scanning Probe Microscopy in Polymers: Introductory Notes

René M. Overney[1,5] and Vladimir V. Tsukruk[2,3]

[1]Department of Chemical Engineering, University of Washington, Seattle, WA 98195
[2]College of Engineering and Applied Sciences, Western Michigan University, Kalamazoo, MI 49008

The purpose of this chapter is to give a brief introduction to *scanning probe microscopy* (SPM) technique, discuss recent developments, and to guide the reader through a variety of recent results presented in this volume. Various SPM techniques including contact, dynamic, force modulation, friction, chemical, electrostatic, adhesion, and thermal modes applied to probing of polymer surfaces are discussed. *Current trends* are summarized including quantitative SPM measurements, probing of nanoscale mechanical properties, studying dynamical surface properties, and exploring local chemical sensing.

Table of contents

[3]Email addresses: René Overney: overney@cheme.washington.edu; Vladimir Tsukruk: vladimir@wmich.edu.

Introduction

Atomic force microscopy (AFM) was invented in 1986 as a logical step in the development of scanning tunneling microscopy (STM), which was introduced in 1982 for studying conductive surfaces *(1)*. The principle of SPM technique is very similar to profilometry, where a hard sharp tip is scanned across the surface and its vertical movements are monitored. As a result of the miniature size of the SPM tip, which is mounted at the bottom end of a cantilever-like spring, it is possible to image the corrugation of the sample (Figure 1a). The AFM technique allowed probing of non-conductive materials that is an obvious advantage for polymeric surfaces.

Since its invention, many AFM applications for characterization of polymeric materials have been developed. The first AFM images of polymeric materials published in 1988 - 1991 ignited a great interest in the polymer community and attracted widespread attention to this emerging experimental technique *(2-4)*. Materials for which new information has been obtained to date range from rigid polymers to soft biological macromolecules, from polymer single crystals to amorphous materials, from liquid crystalline polymers to viscoelastic materials, from molecular monolayers to fibrillar structures, and from single component to multicomponent composite materials (see section *"Polymer Morphology and Structures"* in this volume).

Modern applications of the AFM technique for studying surface morphology typically range from micron to nanometer scales. Heights, depths, and lateral sizes of morphological features, microroughness, in-plane molecular ordering, domain morphology, surface defects, and in-plane orientation of molecules can be measured. Modifications of polymer surfaces are done with the AFM tip including hole-making, formation of texture patterns, drawing figures on a submicron level, initiation of local chemical reactions, and surfaces "cleaning". A new, rapidly developing area of the AFM applications is the probing of local mechanical, viscoelastic, adhesion, and frictional properties of surfaces. Within the past six years, many new operational modes such as frictional, dynamic (non-contact), electrostatic, magnetic, thermal, and chemical force microscopies have been introduced thereby giving birth to a new family of scanning force or scanning probe microscopy (SFM or SPM) techniques.

In this chapter, we summarize basic principles of the SPM technique to guide the reader through a variety of original results and focused reviews presented in this volume and make a general overview on current trends in SPM studies of polymer surfaces. This chapter is an introductory guide for those who are not familiar with the latest developments, trends, modes, concepts, and terminology in the field of SPM technique. The aim of this introduction is to help readers become comfortable with the following collection of articles written for specialists familiar with terminology and basic principles of the SPM applications.

Detailed discussion of STM and SPM technical principles and techniques can be found in a number of recent books, special issues, and reviews *(5-18)*. Earlier observations of soft organic materials have been discussed by Frommer *(12)*. Reneker et al. *(13)* have reviewed SPM applications to polymer crystals and molecular polymer films. Numerous examples of SPM images of various polymer surfaces have been presented by Magonov and Cantow *(6, 14, 15)*. Reviews for organic films have been published by Tsukruk and Reneker *(16)* and DeRose and Leblanc *(17)*.

Basic Principles

Currently, researchers can use the SPM instruments with a wide selection of sophisticated modes and features. A wide selection of scanning modes is available on the modern commercial instruments such as contact, non-contact, in-fluid, force-modulation, friction force, adhesion mapping, and others (see below) *(5, 6)*.

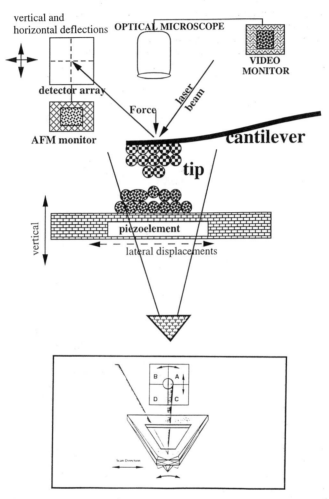

Figure 1. General scheme of SPM instrumentation including possible responses for V-shaped cantilever.

The general scheme of any SPM apparatus includes several major components which allow line-by-line scanning with an atomically sharp tip while monitoring nanometer scale cantilever deflections in vertical and horizontal directions (Figure 1). Precise (within a fraction of a nanometer) 3D movements of either a sample or a cantilever is provided by a tube piezoelement. The SPM tip deflection is monitored by a detection scheme (e. g., array of photodiodes, interferometer scheme, or piezoelectric cantilever) (5). A microfabricated probe consists of a silicon or a silicon nitride cantilever with the integrated pyramidic tip of several microns in height with a tip end radius in the range from 5 to 200 nm. Typical software for 3D image analysis includes a variety of corrections, filtrations, calibrations, and evaluations such as plane-fit, flatten, and high/low frequency filters which can greatly improve visibility of surface morphology details and allow evaluation of the geometrical surface dimensions.

The normal load exerted on the SPM tip varies from 0.01 nN (in a fluid) to 100 μN (stiff short cantilever in humid air) with the typical force range of 0.1 - 100 nN for routine measurements. For a nominal contact area tip/surface of several squared nanometers, this load corresponds to local pressure in the range from several MPas to tens of GPas. Normal acting forces are calculated from a force-distance curve as a total of the adhesive forces (typically 1 - 100 nN for various tip/surfaces pairs in humid air) and spring forces exerted by the deflected cantilever itself. To measure the absolute values of the lateral forces, the torsional deflections of the SPM cantilevers are calibrated according to one of the methods discussed in this volume (see chapter in this volume: Bliznyuk, et al.). A knowledge of spring constants of the SPM cantilevers is critical for the determination of the absolute values of the vertical and torsion forces acting on a tip. Several experimental, analytical, and computational approaches have been recently explored for an estimation of normal (k_n) and torsion (k_t) spring constants of the SPM cantilevers (18-22). Direct experimental measurements of the variations in resonant frequency for the cantilevers loaded with heavy microspheres of different weight is very popular (18). Measurements of resonant frequency for unloaded cantilevers provides the means for the estimation of the actual spring constants if the calibration curves are available (19). Theoretical analytical equations are widely used for the estimation of both k_n and k_t (19-22).

A knowledge of the tip end shape is critical for the recognition of imaging artifacts which can be produced by broken, asymmetric, blunt, or double SPM tips. To estimate tip radii of curvature, R_c, special reference samples with sharp (as compared to the tip end) features should be used. Different reference samples have been proposed including the edges of atomic planes, nanofibres, latex microparticles, whiskers single crystals, and gold nanoparticles (23-25). One example of the SPM image obtained by the scanning of a standard sample with tethered gold nanoparticles by a "blunt" tip is presented in Figure 2. Distinguished triangular features which represent cross-sections of the flattened end of a pyramidic tip are visible on the topographical image. Friction force scanning demonstrates the same triangular features (Figure 2). Bright spikes along the left edge of the triangles are so-called "geometrical friction" resulting from buckling of the SPM tip over the sharp edges during its sliding from left to right.

Software analysis has to be used on-line to provide reasonable quality of recorded images (5, 6). Without discussing in detail all of the capabilities, we briefly summarize here the most important routines. Leveling of a macroscopic tilt is provided by a plane fit. A non-linearity of piezoelement motion is corrected by a proper correction of the applied voltage. High and low pass filters are used to eliminate undesirable frequencies. The off-line analysis capabilities depend upon the software package and the particular application. The most important features that are present in nearly all image analysis packages are cross-section measurements for the

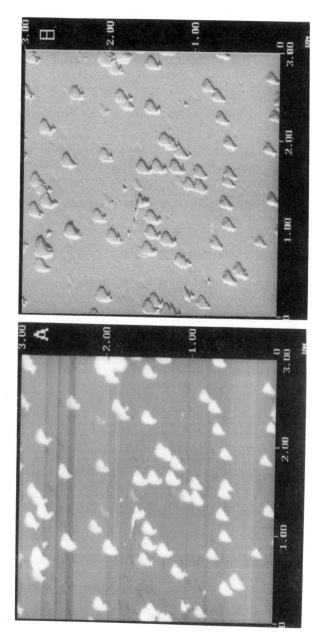

Figure 2. Inverted images of a blunt tip scanned over gold nanoparticles (a) and its friction force image showing "geometrical friction" phenomena (b).

evaluation of heights, lateral sizes, and angles; microroughness measurements (R_a, peak-to-valley distance, rms); the Fourier transformation of topographical images (2D) or selected cross-sections (1D) which provides operators with the means for filtration, suppression of noisy background, emphasis of periodic features, and the averaging of lattice periodicities.

Additional popular features of image analysis software are: 2D autocorrelation functions of images which provide information about statistics of distance distribution; bearing analysis which includes height histograms for averaging of vertical dimensions and the estimation of surface coverage; grain size distribution analysis which provides surface coverage and grain sizes. Spectral properties and fractal dimension of surfaces can be estimated by some packages as well. Usually, image analysis software provides the possibilities for the conversion of the SPM images to one of the standard image formats (e.g., TIFF) that can be used off-line with another image analysis package.

Scan sizes of the SPM images can be as large as 200 μm x 200 μm and as small as 1 nm x 1 nm. The number of pixels usually varies from 64 x 64 to 512 x 512. The range of the surface heights that can be covered by a scanning stage is usually within the range from 0.1 nm to 10 μm. Examples of two reference samples frequently used for piezoelement calibrations are presented in Figure 3. The first image is a rectangular gold-coated grating with a cell size of 10 μm within the 110 μm x 110 μm area. The second micrograph is the 5 nm x 5 nm image of a mica crystalline lattice with an average periodicity of 0.5 nm used for the calibration at a nanometer scale.

Scan rates of the SPM vary from 0.1 to 300 Hz for most applications. Tip sliding velocities range from 10^{-4} μm/sec to 10^4 μm/sec. Specimens can be studied in vacuum, ambient air, fluid or other controlled environments (e. g., elevated temperatures or acidic solutions). Most modern SPM instruments have a built-in optical system that allows analysis of surface textures on an optical scale before the SPM measurements. An important benefit of this combination is the possibility of precise lateral positioning of the SPM tip.

The first commercial SPM instruments became available in 1988 - 1989. Currently, a variety of instruments is available on the market. Very popular microscopes with reliable hardware and sophisticated image analysis software are supplied by Digital Instruments, Inc. Two basic sets, the Nanoscope III and the Dimension 3000, are designed, respectively, as a multimode compact high-resolution tool for research laboratories and a multipurpose universal instrument for a variety of applications and research. The Autoprobe series of the SPM microscopes is manufactured by Park Scientific Instruments, Inc. These highly sophisticated instruments have an open architecture to enable customizing design and have very practical features such as a motorized cantilever holder. Topometrix microscopes (Discover and Accurex) carry all major SPM modes, possess high precision scanning at large scales, and offer a built-in near-field optical microscopy. Burleigh Instruments offers microscopes which are useful for routine measurements and educational purposes as well.

Basic modes of operation

Contact mode and intermittent contact mode. The basic modes of topographical imaging of any SPM instrument are determined by the scanning type (Figure 4). The first operational mode is called contact mode. In this regime, the SPM tip is dragged over the sample surface with a constant velocity and under a constant normal load. The tip is in intimate contact with the surface (point A on a force-distance curve in Figure 5). Scanning can be done in two different fashions: a

8

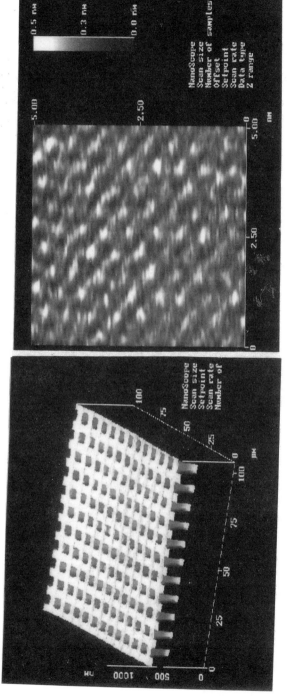

Figure 3. Large-scale 3D grids (left) and atomic scale mica lattice (right) used as reference samples for the distance calibration.

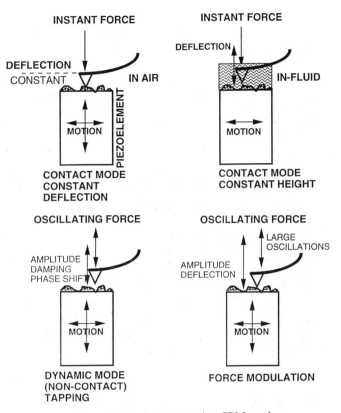

Figure 4. Different scanning SPM modes.

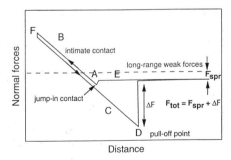

Figure 5. The force-distance curve and different regimes of the tip-surface contact: A: jump-in contact (contact repulsive mode), B: loading part (force modulation mode), C: attrition (contact attraction mode), D: pull-off (adhesion mapping), E: dynamic, non-contact mode; F: maximum loading, the indentation point.

height mode if the cantilever deflection is kept constant by the extending-retracting piezoelement tube, and a deflection mode if the piezotube extension is constant and the cantilever deflection is recorded (Figure 4). The contact mode allows tracking of surface topography with a high precision and also provides the highest lateral resolution of 0.2 - 0.3 nm (down to the true atomic resolution at appropriate conditions *(26)*, see also chapter of Luthi et al. in this volume), but imposes a high local pressure and shear stress on the surface. The small contact area (10^1 - 10^2 nm^2) and significant contribution of the capillary forces are major features of the SPM operations in humid air. Thus, even very low normal forces in the nanoNewton range result in very high normal pressure and shear stresses in the 0.1 - 100 GPa range. Such high local stresses can easily damage soft polymer surfaces as is demonstrated in other chapters in this volume.

Usually, the contact mode can be applied with caution to hard polymers and with great caution to glassy compliant surfaces. The contact mode is necessary for testing of mechanical stability, local elastic response, adhesive forces, and friction properties. However, for high resolution imaging and soft polymers such as rubbers, gels, and elastomers, the contact mode in a fluid is the only choice. Typical fluids used are pure water and alcohol. Fluid selection depends upon polymer dissolution properties. An alternative way to reduce the normal pressure applied to the surfaces is by operating in a dry atmosphere (nitrogen, argon) which eliminates the capillary forces. Both approaches can reduce the local pressure by one or two orders of magnitude to tens of MPa (tens - hundreds atms).

An intermittent contact mode has been introduced to overcome the surface damage problems associated with the high local pressure *(27-30)*. The basic idea of this dynamic mode (frequently called "the tapping mode" for the DI instruments) is replacing the tip-surface contacts with brief approachings of the oscillating tip (Figure 4). The stiff SPM tip oscillates with a high frequency close to its resonant frequency (usually, 200 - 400 kHz). In the vicinity of the surface, weak interactions can significantly change the amplitude of tip oscillations (amplitude detection) and lead to a phase shift (phase imaging) *(29)*. Major factors in the phase contrast mechanism are still being debated but are believed to be a result of viscoelastic response and adhesive forces rather than elastic surface behavior (see below) *(29, 30)*.

The intermittent mode reduces the typical operational forces by at least one order of magnitude (to several nanoNewton or a fraction of nanoNewton). In addition, it virtually eliminates the shear stresses and reduces the time of the tip-surface contact by two orders of magnitude. This type of mode is used for the tracking of surface topography as well as weak forces such as magnetic and electrostatic. The dynamic mode in a fluid is a next step in the force reduction, potentially bringing forces down to tens of pN (local pressure of several atms) *(31)*. Lateral resolution of the intermittent mode is substantially lower than that for the contact mode and is about 1 nm for topography and 10 nm for other properties. However, as has been demonstrated recently, true atomic resolution can be achieved in the truly non-contact mode in a vacuum at special conditions using a new design of electronic feedback (see Luthi et al. chapter in this volume).

Friction force mode. In this mode, the SPM instrument monitors the torsion behavior of the cantilevers caused by the lateral forces acting on the tip end (Figure 1). A typical level of the torsion deflections is within several mrads. The lateral forces are usually associated with the frictional forces between the SPM tip and the polymer surface that originated a popular term "friction force microscopy" (FFM) *(32)*. Obviously, the torsion tip deflections can be caused by a number of independent factors such as natural local slopes of the surfaces and the step edges (so called "geometrical friction", see Figure 2), stick-slip motion, or plowing of soft surfaces by the SPM tip *(33)*. Therefore, the interpretation of the friction force microscopy images should be done with care especially for the soft surfaces and surfaces with sharp features.

In fact, for most surfaces, a major contribution in the torsion signal comes from the shear stresses of probing materials. The corresponding FFM images can be interpreted in terms of the friction force distribution (see Hammerschmidt et al. chapter in this volume) *(34-36)*. For multicomponent materials, the difference in the friction properties can be used for the polymer phase identification. If a proper calibration of the torsional tip deflections is done, a direct determination of the static and dynamic friction coefficients and the shear strength is possible from torsion signal variations (the "friction loop") *(36)*.

Friction measurements for different interacting chemical groups were made and good correlation was observed between the level of the adhesive forces and the friction behavior *(37, 38)*. The role of concurrent intermolecular interactions and surface mechanical responses on the frictional behavior of soft surfaces has been discussed recently (see Vezenov et al. chapter in this volume and below).

Static and dynamic friction can be studied on a molecular scale. Its origin has been found in stick-slip motions of the cantilever spring while scanning under constant contact-load over the sample surface *(39)*. This stick-slip motion is caused by instabilities of the cantilever's spring. The shape and the amplitude of a single stick-slip occurrence are measures for elastic properties of the sample, and the adhesion between sample and tip, respectively. The frequency of the stick slip motion has been found to be periodic for quasistatic sliding, and erratic for faster sliding.

The first SPM experiment that measured wearless friction and elasticity simultaneously on a submicrometer scale found a strong relation between the two sample properties *(40)*. It is important to note that friction is not an intrinsic property of the sample alone. It also contains extrinsic information, such as adhesion forces between sample and tip. If elastic properties dominate the friction response, or the friction signal is adhesion corrected, friction measurements can provide important information about the mechanical surface properties of the sample *(41)*.

Force displacement measurements. Force displacement (FD) measurements use very large amplitude of vertical oscillations. The SPM tip is actually pressed against the surface and then pulled off (Figure 4) *(42)*. If the SPM tip moves laterally, point-by-point, over a surface, a complete distribution of the elastic surface properties can be collected concurrently with the topographical image. The amplitude damping is determined by the elastic surface deformation against a hard tip (the point B on the force-distance curve, Figure 5). The slope of the force-distance curves during direct tip-surface contact is related to Young's modulus of the surface (see below) (Figure 5). Indeed, this technique allows detection of the elastic properties with nanometer lateral resolution for surface with significant local differences in elastic moduli.

In the static mode of operation, the stiffness of the sample can be approximately determined by picturing the sample as a one-dimensional spring (sample spring constant k_s) during a small elastic indentation with a cantilever (spring constant k_l) *(43)*. The effective spring constant k_{sys} (also called stiffness) of the entire system is then given by

$$\frac{1}{k_{sys}} = \frac{1}{k_s} + \frac{1}{k_l}. \tag{1}$$

Based on Hooke's law, the resulting force ΔF on the surface is

$$\Delta F = k_l \Delta z_l = k_s \Delta z_s = k_{sys}(\Delta z_l + \Delta z_s) \tag{2}$$

where Δz_1 and Δz_s are the cantilever deflection and the sample deformation respectively. In order to determine the cantilever deflection in units of a distance it is necessary to calibrate the sensitivity of the cantilever detection scheme. In the setup of a laser beam deflection detector it is the sensitivity of the light-lever/photodiode which has to be determined. Based on equation 1, the detector can be calibrated with the help of a second sample with a stiffness much higher than the stiffness of the cantilever (44). The system spring constant, S_{sys}^{cal}, can be measured with a force-displacement curve, Figure 5. For stiff calibration samples compared to the cantilever spring constant, S_{sys}^{cal} in volt/distance (i.e., the slope of the force-displacement curve in Figure 5) and k_{sys}^{cal} in N/m are related as follows:

$$S_{sys}^{cal} = \frac{1}{\lambda} k_{sys}^{cal} \cong \frac{1}{\lambda} k_1 \Rightarrow \Delta F = \lambda S_{sys}^{cal} \Delta z_1. \tag{3}$$

with λ [N/V] as a calibration factor. Hence, the ratio between the sample indentation and the deflection of the cantilever can be calculated with equations 2 and 3 from the measured slopes of the force-displacement curves as

$$\frac{\Delta z_s}{\Delta z_1} = \left(\frac{S_{sys}}{S_{sys}^{cal}} - 1 \right), \tag{4}$$

and $k_{sys} = \lambda S_{sys}$, is the system stiffness. Thus, the stiffness of the system, k_{sys}, and the stiffness of the sample, k_s, are:

$$k_{sys} = k_{sys}^{cal} \frac{S_{sys}}{S_{sys}^{cal}} \approx k_1 \frac{S_{sys}}{S_{sys}^{cal}} \tag{5a}$$

$$k_s \approx k_1 \left(\frac{S_{sys}^{cal}}{S_{sys}} - 1 \right)^{-1} \tag{5b}$$

The equations 5a and 5b are exact in the limit of an infinitely stiff calibration sample. This calibration technique has been practiced for the determination of the stiffness of polymeric samples in dry nitrogen atmosphere on silicon and ionic crystals, and in various liquid environments on silicon (44).

Dynamic force displacement. Dynamic FD consists of a normal large amplitude saw tooth wave form which is superimposed by a very small (4 nm) amplitude sinusoidal modulation. There is no scanning involved. This technique combines the dc approach of the static FD measurement with a small sinusoidal modulation (ac approach) (44). Its modulation amplitudes are on the nanometer scale. From the measured force signal, the system stiffness and the elastic and viscous component of the modulus can be acquired simultaneously.

This approach has shown to be very effective in measuring inhomogeneities (in the normal direction) and mechanical properties of liquids or melts. It also provides information about the viscosity of liquids near a solid surface. In that respect it is similar to surface forces apparatus with the distinction that the contact area is on the nanometer scale. Recently, dynamic FD measurements have provided viscoelastic information on polymer brushes immersed in good and poor solvents. The static FD measurements were compared to surface forces apparatus measurements and found to be qualitatively consistent within a 10 nm compression regime (44).

Force modulation. In the scanning force modulation technique, the modulation frequency is usually below the resonant frequency of the cantilever and the relaxation frequency of the sample. The sample or the cantilever is modulated in the z-direction (normal direction to the sample surface) at frequencies apart from resonant frequencies of the system, which include the instrument, the sample, the tip and the physical properties of the contact zone. In the scanning mode, the modulation frequency is set above the gain of the feedback-loop. Reasonable modulation frequencies are between 1-20 kHz while scanning, and 10-40 Hz in the static mode (modulation at the same location without scanning). The cantilever response time, the electronic feedback, and system resonance sets the limits in the choice of modulation frequencies.

One problem with the approach discussed above is the unfavorable signal-to-noise ratio due to naturally occurring fluctuations and variations in the system properties. Examples are temperature fluctuations, local contact position and size variations, and local sample imperfections. A static SPM approach for the hardness measurements differs substantially from macroscopic hardness measurements with a Vickers hardness tester. While in the Vickers hardness test a large volume of material is tested, the SPM tip probes only the surface layer.

The amplitude of the oscillating tip is proportional to the elastic properties of the sample and the phase is proportional to the phase shift caused by the sample, which is a measure for the viscoelastic flow (45). Absolute values for viscoelasticity are obtained by normalizing the measurements to a modulation frequency of 10-40 Hz. This is done without scanning to eliminate electronic setup effects which occur at modulation frequencies above 1 kHz. In the elastic regime the Hertzian theory can be applied to relate the z-compliance to Young's modulus, an intrinsic property of the sample (see below). This simplified elastic model does not account for adhesive forces and should therefore not be applied if the 'force-separation curve' indicates variations in adhesion.

An adsorbate-soiled tip can even invert the contrast information. Therefore, measurements must be performed very carefully to avoid frequency-dependent modulation. Environmental conditions, such as humidity, have been found to affect the measurements. As a consequence, caution is advised in reporting absolute values for measurements conducted in air. Relative numbers or comparative absolute numbers collected over a short time are considered more valid.

If the spring constant of the cantilever is very low compared to the sample, the cantilever will bend rather than elastically deform the sample. Only the spring of the cantilever will be measured and not the elastic properties of the sample surface. This is, however, not a serious problem for quantitative measurements on soft polymeric films. Viscoelastic contrast information can be achieved with a cantilever spring constant of 0.01-1.0 N/m on soft films. Reliable absolute measurements demand, however, stiffer cantilevers with spring constants ranging from 0.1 to 10.0 N/m. Yet, wear problems with stiffer cantilevers on deformable films have to be considered. Another method is measurement in the fast modulation regime (GHz regime) where compliance measurements can be conducted with elastically soft probes on hard samples (46).

Young's modulus measurements are achieved by modulating the scanner (or the cantilever) in the surface's normal direction. But there are at least two problems of the experimental set-up: (1) The asymmetry between probing tip and sensor with the sample surface (i.e., the indentation of the tip is not perpendicular to the sample surface), and (2) piezo shearing. Based on such an asymmetric indentation, it has been suggested to simultaneously measure the lateral displacement of the cantilever which is caused by a possible asymmetry or shear process of the piezo (46). By changing in a controlled manner the tilt angle between tip and sample, the lateral piezo motion can be deconvoluted from the lateral modulation signal.

Mapping of the adhesion forces (or a "force volume" mode) is another feature that can be provided by scanning. In this mode, the SPM tip gently touches the surface in each oscillating cycle. Then the pull-off forces are measured and the SPM tip moves to an adjacent point along the scanning line. The surface distribution of the pull-off forces or the adhesive forces (mainly the capillary forces if scanning is in humid air) can be simultaneously collected with the topographical image. Within a reasonable time frame, the number of data points collected can be 128 x 128 providing a spatial resolution better than 30 nm for the typical scan size of 3 μm x 3 μm.

Chemical sensing. It is obvious that SPM operation in the topography mode only rarely cases provides insight into the chemical nature of a multicomponent system. In 1992 Overney et al. introduced the friction force mode as a nano-surface-tool suited to provide chemical distinctions of two-component systems (47). The friction values found on a single component system were compared with the friction values measured on the binary systems. Depending on this correspondence, the material would be identified. In this particular set of experiments, the friction forces were strongly dominated by the mechanical properties of the samples (46).

Lieber's group extended the approach by chemically modifying the cantilever tip and thus enhancing the acting interaction forces (37, 48). This modification is frequently called "chemical force microscopy (CFM)". Because of the particular choice of tip coating material, adhesion forces were considered to dominate the friction forces and, therefore, can be used for identification of different chemical species (chemical sensing). This approach gained popularity among researchers who were interested in the probing of surface chemical composition with a high resolution (see Vezenov et al. chapter in this volume) (37, 48-55). A knowledge of the intermolecular interactions between different chemical groups allows their identification by probing with the SPM tips bearing appropriate chemical groups. Chemical sensing requires the modification of the tip surface by the fabrication of a robust molecular layer with appropriate terminal groups (Figure 6) (37). This technique allows the discrimination of the surface forces related to the intermolecular interactions of different chemical groups. It is also promising for chemical sensing of surfaces with lateral resolution better than 10 nm.

Approaches for the tip modification involve self-assembly to fabricate thiol or silane monolayers by chemisorption onto a gold coated tip or a bare silicon nitride tip. To date, several examples of the modified tips with terminal CH_3, COOH, CH_2OH, CH_2Br, CH_3, SO_3, and NH_2 groups have been demonstrated (37, 56). Selective chemical interactions between these surfaces were tested in air at various humidities, in water, and in solutions with different pH values. Observed variations of interacting energy and friction forces were consistent with a variation of ionic strength of solutions with different pHs. The isoelectric points of surfaces were determined by this technique for the individual chemical groups confined to monolayer surfaces (see chapters written by Vezenov et al. and Tsukruk et al. in this volume).

With recent independent studies in the groups of Colton (57) and Overney (41), it was found that chemical distinctions based on mere friction measurements do not necessarily provide information about the chemical nature of the sample surface only. Both groups showed that the friction signal can significantly vary on chemically homogeneous surfaces just because of local differences in mechanical properties. Therefore, the ability of the material to shear cannot be neglected. Adhesion measurements were also found by the two groups to vary on chemically homogeneous samples because of local variations of the contact area. For instance, differences in grafting density in polyethylene acrylate polystyrene (PEA-xPS) block copolymer film show a significant effect on friction and adhesion despite the fact that the surface is chemically uniform (Figure 7) (41). It was found that the chemically identical surfaces

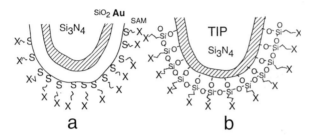

Figure 6. The chemically modified SPM tip: thiol SAM tethered to a gold layer (a) and alkylsilanes assembled on a silicon nitride surface (b).

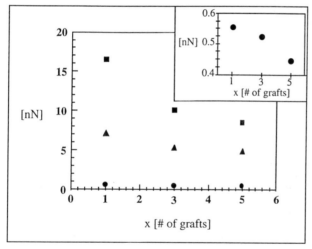

Figure 7. Backbone bending flexibility of PEA-xPS. ■ Adhesion forces provide information about stiffening of graft copolymer surface with increased grafting density. ▲ Absolute lateral forces are adhesion influenced. ● Lateral forces, adhesion corrected. *Inset:* Magnification of adhesion corrected lateral forces. (After (*41*)).

were wetting the tip differently depending on the backbone flexibility of block co-polymers (41).

A chemical distinction of multicomponent surfaces can be very difficult. It is recommended to apply all the SPM modes described in this chapter to identify different microphases. The biggest challenge is the determination of the contact area for chemically modified tips. As long as this value is not established, even relative comparisons of interaction energies can be considered questionable.

Other SPM scanning modes (electrostatic, magnetic, and thermal). Electrostatic force microscopy is based on the detection of the varying capacitance in the vibrating metallic SPM tip that is affected by the surface charge (5, 58). A very similar idea stands behind magnetic force microscopy (5). Depending upon the external field applied and the type of the SPM tip used, the conventional dynamic mode can be converted to electrostatic or magnetic force microscopy. Modification of the scanning conditions is required to eliminate disturbances caused by rough surface topography. Both modes are very important for magnetic and electronic materials and have applications for studying conductive polymers and polymer-metal composites. Lateral resolution of field structures can be as low as tens of nanometers and several isolated charges can be detected.

Studies of the thermal properties of polymer surfaces can be done in the dynamic scanning mode on a SPM instrument equipped with a special thermoregulated stage (49). A very recent development is thermal microscopy which allows the detection of a thermal flux from a surface by probing with a tiny sharp thermopair (60, 61). A polymer sample is placed within a thermocell and a local thermal flux is detected by the nanoprobing thermopair. This design is a microscopic analog of scanning calorimetry which allows testing local thermal conductivity for the selected areas of polymer surfaces. Local glass transition, polymorphic transformations, recrystallization, cure reactions, and melting can be detected for microphase areas of multicomponent polymer surfaces. This mode is still in its infancy but promises exciting developments in studying thermal properties of polymer composites and surfaces (61).

Static probing local micromechanical properties. Analysis of mechanical responses of a polymer surface can be done by studying the force-distance curves while moving the SPM tip toward the surface and back (Figure 4). Different parts of this curve correspond to the different probing modes: scanning in the contact repulsive and attraction modes (A and C), probing of the elastic surface properties in the force modulation mode (point B), and adhesion mapping (point D) (Figure 5). From the slope of the force-distance curve, the point at which the tip is pushed against surface (point B) compliance or elastic modulus can be determined by using one of the models for an elastic contact (see below). The pull-off forces correspond to the adhesive forces between the SPM tip and the surface. Total force applied to the surface is sum of spring forces and adhesion contribution (Figure 5).

The detection of amplitude damping and phase shift in the region E on the force-distance curve (Figure 5) can provide information about the elastic surface properties. The storage and loss moduli and the adhesion properties can be detected (35, 62-66). Indentation of polymer surfaces by applying a high normal load to the tip (the point F on the force-distance curve, Figure 5) can be used for estimation of the local surface hardness (66, 67). Wearing rate and the shear strength can be determined from friction force microscopy data as well (68-70).

Basics of dynamic force approach. The frequency dependence of the viscoelastic properties of a polymer can be approached by various models, such as the

Maxwell model, the *Voigt model* and the *Standard linear solid model*. The *Maxwell model* (damping in series with a spring) predicts Newtonian flow, and the response of a polymer during stress relaxation *(44, 45)*. The classical Hertzian contact theory has been applied in most force modulated SPM studies *(40, 45, 71, 72)*. According to this elastic theory and assuming a sphere-flat configuration, the Young's modulus of the system ($1/E=1/E_1+1/E_2$) elastic constants of both bodies in contact) is given by

$$E = \left(\frac{k_{sys}^3}{6RF_o} \right)^{1/2}$$ (6)

where the equilibrium force is F_o, the contact radius is R ($1/R=1/R_1+1/R_2$, radius of curvatures of both bodies in contact) and the system spring constant, k_{sys} *(40)*. The system spring constant can be determined, (*a*) with the slope of force displacement curves for static elastic measurements (see above), or (*b*) by

$$k_s = k_1 \left(\frac{A_{out}}{A_{in} - A_{out}} \right)$$ (7)

for sinusoidal force modulation measurements. Here, k_1 is the cantilever spring constant, A_{in} is the input modulation amplitude (perturbation) and A_{out} is the response amplitude *(40)*.

The elastic modulus of the sample is calculated with equation 6 along with the knowledge of the cantilever 's normal spring constant and radius of curvature (area of contact). Burnham and Colton demonstrated the applicability of the force-distance method to determine surface mechanical properties *(74)*. However, because of uncertainties of the surface contact area, it is difficult to quantify the results *(75)*.

In a very recent study, the sharp SPM tips were replaced with spherical glass beads with diameters from 4 to 1000 μm *(76)*. Very stiff springs of 4500 N/m were used. Glass bead indentation measurements were performed on polycarbonate (PC) and polystyrene films. A very drastic decrease in the measured modulus of PC was observed over an increase in the sphere's diameter. This is in agreement with the above reviewed comparative study of the SPM and a Vicker's hardness tester *(66)*. The authors of the glass bead indentation work claim the effects of adhesive forces and surface roughness to be responsible for the decrease in the modulus with increased contact area. Static measurements on thin polystyrene films with indentations on the order of more than a tenth of the film thickness show strong contributions from the substrate *(76)*.

Adhesion or long range forces are not considered in the Hertzian theory. Therefore the contact area between non-conforming elastic solids falls to zero when the load is removed *(77)*. Experiments however, like SPM studies, have showed very impressively that measurable competing forces of attraction and repulsion between atoms or molecules in both bodies are acting within a certain separation distance *(71, 78-82)*. Johnson, Kendal and Roberts have considered the aspect of adhesion in the elastic contact regime (JKR model) *(83)*. Hence, the equilibrium force F_0 consist of two loading terms F_{appl} and F_{adh} (see Figure 5) with

$$F_0 = F_{appl} + F_{adh},$$ (8a)

$$F_{adh} = 3\pi RW + \sqrt{6\pi RWL + (3\pi RW)^2},$$ (8b)

where F_{appl} is the applied load, F_{adh} is the load resulting from adhesion *(83)*, and W is the interaction energy per unit area between the sample and probing tip.

By adding a long-range potential to the continuum JKR theory to avoid edge infinities, a very complex and computer intensive theory known as the DMT model

(Derjaguin-Muller-Toporov) was developed (*84*). DMT and JKR theories have been successfully tested in macroscopic experiments (*77*) and also on the microscopic scale with the SPM (*85-87*).

Artifacts in scanning and force modulation measurements. In the case of well ordered monolayers of soft organic films plastic deformations caused by the cantilever tip can be imaged on a consecutive large scale scan (*16, 17, 88*). However, tip induced conformational changes on polymer surfaces are not as easy to confirm. The reason is the scale of deformation. In a Van der Waals crystalline film, for example, single deformations can be expected to occur on a scale of 10 or less square-nanometers leaving a topographically visible trace of deformation. Deformations of polymer surfaces, however, can reach over microns although induced on the nanometer scale (*89*). Hence, friction and surface viscoelasticity measurements have been found to be useful in determining if the cantilever plastically interacted with the polymer sample surface. Topographical imaging alone has been shown to be unconvincing (*41, 89-91*).

The sample's response to force modulations can depend on various variables of operation. The most important ones are the equilibrium load, the amplitude of modulation, and the modulation frequency. As with all variables there is an operational regime which is acceptable. Outside this regime artifacts may occur. Large amplitudes of modulation can also cause unstable equilibrium loads during the scanning process due to feedback problems. Equilibrium loads or amplitudes of operation which exceed the elastic limit of the sample material set the experiment into a regime where quantifying theories are rare and the elastic theories, such as the linear viscoelastic theory, fail.

Questions of gradient of mechanical properties across the polymer film and substrate influence should be carefully addressed in thin film studies. An example of continuous variation of shearing properties for thin polymer films is presented in Figure 8.

A very recent and interesting publication showed that at contact resonances an enhanced contrast force modulation response map can be recorded between the octadecyltriethoxysilane monolayer and mica (*92*). It demonstrates the aspect of resonance force modulation measurements resulting in the reversed contrast. Resonance measurements provide therefore only the information about the existence of heterogeneities in stiffness. They do not identify the stiffer (or softer) areas nor provide absolute or relative values

For additional discussions of nanomechanical probing of polymer surfaces see chapters written by Aime et al., Bliznyuk et al., and Hammerschmidt et al. in this volume.

Limits on Spatial Resolution

Many groups have reported atomic or molecular resolution in friction and topography, (*6, 62*), but only a few have proved true atomic/molecular resolution (*7, 26, 39*) (see chapter written by Luthi et al. in this volume). It is believed that most of the atomic or molecular resolved SPM measurements are not recorded with a single atomic tip. Multiple tip contacts that are commensurable with the sample lattice structure can also provide periodic patterns.

Plastic deformations (i.e., conformational changes) at the surface limit the resolution of SPM measurements. There are four possible origins for plastic deformations: loading pressure, polymer-tip wetting, scan speed, and shear forces. The loading pressure can be reduced by avoiding the formation of capillaries due to humidity and/or by using larger contact radii of curvature. Capillary forces can be diminished in a dry nitrogen gas environment or eliminated in a liquid environment. Cantilever tips of larger radii of curvature can be chosen with the disadvantage of lower imaging resolution. Tip wetting by the polymer sample can be reduced by the choice of (a) the right probing temperature, (b) the time in contact with a particular area, and/or (c) the suitable material for cantilever tip coating. The probing temperature and the time in contact strongly influence the diffusion of the polymers towards the tip. The right tip coating reduces the strength of the interaction. The probing time can be set by the scan speed, and/or by modulation normal to the sample surface. One disadvantage of merely increasing the can speed is reduced feedback control resulting in high shear forces that can cause plastic deformation.

The combination of the mentioned procedures is crucial for reducing plastic deformation and gaining high resolution. The literature is still lacking a comprehensive study of the optimization parameters for defect-free imaging of polymer surfaces, especially for polymers with low glass temperatures.

True atomic/molecular resolution are tested on the ability to detect a single defect site. Examples of defects are lattice steps or single atom (molecule) holes. Atomic (or molecular) moiré pattern can therefore be excluded. An example of such a lattice step defect, measured in air on an ultrathin organic film of 5-(4'-N,N-dihexadecylamino) benzylidene barbituric acid (lipid), is shown in Figure 9a. Overney and Leta have shown that elasticity can be mapped on a molecular scale with a cantilever tip as sharp as the size of a single sample molecule (*63*) (Figure 9b).

SPM Applications in Polymers

Many representative examples of SPM studies of various polymer surfaces ranging from perfectly ordered single crystals to polymer composites with heterogeneous microstructures are presented in this volume. A comprehensive discussion and an extensive bibliography of original publications are presented in chapters devoted to different types of polymeric materials. The first three sections of this volume titled "*Polymer Morphology and Structure*" review surface morphology and microstructures of various polymer surfaces as studied by different SPM modes. The first section includes a review on morphological studies of a classical representative of polymer single crystals, polyethylene, written by Reneker and Chu. Polypropylene crystallization is discussed in chapter written by Stocker and Lotz. Two chapters written by Vancso and Goh, respectively, consider microstructure of polymer and biopolymer fibers.

The second section deals with polymer composites and very compliant materials such as carbon filled polymer blends (LeClere et al.), thermoplastics (Everson), anisotropic polymer films (Biscarini et al.), and biomolecules (Domke et al.). SPM studies on organic molecular films are presented in the third section which includes polymerizable Langmuir monolayers (Takahara et al. and Peltonen et al.) and LB films studied at the air-water interface (Eng et al.).

The fourth section discusses various results on nanoprobing of polymer surface properties. Tsukruk et al. discuss current approaches to quantitative characterization of surface properties. Aime et al. and Hammerschmidt et al. present their results on mechanical probing of deformational and viscoelastic behavior of polymer networks and films. The final section includes several chapters which discuss current trends and modern developments in SPM applications. Resolution limits of SPM technique and true-atomic resolution of crystal lattices are considered in Luthi et al. chapter. Vezenov et al. review CFM applications for various chemically modified surfaces and Tsukruk et al. discuss surface interactions between various chemical groups in different environments (pH and ionic strength). Boland et al. demonstrates examples of nanolithographical works with SPM tips.

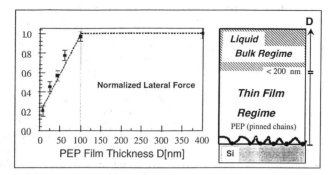

Figure 8. Interfacially confined polyethylene-copropylene system. A continuous change in surface mechanical properties of ultrathin polymer films on silicon is observed. The lateral forces are a measure of the surface stiffness and are decreasing for ultrathin films on silicon. A mechanical boundary layer is formed in closest vicinity to the silicon interface. (From (41)).

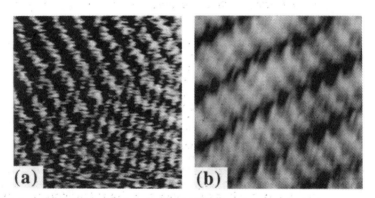

Figure 9. High resolution image of a lipid film surface.

(a) 12×12 nm² scan (in lateral force mode) of a boundary between two domains. The upper limit of the contact area is determined to be on the molecular scale (39).

(b) 3.5×3.5 nm² elasticity scan. The resolution is due to the small contact area on the molecular scale (from (63)).

The following table is intended to provide a brief overview of research domains where the SPM technique has been successfully applied. Here, we briefly summarize some selected references and more complete databases can be found in recent reviews *(6, 12-17, 35)*.

Table. List of SPM applications to polymeric materials.

Surface Morphology	
homopolymers:	Kevlar *(94)*
	polyaniline *(95-97)*
	polycarbonate *(98)*
	poly(epichlorohydrin) *(99)*
	polyethylene *(2, 100-108)*
	poly(hydroxyaniline) *(109)*
	polyimide *(110-112)*
	polyoxymethylene *(113)*
	polypropylene *(114-121)*
	polystyrene *(122-125)*
	polytetrafluoroethylene (Teflon) *(98)*
	polyvyniledenefluoride *(126)*
	polydiacetylenes *(14)*
	polyetheleneterephtalate *(93)*
composites and blends (phase separation)	polypropylene/polyurethane *(127)*
	polystyrene/poly(ethyleneoxide) *(128)*
	polystyrene/poly(methyl methacrylate) *(128)*
	polyethylene-polypropylene *(129)*
	polystyrene-polyvynilmethacrylate *(130)*
	polyethylene-polymethylmethacrylate *(61)*
	polyimides-monomers *(131, 132)*
block copolymers	polyisoprene-polybutadiene *(128, 133-135)*
	polystyrene-polybutadiene *(133)*
	polystyrene-polyparaphenylene *(136)*
	polystyrene-polybutylmethacrylate *(137)*
	polystyrene-polybutadiene-polystyrene *(138)*
	polystyrene-polybutadiene-polymethylmethacrylate *(139)*
	polystyrene/polyethylene-propylene *(90)*
nanoparticle films (latexes, dendrimers)	polystyrene *(122)*
	polybutyl methacrylate *(140, 141)*
	polystyrene-polybutylacrylate *(142)*
	polyamidoimine dendrimers *(143-145)*
	polybenzene dendrimers *(146)*
	grafted polystyrene *(147)*
Mechanical Properties	
indentation	polycarbonate *(66)*
	poly(methyl methacrylate) *(66, 148)*
	polypropylene *(148)*
	poly(sebacic anhydride) *(149)*
friction (lateral forces)	polyoxymethylene *(150)*
	polypropylene *(151)*
	polypyrrole *(152)*

Continued on next page.

Table. List of SPM applications to polymeric materials. *Continued*

Surface Morphology	
	polystyrene/polyethylene propylene (*90*)
	polystyrene/polyethylene oxide (*153*)
scratching, patterning	polycarbonate (*154*)
	poly(methyl methacrylate) (*155*)
	polystyrene (*156, 157*)
shear deformation	poly(oxyphenylene) (*158*)
	polystyrene (*158*)
stretching	polypropylene (*159*)
wetting/dewetting	polyethylene -copropylene (*128*)
	polystyrene/polyethylene-copropylene (*90, 135*)
Surface Treatments	
corona discharge	polypropylene (*151*)
UV irradiation	polyacenaphthalene (*160*)
	polymehtyl methacrylate (*160*)
	polystyrene (*160*)
	polymethyl styrene (*160*)
Langmuir-Blodgett (LB) and Self-Assembled Films	
polymerized LB-films	(*161, 162*)
polymer stabilized LB-films	(*40, 46, 47, 163, 164-166*)
mixed LB films	(*47, 167-169*)
self-assembled monolayers	(*15, 170-174*)
liquid-crystalline LB films	(*175-181*)
Biopolymers (*182*)	
morphology	actin (*183*)
	polyglutamates (*181*)
filaments	neuro filaments (*184*)
	paired helical filaments (*185*)
fibers	collagen (*186-188*)
biodegradation	poly(DL-lactic acid) (*189*)
	DNA (*190*)
crystal growth	lysozyme (*191*)
forces	DNA strands (*192, 193*)
	ligand-proteins (*194*)
adhesion	biotin-streptavidin ligand (*195*)
phase behavior	gelatin (*196*)
stability	polyglutamates (*182*)
surface behavior	enzyme activity (*190*)
	protein motion (197)
	lipids (*198, 199*)
	proteins, DNA (*200*)

Current trends

We refer the reader to particular chapters in this volume for specific and detailed information. Below, we summarize our vision of current trends and prospectives in this new field of polymer science.

Quantitative measurements of friction, elastic, and shearing behavior on a sub-micron scale are crucial for the studying surface properties of multicomponent polymer systems. Promising results are obtained using a combination of topographical, chemical force, force modulation, and friction force microscopy techniques. A major advantage of these methods is the possibility of the local testing of the physical surface properties in relation to the surface topography and chemical composition. Apparently, the focus of SPM studies on polymer surfaces is gradually changing towards *quantification* of the surface measurements rather than simple topographical imaging prevailing in the initial stage of the SPM applications.

Expansion of the family of surface properties being tested by SPM is inevitable and is expected to be a major trend in instrumental development. Within the past seven years, we have watched the coming of new modes for nanoprobing friction, adhesive, electrostatic, magnetic, chemical, and elastic properties. Probing of thermal, viscoelastic, and near field optical properties is under wide implementation currently. We could expect appearance of even more sophisticated highly specialized probing modes focused on, e. g., hydrogen binding or charge-transfer interactions. New methods that increase the rate of surface information collection (such as is known for STM) allows *in situ* observation of surface morphology and microstructure dynamics in a millisecond temporal range (seconds and minutes currently).

Dramatic improvement in lateral resolution could be anticipated as a result of recent developments in the nanotip fabrication and increasing sensitivity of scanning schemes. Microfabricated tips with probe radius down to 5 nm, or even 1 nm (such as nanotubes demonstrated in Ref. 201), can become a routine tool for SPM measurements. Obviously, for compliant polymer surfaces, this transition should be accompanied by significant reduction of applied normal forces. Truly atomic spatial resolution, achievable at special conditions [202], can be used for studying ordered polymer surfaces.

A *"multidimensional" characterization* of polymer surfaces has emerged. Just a few years ago, almost all SPM related publications focused on polymer surface topography and microstructure. Now, discussion of surface topography in conjunction with friction properties, elastic behavior, adhesion, chemical composition, viscoelastic properties, conductive state, and thermal transformations is becoming customary. Such a "multidimensional" collection of experimental data opens exciting opportunities for readdressing the traditional task of "structure-property relationships" for polymer surfaces on the nanometer scale.

A careful design of the SPM tips with controlled *chemical composition of the tip end* provides chemical sensing capabilities. Recent efforts have resulted in substantial progress and the first commercial modified SPM tips have just become available. Apparently, the SPM technique is rapidly moving towards a new level of surface nanoprobing. A key element is a designed nanoprobe with a wide range of controllable properties which will replace a poorly characterized tool with unknown shape, mechanical parameters, and surface chemistry. This development allows quantitative characterization of intermolecular interactions on a submicron scale and opens the door for unambiguous nanomechanical testing of surface properties.

Various combinations of the classical experimental techniques for surface analysis and the SPM technique are used for characterization of polymer surfaces on various scales. Usually, the parameters of surface morphology and microstructure obtained from the SPM method correlate quite well with optical, scanning and transmission electron microscopies and electron and X-ray diffraction for polymers if resolution limits and the nature of data collected are considered. The very local probing capability of the SPM technique provides complementary information that is beyond the possibilities of conventional experimental techniques. The combination of

24

classical surface characterization with the SPM observations produces a pool of quantitative and unambiguous data concerning polymer surfaces.

References

1. Binnig, G.; Quate, C.F.; Gerber, Ch. *Phys. Rev. Lett.* **1986**, *12*, 930.
2. Patil, R.; Kim, S.-J.; Smith, E.; Reneker, D. H.; Weisenhorn, A. L. *Polym. Comm.* **1990**, *31*, 455.
3. Drake, B.; Prater, C. B.; Weisenhorn, A. L.; Gould, S. A.; Albrecht, T. R.; Quate, C.; Cannell, D. S.; Hansma, H. G.; Hansma, P. K. *Science* **1988**, *243*, 1586.
4. Stocker, W.; Bar, G.; Kunz, M.; Moller, M.; Magonov, S. N.; Cantow, H. J. *Polym. Bull.* **1991**, *26*, 215.
5. Sarid, D. *Scanning Force Microscopy*, Oxford University Press, NY, **1991**.
6. Magonov, S.; Whangbo, M.-H. *Surface Analysis with STM and SPM*, VCH, Weinheim, **1996**.
7. Binnig, G. *Ultramicroscopy* **1992**, *42-44*, 7.
8. Hues, S.; Colton, R. J.; Meyer, E.; Guntherodt, H.-J. *MRS Bull.* **1993**, *1*, 41.
9. Burnham, N. A.; Colton, R. J. *J. Vac. Sci. Techn.* **1989**, *A7*, 2906.
10. Burnham, N. A.; Colton, R. J.; Pollock, H. M. *Nanotechnology* **1993**, *4*, 64.
11. Overney, R. M.; Meyer, E. *MRS Bull.* **1993**, *28(5)*, 26.
12. Frommer, J. *Angew. Chem. Int. Ed. Engl.* **1992**, *31*, 1298.
13. Reneker, D. H.; Patil, R.; Kim, S.-J.; Tsukruk, V. V. in: *Polymer Crystallization*, NATO Advanced Series, v. C405, Kluwer Acad. Press, London, **1993**, p. 375
14. Magonov, S. N.; Cantow, H. J. *J. Appl. Polym. Sci.: Appl. Polymer Symp.* **1992**, *51*, 3.
15. Magonov, S. N. *Polym. Sci., Russia,* **1996**, *B38*, 34.
16. Tsukruk, V. V.; Reneker, D. H. *Polymer* **1995**, *36*, 1791.
17. DeRose, J. A.; Leblanc, R. M. *Surf. Sci. Rep.* **1995**, *22*, 75.
18. Cleveland, J. P.; Manne, S.; Bocek, D.; Hansma, P. K. *Rev. Sci. Instrum.* **1994**, *64(2)*, 403.
19. Hazel, J.; Bliznyuk, V. N.; Tsukruk, V. V. *J. Tribolog,* **1997**, submitted.
20. Sader, J. E. *Rev. Sci. Inst.* **1995**, *66(9)*, 4583.
21. Albrecht, T. R.; Akamine, S.; Carver, T. E.; Quate, C. F. *J. Vac. Sci. Technology* **1990**, *A8*, 3386.
22. Butt, H. J.; Siedle, P.; Seifert, K.; Fendler, K.; Seeger, T.; Bamberg, E.; Weisenhorn, A. L.; Goldie, K.; Engle, A. *J. Microscopy* **1993**, *169(1)*, 75.
23. Carpick, R. W.; Agrait, N.; Ogletree, D. F.; Salmeron, M. *Langmuir* **1996**, *12*, 3334.
24. Vesenka, J.; Manne, S.; Giberson, R.; Marsh, T.; Henderson, E. *Biophysical Journal* **1993**, *65(9)*, 992.
25. Goh, M. G.; Juhue, D.; Leung, O. M.; Wang, Y.; Winnik, M. A *Langmuir* **1993**, *9*, 1319.
26. Ohnesorge, F.; Binnig, G. *Science* **1993**, *260*, 1451.
27. Zhong, Q.; Jennis, D.; Kjoller, K.; Elings, V. *Surf. Sci. Lett.* **1993**, *290*, 688.
28. Tamayo, J.; Garcia, R. *Langmuir* **1996**, *12*, 4430.
29. Spatz, J. P.; Sheiko, S.; Moller, M.; Winkler, R. G.; Reineker, P.; Marti, O. *Nanothecnology* **1995**, *6*, 40.
30. Winkler, R. G.; Spatz, J. P.; Sheiko, S.; Moller, M.; Reineker, P.; Marti, O. *Phys. Rev.* **1996**, *B54*, 8908.
31. Hansma, P. K.; Cleveland, J. P.; Radmacher, M.; Walters, D. A.; Hillner, P. E.; Bezanilla, M.; Fritz, M.; Vie, D.; Hansma, H. G.; Prater, C. B.; Massie, J.; Fukunaga, L.; Curley, J.; Ellings, V. *Appl. Phys. Lett.* **1994**, *64*, 1738.

32. Mate, C. M.; McClelland, G. M.; Erlandsson, R.; Chiang, S. *Phys. Rev. Lett.* **1987**, *59*, 1942.
33. Aim, J. P.; Elkaakour, Z.; Gauthier, S.; Mitchel, D.; Bouhacina, T.; Curely, J.; *Surface Science* **1995**, *329*, 149.
34. Lüthi, R.; Meyer, E.; Haefke, H.; Howald, L.; Gutmannsbauer, W.; Güntherodt, H.-L. *Science* **1994**, *266*, 1979.
35. Overney, R. M. *Trends in Polymer Science* **1995**, *3*, 359.
36. Ruan, J.; Bhushan, B. *J. Tribology* **1994**, *116*, 378.
37. Noy, A.; Frisbie, C. D.; Rozsnyai, L. F.; Wrighton, M. S.; Lieber, C. M. *J. Am. Chem. Soc.* **1995**, *117*, 7943.
38. Green, J.-B.; McDermott, M. T.; Porter, M. C.; Siperko, L. M. *J. Phys. Chem.* **1995**, *99*, 10960.
39. Overney, R. M.; Takano, H.; Fujihira, M.; Paulus, W.; Ringsdorf, H. *Phys. Rev. Lett.* **1994**, *72*, 3546.
40. Overney, R. M.; Meyer, E.; Frommer, J.; Guentherodt, H.-J.; Fujihira, M.; Takano, H.; Gotoh, Y. *Langmuir* **1994**, *10*, 1281.
41. Overney, R. M.; Guo, L.; Totsuka, H.; Rafailovich, M.; Sokolov, J.; Schwarz, S. A. *Interfacially Confined Polymeric Systems Studied By Atomic Force Microscopy*; Material Research Society, **1997**, in press.
42. Maivald, P.; Butt, H. J.; Gould, S. A.; Prater, C. B.; Drake, B.; Gurley, J. A.; Ellings, V. B.; Hansma, P. K. *Nanotechnology* **1991**, *2*, 103.
43. Maivald, P.; Butt, H. J.; Gould, S. A. C.; Prater, C. B.; Drake, B.; Gurley, J. A.; Elings, V. B.; Hansma, P. K. *Nanotechnology* **1991**, *2*, 103.
44. Overney, R. M.; Leta, D. P.; Pictroski, C. F.; Rafailovich, M. H.; Liu, Y.; Quinn, J.; Sokolov, J.; Eisenberg, A.; Overney, G. *Phys. Rev. Lett.* **1996**, *76*, 1272.
45. Radmacher, M.; Tillmann, R. W.; Gaub, H. E. *Biophys. J.* **1993**, *64*, 735.
46. Overney, R. M.; Takano, H.; Fujihira, M. *Europhys Lett* **1994**, *26*, 443.
47. Overney, R. M.; Meyer, E.; Frommer, J.; Brodbeck, D.; Luethi, R.; Howald, L.; Guentherodt, H. J.; Fujihira, M.; Takano, H.; Gotoh, Y. *Nature* **1992**, *359*, 133.
48. Frisbie, C. D.; W. Rozsnyai, L.; Noy, A.; Wrington, M. S.; Lieber, C. M. *Science* **1994**, *265*, 2071.
49. T. Nakagawa, K. Ogawa, T. Kurumizawa, and S. Ozaki, *Jpn. J. Appl. Phys.* **1993**, *32*, L294.
50. Berger, C. E.; van der Werf, K. O.; Kooyman, R. P.; de Grooth, B. G.; Greeve, J. *Langmuir* **1995**, *11*, 4188.
51. Sinniah, S. K.; Steel, A. B.; Miller, C. J.; Reutt-Robey, J. E. *J. Amer. Chem. Soc.* **1996**, *118*, 8925.
52. Haugstad, G.; Gladfelter, W. L.; Weberg, E. B.; Weberg, R. T.; Jones, R. R. *Langmuir* **1995**, *11*, 3473.
53. Frommer, J. E. *Thin Solid Films* **1996**, *273*, 112.
54. Tsukruk, V. V.; Bliznyuk, V. N.; Wu, J.; Visser D., *Polymer Prepr.* **1996**, *37(2)*, 575.
55. Schonherr, H.; Vancso, G. J. *Polymer Prepr.* **1996**, *37(2)*, 612.
56. Noy, A.; Vezenov, D. V.; Lieber, C. M. *Annu. Rev. Mater. Sci.* **1997**, *27*, 381.
57. Kiridena, W.; Jain, V.; Kuo, P. K.; Liu, G. *SIA* **1997**, in press
58. Yee, S.; Stratmann, M.; Oeriani, R. A. *J. Electrochem. Soc.* **1991**, *138*, 55.
59. Marti, O.; Hild, S.; Staud, J.; Rosa, A.; Zink, B. in: *Micro and Nanotribology*. Ed. B. Bhushan, NATO ASI Series, Kluwer Press, **1997**, *330*, 455.
60. Hammiche, A.; Pollock, H. M.; Hourston, D. J.; Reading, M.; Song, M. *J. Vacuum Sci. Techn.* in press
61. Hammiche, A.; Pollock, H. M.; Song, M.; Hourston, D. J. *Meas. Sci. Techn.* **1996**, *7*, 142.

26

62. Proc. International Conference on Scanning Tunneling Microscopy (STM'95), S. V. C., USA, July 23-28, 1995) *J. Vac. Sci. Technol.* , **1996**, *B 14.*
63. Overney, R. M.; Leta, D. P. *Tribology Letters* **1996**, *1*, 247.
64. Durig, U; Stadler, A. in: *Micro/Nanotribology and Its Applications*, Bhushan, B., Ed., NATO ASI Series, Kluwer Acad. Publ., Dordrecht, **1997**, p. 61.
65. Hamada, E.; Kaneko, R. *J. Phys. D: Appl. Phys.* **1992**, *25*, A53.
66. Hamada, E.; Kaneko, R., *Ultramicroscopy* **1992**, *42-44*, 184.
67. Kaneko, R.; Hamada, E. *Wear* **1993**, *162/164*, 370.
68. Bhushan, B.; Koinkar, V. N. *Tribology Trans.* **1995**, *38*, 119.
69. Bhushan, B.; Israelachvili, J.; Landman, U. *Nature* **1995**, *374*, 607.
70. Xiao, X.; Hu, J.; Charych, D. H.; Salmeron, M. *Langmuir* **1996**, *12*, 235;
71. Burnham, N. A.; Colton, R. J. In: *Scanning Tunneling Microscopy and Spectroscopy*; Bonnell, D. A., Ed.; V C H Publishers: 220 E 23RD St, Suite 909, New York, 1993.
72. Burnham, N. A.; Kulik, A. J.; Oulevey, F.; Mayencourt, C.; Gourdon, D.; Dupas, E.; Gremaud, G. in: *Micro/Nanotribology and Its Applications*, Bhushan, B., Ed., NATO ASI Series, Kluwer Acad. Publ., Dordrecht, **1997**, p. 421.
73. Salmeron, M.; Neubauer, G.; Folch, A.; Tomitori, M.; Ogletree, D. F.; Sautet, P. *Langmuir* **1993**, *9*, 3600.
74. Burnham, N. A.; Colton, R. J. *J. Vac. Sci. Technol.* A **1989**, *7*, 2906.
75. Hues, S. M.; Draper, C. F.; Colton, R. J. *J. Vac. Sci. Technol.* **1994**, *B12*, 2211.
76. Corcoran, S. G.; Hues, S. M.; Colton, R. J.; Schaefer, D. M.; Draper, C. F.; Meyers, G. F.; DeKoven, B. M.; Webb, S. C. *to be published* **1997**.
77. Johnson, K. L. *Contact Mechanics*; Cambridge University Press, Cambridge, 1985.
78. Blackman, G. S.; Mate, C. M.; Philpott, M. R. *Phys. Rev. Lett.* **1990**, *65*, 2270.
79. Meyer, E.; Heinzelmann, H.; in: *Scanning Tunneling Microscopy II,* Eds. Wiesendanger, R.; Guentherodt, H. J., Springer-Verlag, Berlin, 1992.
80. Hartmann, U. *Adv. Mater.* **1991**, *2*, 594.
81. Weisenhorn, A. L.; Maivald, P.; Butt, H.-J.; Hansma, P. K. *Phys. Rev. B* **1992**, *45*, 11226.
82. Gauthier-Manuel, B. *Europhys. Lett.* **1992**, *17*, 195.
83. Johnson, K. L.; Kendall, K.; Roberts, A. *Proc. Royal Soc. A* **1971**, *324*, 301.
84. Derjaguin, B. V.; Muller, V. M.; Toporov, Y. P. *J. Coll. Interface Sci.* **1975**, *53*, 314.
85. Burnham, N. A.; Dominguez, D. D.; Mowery, R. L.; Colton, R. J. *Phys. Rev. Lett.* **1991**, *64*, 1931.
86. Thomas, R. C.; Houston, J. E.; Crooks, R. M.; Kim, T.; Michalske, T. A. *J. Am. Chem. Soc.* **1995**, *117*, 3830.
87. Meyer, E.; Luethi, R.; Howald, L.; Bammerlin, M.; Guggisberg, M.; Guentherodt, H.-J. *J. Vac. Sci. Technol.* B **1996**, *14*, 1285.
88. Overney, R. M.; Meyer, E.; Frommer, J.; Guentherodt, H. J.; Decher, G.; Reibel, J.; Sohling, U. *Langmuir* **1993**, *9*, 341.
89. Haugstad, G. Presentation at Material Research Society, Boston, 1996.
90. Overney, R. M.; Leta, D. P.; Fetters, L. J.; Liu, Y.; Rafailovich, M. H.; Sokolov, J. *J. Vac. Sci. Technol.* **1996**, *B 14*, 1276.
91. Haugstad, G.; Gladfelter, W. L.; Jones, R. R. *J. Vac. Sci. Technol.* A **1996**, *14*, 1864.
92. Overney, R. M.; Luethi, R.; Haefke, H.; Frommer, J.; Meyer, E.; Guentherodt, H. J.; Hild, S.; Fuhrmann, J. *Appl. Surf. Sci.* **1993**, *64*, 197.
93. Tsukruk, V. V., *Rubber Chem.&Techn.* **1997**, in press

94. Snetivy, D.; Vancso, G. J.; Rutledge, G. C. *Macromolecules* **1992**, *25*, 7037; Dzenis, Yu.; Reneker, D. H.; Tsukruk, V. V.; Patil, R. *Composite Interfaces* **1994**, *2*, 307.
95. Nyffenegger, R.; Gerber, C.; Siegenthaler, H. *Synth. Met.* **1993**, *55*, 402.
96. Avlyanov, J. K.; Josefowicz, J. Y.; MacDiarmid, A. G. *Synth. Met.* **1995**, *73*, 205.
97. Kugler, T.; Rasmusson, J. R.; Osterholm, J. E.; Monkman, A. P.; Salaneck, W. R. *Synthetic Metals* **1996**, *76*, 181.
98. Magonov, S. N.; Kempf, S.; Kimmig, M.; Cantow, H. J. *Polym. Bull.* **1991**, *26*, 715.
99. Singfield, K. L.; Klass, J. M.; Brown, G. R. *Macromolecules* **1995**, *28*, 8006.
100. Wawkuschewski, A.; Cantow, H.-J.; Magonov, S. N. *Polym. Bull.* **1994**, *32*, 235.
101. Gramer, K.; Schneider M.; Mulhaupt, R.; Cantow, H.-J.; Magonov, S. N. *Polym. Bull.* **1994**, *32*, 637.
102. Magonov, S. N.; Qvarnstrom, K.; Elings, V.; Cantow, H. J. *Polym. Bull.* **1991**, *25*, 689.
103. Snetivy, D.; Yang, H.; Vancso, G. J. *J. Mater. Chem.* **1992**, *2*, 891.
104. Annis, B. K.; Reffner, J. R.; Wunderlich, B. *J. Polym. Sci., Part B: Polym. Phys.* **1993**, *31*, 93.
105. Magonov, S. N.; Sheiko, S. S.; Deblieck, R. A. C.; Moller, M. *Macromolecules* **1993**, *26*, 1380.
106. Patil, R.; Reneker, D. H. *Polymer* **1994**, *35*, 1909.
107. Jandt, K. D.; Buhk, M.; Miles, M. J.; Petermann, J. *Polymer* **1994**, *35*, 2458.
108. Eng, L. M.; Jandt, K. D.; Fuchs, H.; Petermann, J. *J. Appl. Phys.* **1994**, *A59*, 145.
109. Bowman, G. E.; Cornelison, D. M.; Caple, G.; Porter, T. L. *J. Vac. Sci. Technol.* **1993**, *A 11*, 2266.
110. Nysten, B.; Roux, J. C.; Flandrois, S.; Daulan, C.; Saadaoui, H. *Phys Rev B-Condensed Matter* **1993**, *48*, 12527.
111. Kim, Y. B.; Kim, H. S.; Choi, J. S.; Matuszczyk, M.; Olin, H.; Buivydas, M.; Rudquist, P. *Mol. Cryst. Liq. Cryst. Sci. Technol.* **1995**, *262*, Sect. A 1377.
112. Patil, R.; Tsukruk, V. V.; Reneker, D. H. *Polymer Bull.* **1992**, *29*, 557.
113. Snetivy, D.; Vancso, G. J. *Polymer* **1992**, *22*, 422.
114. Lotz, B.; Wittmann, J. C.; Stocker, W.; Magonov, S. N.; Cantow, H. J. *Polym. Bull.* **1991**, *26*, 209.
115. Dorset, D. L. *Chemtracts· Macromol. Chem.* **1992**, *3*, 200.
116. Schoenherr, H.; Snetivy, D.; Vancso, G. J. *Polym. Bull.* **1993**, *30*, 567.
117. Snetivy, D.; Vancso, G. J. *Polymer* **1994**, *35*, 461.
118. Snetivy, D.; Guillet, J. E.; Vancso, G. J. *Polymer* **1993**, *34*, 429.
119. Stocker, W.; Schumacher, M.; Graff, S.; Lang, J.; Wittmann, J. C.; Lovinger, A. J.; Lotz, B. *Macromolecules* **1994**, *27*, 6948.
120. Tsukruk, V. V.; Reneker, D. H. *Phys. Rev.*, **1995**, *B51*, 6089.
121. Tsukruk, V. V.; Reneker, D. H. *Macromolecules* **1995**, *28*, 1370.
122. Li, Y.; Lindsay, S. M. *Rev. Sci. Instrum.* **1991**, *62*, 2630.
123. Jandt, K. D.; Eng, L. M.; Petermann, J.; Fuchs, H. *Polymer* **1992**, *33*, 5331.
124. Eppell, S. J.; Zypman, F. R.; Marchant, R. E. *Langmuir* **1993**, *9*, 2281.
125. Kumaki, J.; Nishikawa, Y.; Hashimoto, T. *Journal of the American Chemical Society* **1996**, *118*, 3321; Karim, A.; Tsukruk, V. V.; Douglas, J. F.; Satija, S. K.; Fetters, L. J.; Reneker, D. H.; Foster, M. D. *J. de Phys.* **1995**, *II 5*, 1441.
126. Reifer, D.; Windeit, R.; Kumpf, R. J.; Karbach, A.; Fuchs, H. *Thin Solid Films* **1995**, *264*, 148; Motomatsu, M.; Nie, H.-Y.; Mizutani, W.; Tokumoto, H. *Jpn. J. Appl. Phys., Part 1* **1994**, *33*, 3775.

28

127. Snetivy D.; Vancso, G. J. *Polymer* **1992**, *33*, 432.
128. Schwark, D. W.; Vezie, D. L.; Reffner, J. R.; Thomas, E. L.; Annis, B. K. *J. Mater. Sci. Lett.* **1992**, *11*, 352.
129. Zhao, W.; Rafailovich, M. H.; Sokolov, J.; Fetters, L. J.; Plano, R.; Sanyal, M. K.; Sinha, S. K. *Phys. Rev. Lett.* **1993**, *70*, 1453.
130. Tanaka, K.; Yoon, J.-S.; Takahara, A.; Kajiyama, T. *Macromolecules* **1995**, *28*, 934.
131. Yuba, T.; Yokoyama, S.; Kakimoto, M.; Imai, Y. *Adv. Mater.* **1994**, *6*, 888.
132. Saraf, R. F. *Macromolecules* **1993**, *26*, 3623.
133. Annis, B. K.; Schwark, D. W.; Reffner, J. R.; Thomas, E. L.; Wunderlich, B. *Makromol. Chem.* **1992**, *193*, 2589.
134. Collin, B.; Chatenay, D.; Coulon, G.; Ausserre, D.; Gallot, Y. *Macromolecules* **1992**, *25*, 1621.
135. Liu, Y.; Rafailovich, M. H.; Sokolov, J.; Schwarz, S. A.; Zhong, X.; Eisenberg, A.; Kramer, E. J.; Sauer, B. B.; Satija, S. *Phys. Rev. Lett.* **1994**, *73*, 440.
136. Widawski, G.; Rawiso, M.; Francois, B. *Nature* **1994**, *369*, 387.
137. Collin, B.; Chatenay, D.; Coulon, G.; Ausserre, D.; Gallot, Y. *Macromolecules* **1992**, *25*, 1621.
138. van Dijk and, M. A.; van den Berg, R. *Macromolecules* **1995**, *28*, 6773.
139. Stoker, W.; Beckmann, J.; Stadler, R.; Rabe, J. P. *Macromolecules* **1996**, *29*, 7502.
140. Wang, Y.; Juhue, D.; Winnik, M. A.; Leung, O. M.; Goh, M. C. *Langmuir* **1992**, *8*, 760.
141. Goh, M. C.; Juhue, D.; Leung, O. M.; Wang, Y.; Winnik, M. A. *Langmuir* **1993**, *9*, 1319.
142. Granief, V.; Sartre, A. *Langmuir* **1995**, *11*, 2726.
143. Sheiko, S. S.; Eckert, G.; Ignateva, G.; Musafarov, A. M.; Spickermann, J.; Rader, H. J.; Moller, M. *Makromol. Rapid Commun.* **1996**, *17*, 283.
144. Watanabe, S.; Regen, S. L. *J. Amer. Chem. Soc.* **1994**, *116*, 8855.
145. Tsukruk, V. V.; Bliznyuk, V. N.; Rinderspacher, F. *Langmuir*, **1997**, *13*, 2171.
146. Coen, M. C.; Lorenz, K.; Kressler, J.; Frey, F.; Mulhaupt, R. *Macromolecules* **1996**, *29*, 8069; Hellmann, J.; Hamano, M.; Karthaus, O.; Ijiro, K.; Shimomura, M.; Irie, M. *Langmuir*, **1997**, in press.
147. Sheiko, S. S.; Gauthier, M.; Moller, M. *Macromolecules*, **1997**, *30*, 184.
148. Boschung, E.; Heuberger, M.; Dietler, G. *Appl. Phys. Lett.* **1994**, *64*, 1794.
149. Shakesheff, K. M.; Chen, X.; Davies, M. C.; Domb, A.; Roberts, C. J.; Tendler, S. J. B.; Williams, P. M. *Langmuir* **1995**, *11*, 3921.
150. Nisman, R.; Smith, P.; Vancso, G. J. *Langmuir* **1994**, *10*, 1667.
151. Overney, R. M.; Guentherodt, H. J.; Hild, S. *J Appl Phys* **1994**, *75*, 1401.
152. Li, J.; Wang, E. *Synthet Metal* **1994**, *66*, 67.
153. O'Shea, S. J.; Welland, M. E.; Rayment, T. *Langmuir* **1993**, *9*, 1826.
154. Jung, T. A.; Moser, A.; Hug, H. J.; Brodbeck, D.; Hofer, R.; Hidber, H. R.; Schwarz, U. D. *Ultramicroscopy* **1992**, *42*, 1446.
155. Mamin, H. J.; Rugar, D. *Appl. Phys. Lett.* **1992**, *61*, 1003.
156. Leung, O. M.; Goh, M. C. *Science* **1992**, *267*, 64.
157. Meyers, G. F.; DeKoven, B. M.; Seitz, J. T. *Langmuir* **1992**, *8*, 2330.
158. Yang, A. C. M.; Kunz, M. S.; Wu, T. W. *Mater. Res. Soc. Symp. Proc.* **1993**, *308*, 511.
159. Hild, S.; Gutmannsbauer, W.; Luthi, R.; Haefke, H.; Guntherodt, H. J. *Helv. Phys. Acta* **1994**, *67*, 759.

160. Goh, M. C. *Advances In Chemical Physics*, v. XCI, Eds. Prigogine, I.; Stuart, A., John Wiley, 1995, p.1.
161. Arisawa, S.; Fujii, T.; Okane, T.; Yamamoto, R. *Appl. Surf. Sci.* **1992**, *60-61*, 321.
162. Peltonen, J. P. K.; He, P.; Rosenholm, J. B. *J. Am. Chem. Soc.* **1992**, *114*, 7637.
163. Overney, R. M.; Takano, H.; Fujihira, M.; Meyer, E.; Guntherodt, H. J. *Thin Solid Films* **1994**, *240*, 105.
164. Overney, R. M.; Bonner, T.; Meyer, E.; Ruetschi, M.; Luethi, R.; Howald, L.; Frommer, J.; Guentherodt, H.-J.; Fujihira, M.; Takano, H. *J. Vac. Sci. Technol.* *B* **1994**, *12*, 2227.
165. Chi, L. F.; Anders, M.; Fuchs, H.; Johnston, R. R.; Ringsdorf, H. *Science* **1993**, *259*, 213.
166. Chi, L. F.; Eng, L. M.; Graf, K.; Fuchs, H. *Langmuir* **1992**, *8*, 2255.
167. Tsukruk, V. V.; Einloth, T. L.; Van Esbroeck, H.; Frank, C. W. *Supramol. Sciences*, **1995**, *2*, 219.
168. Tsukruk, V. V.; Bliznyuk, V. N.; Reneker, D. H.; Kirstein, S.; Möhwald, H. *Thin Solid Films*, **1994**, *244*, 763.
169. Tsukruk, V. V.; Bliznyuk, V. N.; Reneker, D. H., *Thin Solid Films* **1994**, *244*, 745.
170. Overney, R. M.; Meyer, E.; Frommer, J.; Güntherot, H. J.; Decher, G.; Reibel, J.; Sohling, U. *Langmuir* **1993**, *9*, 341.
171. Schwartz, D. K.; Viswanathan, R.; Zasadzinski, J. A. *Langmuir* **1993**, *9*, 1384.
172. Ge, S.; Takahara, A.; Kajiyama, T. *Langmuir*, **1995**, *11*, 1341.
173. Tsukruk, V. V.; Lander, L. M.; Brittain, W. J. *Langmuir* **1994**, *10*, 996.
174. Tsukruk, V. V.; Everson, M. P.; Lander, L. M.; Brittain, W. J. *Langmuir* **1996**, *12*, 3905.
175. Josefowicz, J. Y.; Maliszewskyj, N. C.; Idziak, S. H.; Heiney, P. A.; McCauley, J.; Smith, A. B. *Science* **1993**, *260*, 323.
176. Maliszewskyj, N.; Heiney, P.; Josefowicz, J. *Langmuir* **1995**, *11*, 1666.
177. Tsukruk, V. V.; Reneker, D. H.; Bengs, H.; Ringsdorf, H. *Langmuir* **1993**, *9*, 2141.
178. Tsukruk, V. V.; Bengs, H.; Ringsdorf, H., *Langmuir* **1996**, *12*, 754.
179. Tsukruk, V. V.; Janietz, D *Langmuir* **1996**, *12*, 2825.
180. Tsukruk, V. V.; Reneker, D. H.; Foster, M.; Reneker, D. H.; Schmidt, A.; Wu, H.; Knoll, W. *Macromolecules* **1994**, *27*, 1274.
181. Tsukruk, V. V.; Reneker, D. H.; Foster, M.; Reneker, D. H.; Schmidt, A.; Wu, H.; Knoll, W. *Langmuir* **1993**, *9*, 3538.
182. Haugstad, G. *Trends in Polymer Science* **1995**, *3*, 353.
183. Manne, S.; Hansma, P. K.; Massie, J.; Elings, V. B.; Gewirth, A. A. *Science* **1991**, *251*, 183.
184. Karrasch, S.; Heins, S.; Aebi, U.; Engel, A. *J. Vac. Sci. Techn. B* **1994**, *12*, 1474.
185. Pollanen, M. S. *Nanotechnology* **1995**, *6*, 101.
186. Chernoff, A. G.; Chernoff, D. A. *J. Vac. Sci. Techn. A* **1992**, *10*, 596.
187. Baselt, D. R.; Revel, J. P.; Baldeschwieler, J. D. *Biophys. J.* **1993**, *65*, 2644.
188. Gale, M.; Markiewicz, P.; Pollanen, M. S.; Goh, M. C. *Biophys. J.* **1995**, *68*, 2124.
189. Shakesheff, K. M.; Davies, M. C.; Roberts, C. J.; Tendler, S. J. B.; Shard, A. G.; Domb, A. *Langmuir* **1994**, *10*, 4417.
190. Radmacher, M.; Fritz, M.; Hansma, H. G.; Hansma, P. K. *Science* **1994**, *265*, 1577.

191. Durbin, S. D.; Carlson, W. E. *J. Crystal Growth* **1992**, *122*, 71.
192. Lee, G. U.; Chrisey, L. A.; Colton, R. J. *Science* **1994**, *266*, 771.
193. Kurihara, K.; Takashi, A.; Nakshima N. *Langmuir* **1996**, *12*, 4053.
194. Stuart, J. K.; Hlady, V. *Langmuir* **1995**, *11*, 1368.
195. Haugstad, G.; Gladfelter, W. L.; Weberg, E. B.; Weberg, R. T.; Weatherill, T. D. *Langmuir* **1994**, *10*, 4295.
196. Hild, S.; Gutmannsbauer, W.; Luethi, R.; Fuhrmann, J.; Guentherodt, H.-J. *Journal of Polymer Science, Part B* **1995**, *34*, 1953.
197. Thomson, N. H.; Fritz, M.; Radmacher, M.; Cleveland, J. P.; Schmidt, C. F.; Hansma, P. K. *Bioplys. J.*, in press
198. Birdi, K. S.; Vu, D. T. *Langmuir* **1994**, *10*, 623.
199. Radler, J.; Radmacher, M.; Gaub, H. E. *Langmuir* **1994**, *10*, 3111.
200. Hansma, H. G. *J. Vacuum Sci.&Techn.*, **1996**, *B14*, 1390.
201. Dai, H.; Hafner, J. H.; Rinsler, A. G.; Colbert, D. T.; Smalley, R. E. *Nature*, **1996**, *384*, 147.
202. Overney, R. M.; Takano, H.; Fujihira, M.; Overney, G.; Paulus, W.; Ringsdorf, H. In *Forces in Scanning Probe Methods*; Guentherodt, H.-J., Anselmetti, D., Meyer, E., Eds.; Kluwer Academic Publishers: London, **1995**; Vol. 286.

POLYMER MORPHOLOGY AND STRUCTURE: COMPOSITES AND BIOPOLYMERS

Chapter 2

Morphology, Folding, and the Structure of Polymer Single Crystals

D. H. Reneker and I. Chun

The Maurice Morton Institute of Polymer Science, University of Akron, Akron, OH 44325–3909

Observations made, with the atomic force microscope (AFM), of the fold surfaces of lamellar crystals of polyethylene and other polymers, and of the lateral surfaces of polyethylene fibers, films, and fractured lamellae, are reviewed. New AFM observations of fold surfaces are presented. The images of the fold surface are generally noisy, difficult to obtain, and ambiguous. Folds are most strongly connected to the interior of the crystal, and much more loosely connected to their neighbors, a circumstance that permits nanometer scale roughness to occur, both during growth and during annealing. The limitations of contemporary tips are such that roughness can confuse the molecular details observed on the fold surfaces, while the same tip can produce more meaningful molecular scale images of the locally smooth lateral surfaces.

Since the recognition, around 1957, that lamellar crystals grow when polyethylene is precipitated from solution, folded chain lamellar crystals have become a paradigm for the morphology of crystalline polymers. Lamellar crystals have joined other enduring paradigms for the structure of polymers, which include the random coil, the fringed micelle, and the extended chain fiber. Each of these paradigms may be appropriate for the description of the morphology of a particular polymer, depending upon the way the polymer is solidified.

There has been much speculation and argument about the possible arrangements of the molecular segments at the fold surface. The 1979 Faraday Discussions of the Chemical Society (*1*) were devoted to this question. The fold surfaces, that is, the surfaces at which the molecules fold and re-enter the lamella, are special because each fold is strongly connected to the interior but only weakly connected to its neighbors on the surface. The strong chain that extends 10 nm or more into the interior is very stiff when it is the planar zig-zag conformation, and quite flexible when it is not. The flexibility, arising from rotations around bonds, makes it possible to create crystallographic defects that rotate or translate the parts of the chain through which the defect moves. The conformation at each fold on the surface is influenced by the way the attached planar zig-zag segments, called stems, fit into the interior of the crystal, as communicated by the motions of these defects.

Our interest includes and extends beyond the need for a better understanding of the mechanical properties of polymers, to questions that arise in the design of nanoscale machines (*2*). From this viewpoint, a linear polyethylene molecule is a mechanism that translates and rotates through its neighbors in a crystal, using a well defined set of repetitive motions. The molecules can be caused to change shape and transport a segment bearing a reactive or interactive group to a particular place in a potentially useful way.

Some Relevant AFM Observations. The advent of the Atomic Force Microscope (AFM) in the 1980's, with the demonstrated ability to observe the shape of molecular segments in the size range expected for the folds, offered a new way to directly observe the structure of the fold surface. The goal of obtaining clear, meaningful and reproducible images of individual folds in the array of folds that form the surface proved to be elusive, although the ability to make many quantitative measurements of morphological parameters of folded chain lamellar crystals of polyethylene was demonstrated. For example, measurement of the changes in lamellar thickness can be performed on areas only a few tens of nm in diameter with a sensitivity that surpasses all previous methods. The AFM, and other scanning probe methods will become more and more rewarding for elucidating the properties and behavior of nanoscale regions of lamellar crystals of polymers as expected improvements in the instruments occur.

Annealing. Changes in the thickness of lamellar crystals that occur during annealing, that is, during heating to a temperature below the temperature at which the crystal will melt, produce several characteristic changes in the morphology of the lamellae which can be observed in detail. These changes are produced by diffusive motion of the polymer chains through the crystal lattice of the lamellar crystal. There is evidence that the diffusion is mediated by the propagation of crystallographic defects which can move into and through the folds. The mobile defects provide a mechanism, which can be modeled in atomistic detail, that connects the geometry and position of folds with the position of chains and defects inside the crystal. The lamellae become thicker while they retain their crystallographic orientation. More detailed observations of individual folds at the fold surface, with improved scanning probe microscopes, are expected to provide information needed to corroborate and further develop the defect model of chain motion.

In areas where one lamella is lying on other lamellae, cooperative thickening occurs in adjacent layers. The annealed multilayer crystals show deeper, more dramatic corrugations than annealed monolayer crystals. This suggests that the thick regions of the top layer ride on the top of the thick regions of the underlying layer. There is also evidence that stems from one lamella penetrate into the other during annealing. Reneker and Chun (*3*) (see Figure 3 of this reference) show parts of two overlapping crystals that were annealed at a temperature of 132°C. Holes with thickened rims formed in the regions where the crystal is only one lamella thick, and widely spaced (130 nm) elongated patches formed in the region where the crystals overlap. At the highest temperature the peaks grew to almost twice their original thickness, and the separation between them increased. Areas of the mica substrate some 10's of nanometers across were exposed as the polyethylene molecules were incorporated into regions that are twice the thickness of the original lamella. Similar evidence is shown in electron micrographs obtained by Roe, et al. (*4*).

These changes in morphology during annealing require a lot of motion of molecules at the fold surface. As a lamella thickens, the segments that are originally at the fold surface move into the interior and are replaced by segments that were originally inside the lamella.

Mobile Crystal Defects That Move Polymer Chains. A fundamental interest in examining the positions of folds on a fold surface arises from the possibility of observing the changes produced by crystal defect mechanisms that move the polymer chain during annealing. When a folded chain lamellar crystal of polyethylene is annealed, the polymer molecules rearrange themselves to lower their energy by increasing the stem length and reducing the number of folds. Statton and Geil (*5*) observed this process by measuring, with small angle x-ray diffraction, the change in the thickness of lamellar crystals in a sedimented mat of crystals, as the mat was heated. Thin regions become thinner as the molecules move in a way that allows the longest stems to become longer. To accomplish this, the chain must move through the crystal lattice, since the thickening process occurs at temperatures below the melting temperature.

Reneker (*6*) described a crystallographic defect, now called a dispiration, which can produce the motions of the chain required to increase the crystal thickness. The dispiration, as it diffuses through a segment of the chain, moves the segment forward by one half a repeat unit, and rotates the segment by 180° around its axis. Only small, elastic deformations of adjacent molecules are required. Dispirations can move into a fold and raise it above the fold surface, or they can be emitted from a fold and diffuse away along the chain, thereby lowering the fold. The passage of a succession of dispirations through the chains in a lamellar crystal can accomplish the lengthening of the stems that occurs when the crystal is heated. A fold is relatively permanent, but its internal conformation changes when it interacts with a dispiration. A fold disappears when a chain end moves through the fold. The dispiration is part of a family, (*7-8*) of small crystallographic defects that can move a polyethylene chain, in several ways, inside a crystal. The dispiration is the defect most suited to accomplish the changes that occur during annealing of crystalline polyethylene.

Folds, stems, and crystallographic defects, in lamellar crystals of polymers, provide detailed mechanisms for balancing the thermodynamical costs of irregularities at a fold surface (often called roughness) against structures in the interior of lamellar crystals. These mechanisms are amenable to molecular dynamic modeling that is kinematically detailed, and specific to the particular chemical structure of the polymer molecule. Models involving folds, stems, and defects provide a framework for a detailed and molecule specific analysis of molecular motions that occur in polymer crystals. At the same time, the AFM and related scanning probe methods promise to provide detailed information needed about the movements of the folds, to corroborate such models, and to utilize crystallographic defects to improve our understanding of the connections between nanoscale and microscale properties of polymers.

Pressure on a fold can cause an interstitial type dispiration to be created in one of the stems attached to the fold. Once created, the dispiration (which carries an extra CH_2 and 180° of twist) can diffuse in a random walk way, to eventually enter the fold at the other end of the stem and increase the protrusion of that fold as the extra CH_2 group is incorporated into the fold. At the same time, the conformation of the bonds involved in the fold changes to accomodate the twist carried by the dispiration. Movement of other dispirations and some related phenomena can produce increases in the lengths of the stems by a factor of two or more. The material to increase the length of the stems comes from reducing the number of stems, which corresponds to decreasing the width of the crystal. The AFM is, in principle, capable of observing such changes as they occur. Observations of this sort would produce a wealth of new information.

Folded Chain Lamellar Crystals. Many excellent reviews of lamellar crystals of polyethylene and other polymers exist. A.W. Keller (9) provided an interesting and readable historical perspective. He noted that chain folding occurs because it is the fastest way for a long chain to be incorporated into a crystal. The length of the stems is mainly determined by the crystallization conditions, and is inversely related to the supercooling.

P. H. Geil (10) provided a comprehensive summary of the early research on lamellar crystals. D. C. Bassett (11) summarized the morphology and growth of lamellar crystals up to 1981. B. Wunderlich (12) dealt with the polymer morphology in a comprehensive way in his textbook. L. Mandelkern (13) provided a careful statement, of the problems recognized by those who favored larger random deviations from the idealized perfect crystals, which was updated in a later paper (14). F. Khoury and E. Passaglia (15) discussed broader aspects of the morphology of crystalline polymers, of which the morphology of lamellar polymer crystals forms an important part. J. D. Hoffman, G. T. Davis, and J. I. Lauritzen (16) described the growth and morphology of lamellar crystals.

There are several reliable pieces of morphological information that indicate a high degree of order in the positions of the folds on polyethylene crystals. Paraffin crystals grow epitaxially on the fold surfaces of polyethylene crystals (17) in an ordered way that could not occur unless the fold surfaces have some regular order. Polyethylene crystals grown from solution generally develop a hollow pyramidal habit as a consequence of the fact that the folds are bulky and can fit together better if the folds are tangent to a plane that is not perpendicular to the stems. Reneker and Geil (18), and Keller (19) observed that the facets formed by the fold surfaces are crystallographic planes with small Miller indices. The conclusions drawn from the epitaxy and from the fold surfaces of pyramidal crystals are all based on the average arrangement of the folds. A small number of localized departures from this average do not necessarily prevent epitaxy of paraffin crystals or the development of the hollow pyramidal habit.

Morphological Observations of Lamellar Crystals. Here we simply summarize, with references to a few figures and discussion in other publications. Patil and Reneker (20) showed several pyramidal lamellar crystals of polyethylene at both low and high magnification. (The scale markers on Figures 10, 11, 12, and 13 of (20) should all be labeled 5 nm.). Reneker, Patil, Kim, and Tsukruk (21) showed a short segment of a fold domain boundary surrounded by patches. It is characteristic for the patches along the fold domain boundaries in collapsed pyramidal crystals of polyethylene to be a nanometer or so higher than those in the surrounding material. Reneker, et al. (22) showed AFM images of single crystals, step changes in lamellar thickness due to an increase in cooling rate, fold domain boundaries and patches, growth spirals, and a lamellar crystal that thickened during annealing, with the appearance of large holes from which the material for thickening was removed. Patil, Kim, Smith, Reneker, and Weisenhorn (23) showed growth spirals on dendritic crystals of polyethylene.

In the diagram of a hollow pyramidal crystal shown in Figure 1, the molecules stand parallel to the axis labeled **c**. The top and bottom planes of the region with corners at **a**, **b**, and **c** are called the fold surfaces, to which the folds are tangent in an ideal crystal. The narrow vertical surface between **a** and **b** is called a fold plane. A part of the crystal is cut away in the diagram to show the shape. Hollow pyramidal crystals deform in several characteristic ways as they collapse onto a flat substrate while the solvent from which they grew is evaporated. Folded pleats along the shorter axis of the lozenge, tears near the ends of the shorter axis of the lozenge, wrinkles along the (310) crystal planes, and slip between fold planes have all been observed. Collapse of areas near the growth faces can occur by rotation of one of the fold surfaces of a crystal into the plane of the substrate. Contact forces attach the lamella to the substrate and hold it while the remote parts of the crystal collapse by one or more of the above mechanisms. The AFM makes it possible to compare undeformed areas with other parts of the crystal which are obviously deformed.

AFM Images of Fold Surfaces on Polyethylene and Other Polymers. R. Patil and D. H. Reneker (20) examined the fold surfaces of lamellar crystals of linear polyethylene grown by a self seeding procedure. High spots observed on this surface formed a two dimensional array. The spacings and angles of the array were consistent with those expected for folds viewed from a direction perpendicular to the fold surface of a pyramidal crystal in which each fold plane is displaced from the preceding one by 1.5 repeat units in the chain axis direction. Other samples prepared in nominally the same way showed only disordered images of the fold surfaces of pyramidal crystals.

Nie, Motomatsu, et al. (24) obtained a periodic image on a square lattice with a lattice spacing near 0.6 nm for polyethylene oxide in a phase-separated polystyrene-polyethylene oxide blend that was spin-coated onto cleaved mica. The polyethylene oxide lattice dimensions in the plane perpendicular to the chains, determined from x-ray diffraction, are 0.656 by 0.652 nm. This periodic image was clearer in regions that were first scanned many times with a tip force of around 50 nN and then imaged with a much smaller force. Motomatsu, et al. (25) used scanning force microscopy to obtain an image of the top layer of a growth spiral of polyethylene oxide that grew from the melt at a temperature of 65°C. A square array of spots with a spacing of 0.74 nm was observed, which is not too far from the value of 0.65 nm expected from crystallographic data for folds on polyethylene oxide. The stems of the polyethylene oxide molecules are perpendicular to the fold surface in lamellar crystals (26), as has been shown by electron microscopy and electron diffraction observations. Polyethylene oxide crystals are flat, with no indication of a hollow pyramidal habit.

Stocker, et al. (27) resolved the folds at the surface of crystallized cycloparaffins, and thereby established that it is within the capability of a typical atomic force microscope to produce recognizable images of fold surfaces.

A group from the Universities of Groningen and Konstanz (28) reported that in their examination of the fold surfaces of polyethylene single crystals, a regular pattern on the molecular scale was never clearly resolved, even though their instrument resolved the mica lattice, and produced images of polytetrafluoroethylene molecules deposited at the surface of highly oriented thin films.

Kajiyama, et al. (29) and Takahara, et al. (30) used frictional force scanning probe microscopy and AFM to measure the frictional force in different directions on the fold surfaces of polyethylene single crystals grown from polyethylene molecules with different molecular weights. The frictional force observed varied both with molecular weight and with the direction in the fold surface in which the probe moved. The anisotropy of the frictional force was higher for the lower molecular weight samples, which were presumed to have more regular folds.

Lateral Surfaces of Oriented Films, Fibers, and Fractured Lamellae. Lin and Meier (*31*) showed an image of a folded chain on surface of a sample of highly oriented polyethylene chains. This fold is not part of a fold surface, and was presumably buried in the crystal or created on the surface when the sample was prepared. Their image shows both the separation between parallel polyethylene chains and some aspects of the periodic features along the chains, proving that tips now available can resolve features on a lateral surface that are much smaller than those expected on a fold surface.

Sumpter, et al. (*32*) used molecular dynamics calculations to study atomistic details of the interaction of an atomic force microscope probe with the lateral surface of a polyethylene fiber. The calculations led to the suggestion that the optimum working range for the tip force is 0.01 to 10 nN, to achieve high sensitivity, and at the same time, minimize deformation of the fiber by the tip. Sumpter's calculations showed that after removing a probe which had been forced into the polymer fiber, the polymer returned, essentially, to its original structure. Sumpter's calculations show that tip interactions with a polymer surface are complex even for tips with idealized shapes, and yet some molecular scale features are captured in the AFM image.

A fracture surface of pressure-crystallized extended-chain polyethylene with its normal perpendicular to the fiber axis was successfully imaged by Annis, et al. (*33*) This surface is the lateral surface of a lamellar crystal grown from the melt. Periodic variations in elevation resembling chains with spacings roughly appropriate to polyethylene were observed, but the orientation of the periodic elevations was perpendicular to the expected chain direction.

Wawkuschewski et. al. (*34*) obtained images of the surface of a stretched polyethylene tape, imaged with the tip and the sample immersed in water. At forces around 1 nN, slender fibrils were seen. Increasing the force to 10 nN revealed transverse striations on each of the fibrils, with 10 to 30 nm spacings between the striations, corresponding to the long period observed with small angle x-ray scattering from oriented polyethylene samples. At higher magnification, parallel polyethylene molecules were observed in the fibrils.

Magonov, et al. (*35*) examined microtomed surfaces of cold-extruded polyethylene. The cut surfaces had their normal direction perpendicular to the extrusion direction. Microfibrils with diameters of 20 to 90 nm were observed. Rows of raised spots were interpreted as individual molecules. Molecules appeared and disappeared as they ran in and out of the surface layer. The 0.25 nm spacing of the polyethylene repeat unit along the chain was observed.

Vancso (*36*) used a variation of atomic force microscopy called lateral force mode microscopy to obtain, molecular scale images of the surface of poly-(tetrafluoroethylene) films obtained by pressing and sliding a block of the polymer on a clean, heated glass surface. Both the lateral force mode and the contact mode produced images in which the 0.578 nm spacing between the poly(tetrafluoroethylene) molecules was resolved, when the scanning direction was across the axes of the molecules. When the scanning direction was along the axis of the oriented molecules, the parallel molecules were resolved in the contact force mode, but a flat, featureless image was obtained in the lateral force mode.

Experimental

The polymer used was Standard Reference Material 1475, Linear Polyethylene, ($M_w \approx$ 52,000, $M_w/M_n \approx 2.90$), obtained from the National Institute of Standards and

Technology, Gaithersburg, MD 20899, USA. The lamellar crystals were prepared by first completely dissolving about 6 mg of polyethylene in 20 ml of xylene (about 0.03% concentration) at 125°C. Large, complicated crystals were grown by cooling this solution to room temperature. The milky suspension was then reheated by placing it in an oil bath held at a temperature of 98 °C. As soon as the mixture became transparent, which occurred as the crystals separated into fragments smaller than the wavelength of light, the re-dissolved solution was cooled to room temperature by removing the test tube from the 98°C bath. The suspension of crystals in xylene was sprayed onto freshly cleaved mica, dried, and annealed. Annealing was performed in air at 111°C for 30 minutes. The AFM images were obtained with a Nanoscope II atomic force microscope manufactured by Digital Instruments, of Santa Barbara, CA.

As-Prepared Lamellar Crystals. Crystals of polyethylene, prepared as described above, were covered with patches 15 to 40 nm in diameter that rose to a summit about one nanometer higher than the average plane of the surface. Patches in regions which appeared to be deformed during collapse of the pyramid were larger and elongated. The peak to valley height of the larger patches was about 1.5 nm. The highest patches on unannealed crystals occur (*21*) (see Figure 1 of this reference) along the fold domain boundaries. Figure 2 of this same reference shows polyimide chains in a polyimide fiber. Also, patches in the vicinity of fold domain boundaries between (200) and (110) types of fold domains are shown (Figure 3) in a paper (*22*) by Reneker, et al.

Figure 2 shows a high magnification image of the fold surface obtained from the crystals prepared as described above. The image shown was produced by filtering the data with a filter that removed spatial frequencies over 2.5 cycles per nanometer. This filter emphasizes information about the height, spacing, separation, and general shape of folds. The filter removed noise that corresponds to variations in height that occur in regions smaller than the spacing between chains in polyethylene crystals.

Similar images, one of which is shown by Magonov and Reneker (*37*), were obtained from the surfaces of several other lamellar crystals in the same preparation. In all the images, the high spots are in rows with directions and spacings that are clearly evident, but not perfectly aligned on a regular lattice that can be easily related to the positions of chains in a polyethylene crystal. The average directions and spacings of the rows in these images are affected by the angle between the normal to the fold surface and the axis perpendicular to the scanned plane of the atomic force microscope, which angle varied because these crystals sometimes had growth spirals which caused some regions of the lamella to collapse with their normal directions significantly different than the normal direction of the mica substrate.

Annealed Lamellar Crystals. Figure 3a is a high magnification image of the fold surface of a crystal annealed at a temperature of 111°C for 30 minutes, which had increased in thickness and developed large scale roughness. Figure 3b shows the calculated Fourier transform of the data, with a superimposed square that shows the low-pass filter that removed frequencies over 2.5 cycles per nanometer. The scale shown along the abscissa also applies to the ordinate. The same filter was used to produce Figure 2. Figure 3c is the image obtained after the filtered diffraction pattern is transformed back to an image. Over the 10 by 10 nm area shown, the image of the surface is similar to the image of the fold surface of the as-prepared crystal shown in Figure 2. In areas of the fold surface 10 nm across, the roughness of the fold surface, observed with the AFM, is essentially the same before and after annealing.

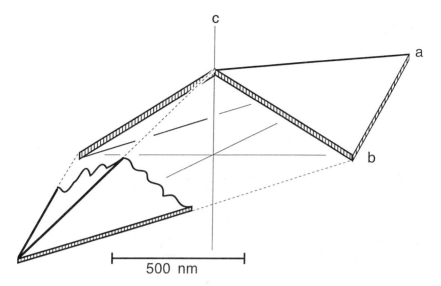

Figure 1 Diagram of a hollow pyramidal crystal of polyethylene.

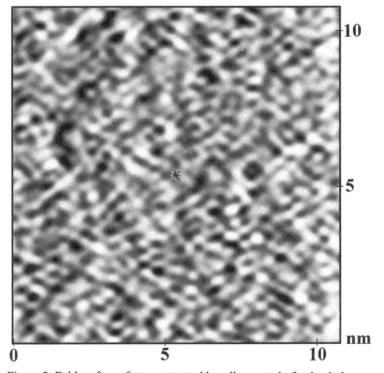

Figure 2 Fold surface of an as-prepared lamellar crystal of polyethylene.

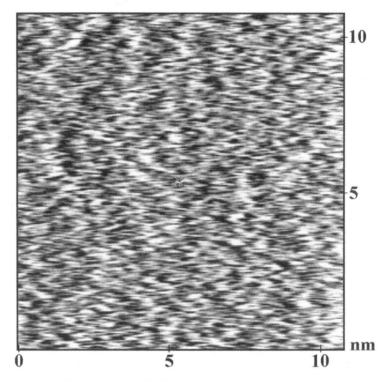

Figure 3a Annealed lamellar crystal: Observed image.

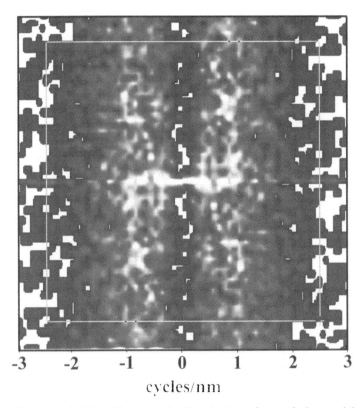

-3   -2   -1   0   1   2   3

cycles/nm

Figure 3b Annealed lamellar crystal: Fourier transform of observed image, showing the boundary of the low-pass filter as a white line.

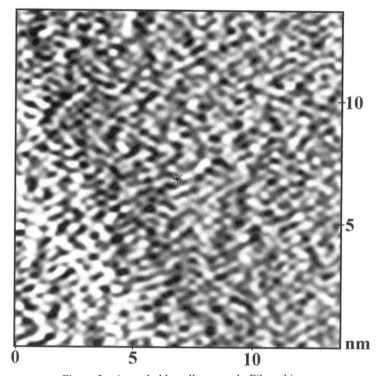

Figure 3c Annealed lamellar crystal: Filtered image.

Figures 4a and 4b show a comparison of the "surface" view of the as prepared crystal in Figure 2 and the annealed crystal in Figure 3. The roughness and general appearance of the two surfaces before and after annealing are essentially the same. A tendency for a regular lattice spacing and for the alignment of the high points into rows is apparent in Figures 2 and 3c.

Discussion

There are many examples of the observation of the ends of paraffin molecules, which are smaller than a fold, folds at the surface of crystalline cyclo-alkanes, and molecules on the sides of oriented fibrils of polyethylene to prove that the AFM should resolve chain folds in polyethylene, if the fold surface is reasonably perfect and flat. The folds on a hollow pyramid should be easier to resolve since the distance between the folds in successive planes is even larger than in the flat crystals. The resolution of the AFM is high enough to resolve the ends of paraffin chain in the n-alkane with 32 carbon atoms (dotriacontane). Side by side chains were resolved in fibers or oriented films of polyethylene, a copolymer of vinylidene fluoride and tetrafluoroethylene, and polyimides.

Yet, what is most often observed can be interpreted as a more or less disordered pattern of the high spots on the fold surface. The filtered image of recently prepared crystals, shows an array of high spots that may indicate the position of folds, although neither the spacings or angles have a simple relation to those expected from a flat, orderly fold surface of a crystal of polyethylene.

The future goal of AFM observations becomes to measure not only the position of each fold in its fold surface, but also to determine how far above or below the position of the median fold surface each fold actually is. This is a different kind of problem than has been successfully dealt with by the image interpretation algorithms now available for atomic force microscopy.

A relatively simple method used by Patil and Reneker (20), for analyzing such data consisted of rotating the expected positions of the folds on a fold surface of a hollow pyramidal crystal into the orientation in which they would be seen by the tip of the AFM. The spacings between rows of folds, and the angles between rows were calculated from this viewpoint, and compared with thespacings and angles observed in the AFM image of a small, flat area on a collapsed hollow pyramidal crystal. This strategy can be successful only when almost every fold is tangent to the fold surface. If some folds stick up a few repeat units, or are at the bottoms of holes that are only a few repeat units deep, the AFM will see these folds projected to a "wrong" position in the image plane.

The straight stems inside the lamellar crystal diagrammed in Figure 1 are parallel to the c-axis of the crystal. The stems are inclined with respect to the normal to the fold surface. It is reasonable to hypothesize that the roughness is created by folds that protrude from the fold surface in the direction of the chain axis. Such folds have a component of their protrusion both in the plane and normal to the plane. Folds that protrude more than one or two repeat units are expected to collapse onto the surface. Additional roughness can come from folds that sit below the level of the fold surface. All these possibilities add to the difficulty of observing the folds with the AFM.

The fact that the stems are inclined with respect to the fold surface means that a fold protruding from (or sitting below) the fold surface in a direction parallel to the stem, has components both along and perpendicular to the axis of the tip. The height,

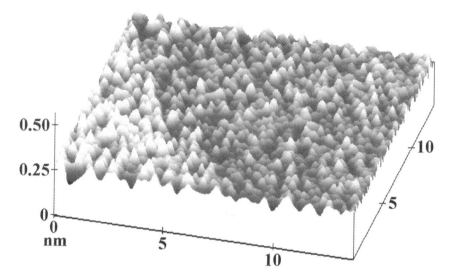

Figure 4a "Surface view" of lamellar crystal: As-prepared crystal.

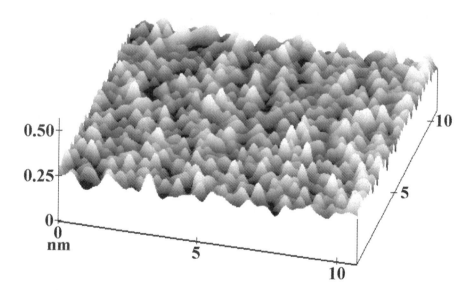

Figure 4b "Surface view" of lamellar crystal: Annealed crystal.

measured with an ideally sharp tip, is less than the amount the fold protrudes in the direction of the stem axis. The projection of the high point of the fold onto the AFM image is displaced from the lattice point at which the stem emerges from the fold surface. The high points observed on the image therefore have a variable relationship to the crystal lattice in the interior of the lamella. The high spots will not necessarily line up along crystallographic directions. The measured spacings will not be the geometrical projection of the lattice spacing into the fold surface.

Figure 5 shows a diagram of the molecular stems in a lamella, inclined with respect to the normal to the fold surfaces. The folds are displaced, from the fold surface, by distances of around 1 nanometer, in the direction of the chain stems. An idealized tip with a radius of 2 nanometers is also shown at the same scale in Figure 5.

Patches. For lamellar crystals produced by self seeding procedures, a typical surface feature, which we call patches, occurs in AFM images. A patch is 15 to 40 nm in diameter, has an irregular rounded shape, and it rises gradually to a summit near its center, which is from 0.25 to 1 nanometer above the valleys between adjacent patches.

The part of the lamella under a typical patch contains a few hundred typical polymer molecules. The amplitude of the patches in the direction perpendicular to the fold surface is less than one-tenth the thickness of the lamellar crystal being observed. It is a common practice to show the AFM images with a much higher magnification in the direction normal to the plane of the image than the magnification in the plane of the image. This practice makes the patches appear as mounds or cones with rounded tips and emphasizes that patches are a recurrent and outstanding feature of AFM images of the fold surfaces of the lamellar crystals of polyethylene.

A patch may be a realistic image of the shape of the surface, but because of the unkown shape of the tip, a patch may instead be the result of a sharp protuberance on the fold surface that is "dilated", by the tip, as described by Villarrubia (38), so that the image is closer to the shape of the tip than to a fold protuding from the sample. In either case, the image of a patch reveals an elevation on the fold surface. The ambiguities arising from the unknown tip shape affect the shape of the slopes leading up to the highest part of the patch. Evidence about patches from electron microscopy is ambiguous because the material used to shadow the crystals often has a patch-like appearance on both the crystal and the substrate. In the discussion elsewhere in this paper, it is assumed that the shape of the patches is a good representation of the shape of a feature of the fold surface.

Smaller patches are also observed, on the slopes of the larger patches or sometimes on flat areas between the larger patches. The AFM images are consistent with the assumption that the height of adjacent folds, as measured with the AFM, varies stochastically in such a way that the patches rise gradually to a summit without localized steps.

Preliminary observations suggest that in annealing experiments the patches grow higher and the valleys grow deeper while the patches maintain about the same diameter. This supports the identification of the patches with the "coherently annealing volume" postulated by Sanchez, et al. (39). Patches undergo changes during annealing which include the development of aligned rows of patches, the incorporation of smaller patches into larger ones, and the growth in thickness of the larger patches, with the simultaneous formation of valleys and holes. Reneker and Chun (3) show the patches on a crystal annealed at 115°C in their Figure 1, and patches arranged in criss-crossing rows in their Figure 2. In their Figure 3, they show the difference between an annealed single lamella lying on mica and a lamella lying on top of another lamella.

Figure 5 Molecular stems in a lamella with rough surface adjacent to an idealized tip at the same scale. Commercially available tips are thought to have a radius of 20 nm, although little is known about the shape within one nanometer of the part of the tip that contacts the sample.

Roughness. We use the term "roughness" in connection with the morphology of the fold surface. The roughening can be described as a consequence of folds protruding above the fold surface, or embedded in the crystal, just inside the fold surface as diagrammed in Figure 5.

The observed growth of lamellar crystals of polyethylene as hollow pyramids is used to support the hypothesis that the fold surfaces are orderly and smooth. The orderly arrangement of bulky folds is suggested to be the primary cause that leads to the growth of hollow pyramids. The growth of hollow pyramidal crystals occurs according to rules that guide the placement of adjacent folds. Regular "stepping down" of adjacent fold planes, as shown on page 133 of Geil (*10*), or of adjacent folds in the same fold plane, as shown on page 275 of Frank(*9*), may occur. In both of these cases, the folds are all tangent to the fold surface as the crystal grows.

A different rule, that permits some disorder but requires enough orderly arrangement of folds to create hollow pyramids is required. Such a rule must average over disordered regions and maintain the average spacing between folds required for the construction of a hollow pyramid.

An alternative possibility is that the hollow pyramids grow with orderly and smooth fold surfaces, but that during cooling to room temperature, collapse of the pyramid onto a flat mica substrate, and the other steps in preparing the crystal for observation, dispirations and other crystallographic defects move in such a way that the smooth fold surface is roughened. Since changes of just this nature are responsible for the increase in lamellar thickness that occurs during annealing, observations of changes in the folds with temperature with an atomic force microscope equipped with a hot stage could shed much light on these processes.

The observed epitaxial growth of paraffin crystals on the fold surfaces also requires that the folds have an orderly arrangement. It is easy to see how a linear molecule could span a localized disordered region, so epitaxy does not imply that the folds have perfect order on the fold surface.

The theoretical models which were developed to describe the growth of lamellar crystals of polymers offer reasons that the fold surfaces may be rough on a molecular scale. These models support the possibility that the fold surfaces are rough even as the crystals grow. Frank and Tosi (*40*) considered a distribution of stem lengths within each lamellar crystal, which would result in rough fold surfaces, since some stems would end at different levels than their neighbors. In their formulation, the average stem length converged to a value that depended on the amount of supercooling and did not depend upon the stem length in the underlying fold plane upon which the molecule was growing. Lauritzen and Passaglia (*41*) allowed the lengths of stems to vary but constrained the average stem length to be the same as that of the chain folded lamella upon which the growth occurred. The standard deviation for the stem length distribution that they calculated was around one nanometer, close to the observed height of the patches and somewhat more than the roughness of a fold surface in a typical 10 nanometer square.

AFM Tips. The tip shape for the AFM has attracted widespread interest. Tips with known shapes in the region within one nanometer of the apex are not readily available. Algorithms that correct images for the non-ideal shape of real tips do not work well without knowledge of the shape of the tip near its apex, where multiple protrusions in different directions are likely to exist, particularly at the atomic scale. Both elastic and plastic indentation of the sample by the tip can easily occur (*32*). Real tips for the

AFM cannot determine the complete shape of a molecular segment because even an atomically sharp tip is too large to explore the interior parts of a molecule.

Tips that can resolve intermolecular distances on the lateral surfaces of highly oriented polyethylene produce images showing more disorder in the arrangements of folds on the surface, although the folds must have structural features that are somewhat larger than the distance between adjacent polyethylene molecules. Figure 5 shows how a tip with a diameter of two nanometers interacts with a rough fold surface. Commercial tips are said, on the basis of transmission electron micrographs, to have a tip radius of 20 nm, but their performance in resolving smaller features on smoother surfaces indicates that there are smaller protrusions that interact with the surface. It is clear that if the tip diameter in Figure 5 were increased by a factor of ten, a one nanometer protrusion on the fold surface would shield a relatively large surrounding area from observation.

Efforts are being made to use a fullerene tube as a tip (42). The sealed tip of a fullerene tube is nearly spherical with a diameter of about 1.38 nm. Each of the carbon atoms at the end forms an apex which protrudes slightly. On a very smooth surface, a favorably oriented tube could resolve structures that protrude no more than about 0.1 nm. Higher protrusions on the sample would encounter other parts of the tip and produce a complicated image. Yet, such a tip would be a great improvement over contemporary tips, because corrections based on the known shape could be made.

The indentation of a tip with a spherical radius of 5 to 30 nm can be calculated by methods of continuum mechanics. The diameter of the indented area of the sample, for forces that are often encountered, can be many nanometers. The fact the features with atomic size dimensions are seen, can be rationalized in several different ways: Very small forces reduce the size of the indented area; Periodic features on the tip can interact with periodic features on the sample to produce periodic features in the images; A small protrusion on the tip may detect a reduced but measurable effect from equally small protrusions in a nearly flat sample, even though both the tip and the sample are elastically deformed. The very real possibility that the tip may have several points that touch different parts of the sample at different times is too complicated to say much about, although such a tip could produce images with atomic scale features.

The sizes of features on a surface, such as steps, protrusions, tips, holes, and crevices, may be used to place an upper bound on the size of an AFM tip. The tip shape can be deduced from the image of a sample surface that contains sharp features of this sort. Villarrubia (40) showed that the deduced tip shape can be a good estimate of the actual tip shape, and the deduced shape can be used to improve the accuracy of measurement on the parts of the surface which are accessible to the tip.

The tips used in this work do a good job of locating the highest points on a reasonably flat surface. But as soon as the tip moves down from the highest point, the finite radius of the tip begins to dilate the image of the highest point. This may not be disastrous, particulary at lower magnification, but if another part of the tip touches another molecular segment as the tip slides down the slope of the first point, the possibilities for interpreting the tip height in terms of surface morphology rapidly become very complex and ambiguous. This is the problem encountered in this attempt to observe the folds on polymer crystals.

A crystallographic array of atomic scale tips could be forced into contact with the surface of a crystal and still produce a periodic image as the array moved over the periodic lattice of the crystal. The periodic interaction is likely to include relatively large motions of some atoms away from the lattice points of the crystal.

Summary

Lamellar crystals of polymers, which were studied intensively by many researchers since 1957, are idealized as having planar zig-zag stems on a three dimensional translationally symmetric crystal lattice. Chain ends, twists, crystallographic defects, growth spirals, hollow pyramidal habits, and other morphological features are also needed to explain the physical properties of polyethylene crystals. The AFM, with its high sensitivity to distances perpendicular to the surface of a lamellar crystal, provides a new way to look at the surfaces of these crystals. It has the capability of imaging, at the atomic scale, details of these features of the crystals that were never seen so clearly with other microscopic techniques.

An image of very well ordered folds obtained was reported (20). Images of fold surfaces obtained subsequently on crystals prepared at a different time, indicate that the fold surfaces are more complicated. The crystals used in the earliest experiments appear to have fold surfaces that are significantly smoother than the fold surfaces of crystals prepared later.

Correlation of AFM measurements with sample preparation conditions may lead to the production of crystals with smooth fold surfaces. The resulting information would be helpful for improving theories of the growth of lamellar crystals where roughness of the fold surface is used as a parameter, and for observing the interactions between crystal defects and the fold surface.

Both new and familiar morphological features of the crystals were seen in the images from the AFM. In most of the materials examined, flat fold surfaces that contained regular arrays of folds all at the same level were not found. We conclude that the fold surfaces have a limited roughness that does not interfere with the growth of hollow pyramidal crystals or with the epitaxy of crystals of evaporated polyethylene oligomers.

If the fold surface is rough, chain tilt complicates the interpretation of the data obtained with an AFM probe with its axis perpendicular to the fold surface, even if the tip has an atomic radius of curvature, because the distance a fold protrudes above the fold surface, in the direction of the chain axis, has components both in the fold surface and perpendicular to it. Also, the highest point the tip sees on a fold when the tip axis is parallel to the chain axis is different from that seen when the tip axis is perpendicular to the fold surface.

The roughness of the fold surface occurs on a lateral scale of a few nanometers, and on a similar scale in the direction of the axis of the stem. Little is known about the details of the tip geometry on this scale, but it is reasonable to suppose that the tip is blunt and complicated, and therefore cannot follow all relevant details. Perhaps better tips will come in the near future, along with algorithms for processing the images to remove some of the effects of the finite size and the complicated shape of each tip.

AFM observations suggest the presence of vacancy-like holes trapped between the rough fold surfaces of adjacent lamellae. X-ray long period and density measurements could be fitted to a vacancy model rather than to the conventional amorphous layer model of the interlamellar region. If the absent mass represented by the vacancy is interpreted, not as a vacancy, but as an indication of the presence of amorphous polymer with a density about 4/5 the density of crystalline material, then the volume of amorphous material predicted from density data will be about 4 times larger than the volume of the vacancies. In other words, the usual crystal and amorphous phase interpretation of the density measurements predicts a much thicker disordered region between lamellae than actually occurs if vacancies are the primary cause of the reduced density.

Conclusions

The well known arguments that favor regular folding are carefully considered. These arguments conclude that the folds are orderly, at least to a degree high enough to create surfaces that cause nucleation of epitaxially oriented crystals of evaporated polyethylene oligomers, and to guide the building of crystals shaped like hollow pyramids.

Polymers that crystallize with their stems perpendicular to the lamella may yield new information about the behavior of folds more easily and sooner than polyethylene. Motomatsu's work on polyethylene oxide (24-25) is an example of the observation of a regular array of folds.

The atomic force microscope works with high resolution on atomically flat surfaces, but, because of the unknown shape of the tips, many ambiguities appear in the data when surfaces have deviations up and down of a few tenths of a nanometer. Features corresponding to single polymer molecules on the lateral surfaces of polymer crystals, oriented films, and fibers are easier to recognize in the AFM images than folds, even though the folds are bigger and the folds are more widely separated than the molecular features on the lateral surfaces. This indicates that the lateral surfaces are smoother.

The prospects are good for observing interesting and useful nanostructures made by attaching molecular structures with a polyethylene tail to the fold surface by incorporating the tail into the lamellar crystal. Lamellar polymer crystals may become an interesting structural element for designers of nanoscale machinery (2).

AFM and other scanning probe microscopy methods promise the means to observe crystallographic defects in polyethylene as the defects move into and out of the folds. Changes in lamellar thickness occur at in a few minutes at temperatures over 100°C for polyethylene. Motion of folds due to crystal defects is expected to occur much more slowly at room temperature. It is reasonable to hope, that in the near future, changes in the fold positions can be measured in real time as the crystal is thickening on a hot stage.

Better tips, with sharper apices and with the shape known in atomic detail, would bring large benefits to examination of polymeric structures, and in many other uses of the AFM.

Acknowledgements

We are grateful for an unrestricted grant from the Dupont company which made the recent experiments reported here possible.

References

1. Young, D. A.; Ed.; *Faraday discussions of the Chemical Society;* Number 68; Faraday Division, Royal Society of Chemistry: London, England, 1979.
2. Reneker, D. H.; Chun, I. *Nanotechnology* **1996**, *7*, pp. 216-223.
3. Reneker, D. H.; Chun, I. *Polym. Prepr.* **1996**, *37*, pp. 546-547.
4. Roe, R. J.; Gieniewski, C.; Vadimsky, R. G. *J. Polym. Sci., Polym. Phys. Ed.* **1973**, *11*, pp. 1653-1670.
5. Statton, W. O.; Geil, P. H. *J. Appl. Polym. Sci.* **1960**, *3*, pp. 357-361.

51

6.   Reneker, D. H. *J. Polym. Sci.* **1962**, *S 59*, pp. 539-541.
7.   Reneker, D. H.; Mazur, J. *Polymer* **1988**, *29*, pp. 3-13.
8.   Reneker, D. H.; Mazur, J. In *Computer Simulation of Polymers*; Roe, J., Ed.; Prentice Hall: **1990**, pp. 332-340.
9.   Keller, A. W. In *Sir Charles Frank, OBE, FRS*, Chambers, R. G.; Enderby, J. E.; Keller; A. Lang R.; Steeds, J. W., Ed.; Adam Hilger: Bristol, Philadelphia, New York, **1991**, pp. 265-321.
10.   Geil, P. H. *Polymer Single Crystals*; John Wiley & Sons, Inc.: New York, **1963**.
11.   Bassett, D. C. *Principles of Polymer Morphology*; Cambridge University Press: Cambridge, **1981**.
12.   Wunderlich, B. *Macromolecular Physics*; Academic Press: New York, **1973**; Vol. 1.
13.   Mandelkern, L. *Crystallization of Polymers*; McGraw-Hill: New York, **1964**
14.   Mandelkern, L. *Polym. J. (Tokyo)* **1985**, *17*, pp. 337-350.
15.   Khoury, F.; Passaglia, E. In *Treatise on Solid State Chemistry*; Hannay, N. B., Ed.; Plenum Press: New York, **1976**, Vol. 3; pp. 335-496.
16.   Hoffman, J. D.; Davis, G. T.; Lauritzen, J. I., Jr. In *Treatise on Solid State Chemistry*; Hannay, N. B., Ed.; Plenum Press: New York, **1976**, Vol. 3; pp. 497-614.
17.   Wittman, J. C.; Lotz, B. *J. Polym. Sci., Polym. Phy. Ed.* **1985**, *23*, pp. 205-226.
18.   Reneker, D. H.; Geil, P. H. *J. Appl. Phys.* **1960**, *31*, pp. 1916-1925.
19.   Keller, A. W. *Rep. Prog. Phys.* **1968**, *32*, pp. 623-704.
20.   Patil, R.; Reneker, D. H. *Polymer* **1994**, *35*, pp. 1909-1914.
21.   Reneker, D. H.; Patil. R.; Kim. S. J.; Tsukruk, V. In *Proceedings of the 51st Annnual Meeting of the Microscopy Society of America*; Bailey, G. W.; Rieder, C. L., Ed.; San Francisco Press: **1993**, pp. 874-875.
22.   Reneker, D. H.; Patil, R.; Kim, S. J.; Tsukruk, V. In *Crystallization of Polymers*; Dosière, M., Ed.; NATO ASI series C, Mathematical and physical sciences; Kluwer Academic Publishers: Dordrecht, Boston, London, 1993, Vol. 405; pp. 357-373 and 398-400.
23.   Patil, R.; Kim, S. J.; Smith, E.; Reneker, D. H.; Weisenhorn, A. L. *Polymer* **1990**, *31*, pp. 455-458.
24.   Nie, H.-Y.; Motomatsu, M.; Mizutani, W.; Tokumoto, H. *J. Vac. Sci. Technol. B* **1995**, *13*, pp. 1163-1166.
25.   Motomatsu, M.; Nie, H.-Y.; Mizutani, W.; Tokumoto, H. *Polymer* **1996**, *37*, pp. 183-185.
26.   Chen, J.; Cheng, S. Z. D.; Wu, S.; Lotz, B.; Wittman, J.-C. *J. Polym. Sci. B: Polym. Phys.* **1995**, *33*, pp. 1851-1855.
27.   Stocker, W.; Bar, G.; Kunz, M.; Magonov, S. N.; Cantow, H.-J. *Polym. Bull.* **1991**, *26*, pp. 689-695.
28.   Grim, P. C. M.; Brouwer, H. J.; Seyger, R. M.; Oostergetel, G. T.; Bergsman-Schutter, W. G.; Arnberg, A. C.; Güthner, P.; Dransfeld, K.; Hadziioannou, G. *Makromol. Chem. Macromol. Symp.* **1992**, *62*, pp. 141-155.
29.   Kajiyama, T.; Ohki, I.; Takahara, A. *Macromolecules* **1995**, *28*, pp. 4768-4770.
30.   Takahara, A.; Ohki, I.; Kajiyama, T., *Polym. Prepr.* **1996**, *37*, pp. 577-578.
31.   Lin, F.; Meier, D. J. *Langmuir* **1994**, *10*, pp. 1660-1662.
32.   Sumpter, B. G.; Getino, C.; Noid, D. W. *J. Chem. Phys.* **1992**, *96*, pp 7072-7085.

33. Annis, B. K.; Noid, D. W.; Sumpter, B. G.; Reffner, J. R.; Wunderlich, B. *Makromol. Chem. Rapid Commun.* **1992**, *13*, pp. 169-172.
34. Wawkuschewski, A.; Cantow, H.-J.; Magonov, S. N.; Sheiko, S.; Möller, M. *Polym. Bull.* **1993**, *31*, pp. 693-698.
35. Magonov, S. N.; Quarnstrom, K.; Elings, V.; Cantow, H. J. *Polym Bull.* **1991**, *25*, pp. 689-695.
36. Vancso, G. J.; Förster, S.; Leist, H. *Macromolecules* **1996**, *29*, pp. 2158-2162.
37. Magonov, S. N.; Reneker, D. H. In *Annu. Rev. Mater. Sci.*; Annual Reviews: Palo Alto, **1997**, Vol. 27; pp. 175-222.
38. Villarrubia, J. S. *Surf. Sci.* **1994**, *321*, pp. 287-301.
39. Sanchez, I. C.; Peterlin, A.; Eby, E. K.; McCrackin, F. L. *J. Appl. Phys.* **1974**, *45*, pp. 4216-4219.
40. Frank, F. C.; Tosi, M. *Proc. R. Soc. London Ser. A*: **1961**, *263*, pp. 323-329.
41. Lauritzen, J. I., Jr.; Passaglia, E. *J. Res. Nat. Bur. Stand. Sect. A*: **1967**, *71A*, pp. 261-275.
42. Thess, A.; Lee, R.; Nikolaev, P.; Dai, H.; Petit, P.; Robert, J.; Xu, C.; Lee, Y. H.; Kim, S. G.; Rinzler, A. G.; Colbert, D. T.; Scuseria, G. E.; Tomanek, D.; Fischer, J. E.; Smalley, R. E. *Science* **1996**, *273*, pp. 483-487.

Chapter 3

Atomic Force Investigations of Epitaxially Crystallized Tactic Poly(propylenes)

Wolfgang Stocker[1], Andrew J. Lovinger[2], Martina Schumacher[3], Sabine Graff[3], Jean-Claude Wittmann[3], and Bernard Lotz[3]

[1]Humbolt Universität zu Berlin, Institut für Physik, Invalidenstrasse 110, D-10115 Berlin, Germany
[2]Lucent Technologies, Bell Laboratories, Murray Hill NJ 07974
[3]Institut Charles Sadron, CNRS—Université Louis Pasteur, 6, rue Boussingault, 67083 Strasbourg, France

Atomic Force Microscopy (AFM) examination of epitaxially crystallized isotactic and syndiotactic polypropylenes (iPP and sPP) has been performed with methyl group resolution. The control of the crystalline morphology and its analysis with electron diffraction and AFM techniques makes it possible to gain unmatched insights into various aspects of the structure of the contact face. As reviewed in this paper, the combination of sample preparation and investigation techniques has made it possible (a) to reveal the exact nature of the contact face in αiPP, (b) to visualize the left and right hands of helices in sPP and (c) to have a first direct space visualization of the frustrated nature of βiPP.

Atomic Force Microscopy of polymers with submolecular resolution is, in principle, possible since methyl group resolution can be achieved. At present however, this level of resolution is only confirmed when *recognizable patterns* of submolecular features can be imaged. This condition excludes of course a submolecular analysis of amorphous polymers, in particular when long, flexible side-chains are involved. It also explains the difficulty of imaging the "amorphous" fold surface of a linear polymer such as polyethylene (1), even though the folding of chains along well known crystallographic growth planes introduces some degree of orientational order, as revealed for example by polymer decoration techniques (2).

Crystalline polymers provide a priori more appropriate samples, since AFM surface patterns can be compared with the crystal structure determined by diffraction techniques. However, a major difficulty is linked to the fact that most orientation procedures yield a fiber orientation, which leaves an ambiguity about the nature of the exposed crystallographic plane, and therefore about the anticipated surface pattern.

In this paper, we review three projects in which we have succeeded in overcoming the above mentioned difficulties and ambiguities by an appropriate choice of polymers and sample preparation conditions. The polymers are isotactic and syndiotactic polypropylenes (iPP and sPP). Since the main chains bear only methyl groups, there is no side-chain conformational freedom: the methyl groups position is fully determined by the main chain conformation, and is therefore unlikely to be modified in the exposed face (or, for that matter, by effect of the AFM tip). Furthermore, both polymers exhibit conformational (for sPP) and structural (for both iPP and sPP) polymorphism: three crystalline polymorphs are known for both systems, which involve different helix geometries (for sPP) (3), or rest on a single three-fold helix geometry (for iPP) (4).

The sample preparation makes extensive use of epitaxial crystallization on low molecular weight organic substrates (5). Selection of appropriate substrates makes it possible to induce various polymorphs of crystalline polymers and, in all cases, to expose well defined crystallographic contact planes. These planes are of the type ($hk0$), i.e. the polymer chains or helices lie flat in the contact plane: AFM explores the lateral organization of the polymer helices, which provides (among others) a possibility to probe the helical hand. Finally, the structure of the film (lamellar, crystal and nature of the contact face) is determined by electron microscopy and electron diffraction. These data make it possible to predict and compare the AFM images against anticipated methyl group patterns. Conversely, AFM can provide insights into the structure of the contact plane that are not accessible to electron diffraction analysis.

The complementary contributions of electron diffraction and AFM analysis are illustrated here for three studies, presented in their chronological order: (a) the analysis of the exact nature of the contact face of iPP in its α and γ phases (b) the observation of right and left helical hands of syndiotactic polypropylene and (c) the AFM recognition of the frustrated nature of the β phase of iPP.

Experimental procedures.

The sample preparation procedures are an essential ingredient in the present investigations. First, a thin film of polymer is deposited on a glass slide by evaporation of a dilute solution in a suitable solvent. The film thickness is limited to the 10-30 nm range, since it will also be used for transmission electron microscopy. Single crystals of a suitable low molecular weight (Mw) organic material (prepared in a different experiment) are then deposited on the polymer film with their mother liquor, and the latter is left to evaporate. The whole sample is then heated in order to melt the polymer but not the low Mw crystal substrate, and cooled quickly in the lower part of the crystallization range of the polymer to induce epitaxial crystallization. The organic substrate is then dissolved, leaving the contact face exposed. The resulting thin film is suitable for AFM examination without any further processing. This examination is actually facilitated by the fact that the polymer film is nearly molecularly flat (as a result of growth in contact with the low Mw crystal) and that appropriate areas are recognizable by the imprint left by the latter crystal in the polymer film (cf. Figure 6a): suitable areas are selected with the binocular attachment of the AFM. The same film can be used for parallel electron microscopy (EM) and electron diffraction (ED) investigations. For this purpose, it is shadowed (if desired), coated with a carbon film, floated off on water and mounted on EM grids.

In our various investigations, we have used samples in the $\approx 10^5$ Mw range (these characteristics are not essential in the present study) provided by Elf-Atochem (iPP), Exxon Chemical International (iPP and sPP) and Montell (sPP). Examination techniques included standard Philips CM 12 electron microscope and Nanoscope III AFM instrument. The latter is equiped with a liquid cell attachment: resolution needed to achieve methyl group imaging is best obtained when the samples are examined in a liquid medium (cf. ref. 6 a and b, and ref. 10).

The contact faces of iPP in its α and γ phases.

Investigation of the contact faces of αiPP (Figure 1) provided the most illustrative reasons for the need to combine AFM and electron microscopy studies (6). Isotactic polypropylene crystallizes epitaxially on benzoic and nicotinic acid and forms a kind of "cross-hatched" lamellar morphology reminiscent of the so-called "quadrites" grown from dilute solution (7) (Figure 2a). An analysis by electron diffraction, combined with the known crystal structure of the α phase, makes it possible to describe in *almost* every detail the film structure. The iPP contact face is the lateral (010) or *ac* face. Taking advantage of an earlier analysis of the homoepitaxy of αiPP (8), it is further possible to determine the azimuthal setting of right- and left-handed helices in the two sets of lamellae giving rise to the cross-hatched morphology. Indeed, it is known that successive *ac* crystallographic planes are made of right-handed and of left-handed helices only. Furthermore, the azimuthal setting of these helices on their axis differs, which means that alternate *ac* planes differ both by the hand of their constituting helices (i.e. right or left) *and by the density of methyl groups exposed in the contact face*. Diffraction techniques *cannot, and do not* give access to the exact nature of the *ac* contact face, i.e. they leave an ambiguity in the structural analysis of the epitaxy. AFM *can* solve this ambiguity by probing the density of methyl groups in the contact face. In short, the argument sums up to the following dilemma: is the pattern of methyl groups in the lozenge (nearly square) shaped *ac* contact face similar to the "five" face of dice (Figure 1b), or to the "four" face (Figure 1c)?

The AFM answer is unambiguous: the *ac* contact face is the "four" face, i.e. with the lowest density of methyl groups (Figure 2b). The images display indeed a 0.65 nm-side lozenge-shaped pattern of methyl groups (diameter: 0.4 nm) which matches perfectly one of the two patterns of *ac* faces. AFM therefore helps to *discriminate* between two possible contact faces (6).

When taking into account the crystal structure of αiPP and the lamellar morphology, analysis of the fairly "simple" pattern of Figure 2b can be taken well beyond this discriminative step: it is indeed possible to determine the actual hand of the exposed helices through a two-step reasoning paralleling that used to analyze the quadrite structure (cf. above): (a) Knowing (from smaller scale images) that the lamellar orientation is at 11 o'clock in the imaged region, the helical stems (normal to the lamellar surface) are oriented at $\approx$ 2 o'clock: rows of methyl groups oriented in that direction belong therefore to the same helix. (b) The actual helical hand is deduced from the relative heights of the various helices, i.e. from the lozenge geometry of the methyl group pattern. When only one

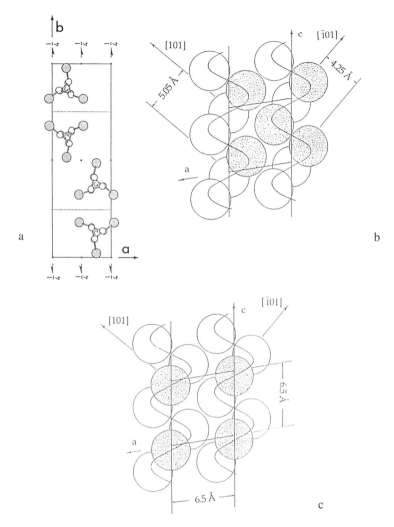

Figure 1. (a) Crystal structure of the α phase of isotactic polypropylene. Note the existence of layers of isochiral helices parallel to the *ac* plane (successive layers being antichiral), and the different azimuthal settings of chains in successive planes. The methyl groups which may be imaged by AFM are shaded. (b, c) Pattern of exposed methyl groups (shaded) illustrating their different densities in successive (010) planes. The patterns resemble the "five" and "four" faces of dice and correspond to exposed right-handed (b) and left-handed helices (c), respectively. The present faces are observed when seen along the -*b* axis orientation. If seen along the +*b* orientation, the patterns would be symmetrical (relative to a vertical axis) and the hands of helices reversed. Reproduced from reference 6.

a

Figure 2. (a) "Cross-hatched" morphology of lamellar crystals of αiPP standing on edge, obtained by epitaxial crystallization on benzoic acid. Electron micrograph, Pt/C shadowing. Scale bar: 1 μm (b) Fourier-filtered AFM image of the (010) contact face of αiPP illustrating the methyl groups (dark) on a lozenge-shaped pattern of 0.65 nm side. Imaging conditions: in air. Size of imaged area: $2*2$ nm^2. Reproduced from reference 6.

Continued on next page.

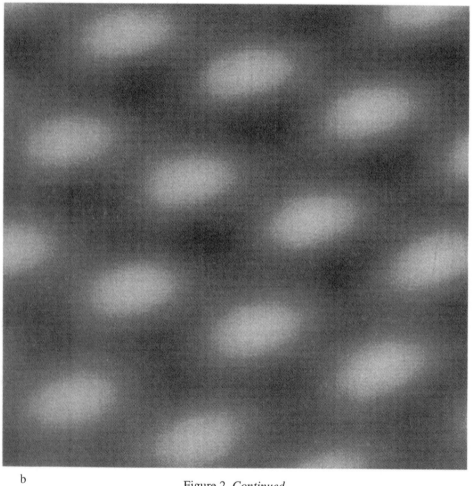

b

Figure 2. *Continued.*

methyl group of the helix is exposed in the *ac* face, the underlying helical path (buried as far as probing by AFM is concerned) is parallel to the *long* axis of the lozenge. Combining this information with the known helix axis orientation, it follows that the helices imaged in Figure 2b are *right-handed*.

Note further that *identical* AFM images are obtained when the probed contact face is located in the differently oriented set of lamellae. However, the lamellar orientation is then at 1 o'clock, the helix axis at 10 o'clock, and the helices *left-handed*. This analysis indicates that the two lamellar orientations generated by the epitaxial growth (Figure 2a) rest on a *single* polymer-substrate *epitaxial relationship*, but involve contact layers made of *helices of opposite hand*. In other words, the two lamellar orientations reflect the symmetric tilts (relative to the helix axis) of the helical paths in antichiral helices.

A similar epitaxial relationship has been established for the γ phase of iPP which is built up with layers structurally identical to those of the α phase: the same pattern of methyl groups is indeed observed. However, two non-parallel chain orientations exist in the γ phase unit-cell (9). With reference to Figure 2b, they are parallel to the two rows of methyl groups at 2 o'clock and 10 o'clock, respectively. The lamellar surface, symmetrically oriented with respect to the two chain orientations, is at 12 o'clock. This different lamellar orientation thus provides an "AFM marker" which helps identify the γ phase, but its symmetric orientation with respect to the rows of methyl groups does not allow identification of the underlying helix axis orientation, and therefore precludes determination of the actual hand of exposed helices which was performed for αiPP (6).

Helical hand in syndiotactic polypropylene.

The issue of identifying helical hand has been further developed in a subsequent work with syndiotactic polypropylene (10) which is based on the same combination of electron diffraction analysis and AFM probing of epitaxially crystallized films. Whereas in the iPP films, helical hands could be *deduced* on the basis of a "crystallographic" reasoning, epitaxially crystallized sPP provides an opportunity to *visualize* in real space the helical hand of constituent helices.

An earlier extensive study of sPP single crystals (11) had demonstrated that, contrary to initial belief, the stable structure of sPP is based on a fully antichiral packing mode. The unit-cell is orthorhombic with parameters $a = 1.44$ nm, $b = 1.12$ nm, $c = 0.74$ nm, and antichiral helices alternate in both the *ac* and *bc* planes. The latter plane is quite appropriate to reveal helical hand for both conformational and geometric reasons. Indeed, due to the *ttgg* conformation of the chain, successive methyl groups project in opposite directions, and are parallel and underline the (mostly buried) main chain helical path. The helix may be visualized as a rectangular staircase with, in the *bc* plane, a flight of three "steps" tilted at 45° to the helix axis. These steps are a CH_3, a (main chain) CH_2 and a CH_3 which build up a pseudo n-pentane unit (Figure 3). Also, the *bc* face is structurally ideal for AFM investigations since it is flat (same *z* coordinates for both CH_3s and CH_2 groups) and should display a characteristic pattern of interlocking flights of pseudo-*n*-pentane units tilted along the *b* axis alternatively at + and at - 45° to the *c* (helix) axis (Figure 3).

Suitable samples of sPP with the exposed *bc* plane were produced by using crystals of *p*-terphenyl as the substrate, and the crystal structure and nature of the contact plane confirmed by electron diffraction. AFM imaging in a liquid medium does indeed reveal (even in unfiltered images, cf. Figure 4a) the expected pattern generated by the tilted *n*-pentane units. Relative orientation of the units and scan direction is more appropriate to reveal the left-handed helices in Figure 4a, but right-handed helices are also clearly visible in Fourier filtered images (Figure 4b).

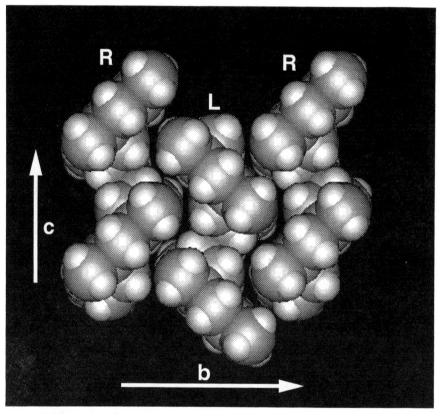

Figure 3. Syndiotactic polypropylene helix conformation (of type *ttgg*) and interlocked packing of right-handed (R) and left-handed (L) helices in the (100) or *bc* plane of the stable crystal modification. Note the prominence of the helical path (tilted at $\approx 45°$ to the helix axis (vertical)), which results from alignment of two successive methyl side-chains and the middle backbone CH_2.

a

b

Figure 4. Unfiltered (a) and Fourier-filtered (b) images of the (100) contact plane of syndiotactic polypropylene. Helix axes are tilted at ≈ one o'clock. Note the better resolution of left handed helices, as a result of more appropriate orientation relative to scanning direction. Right-handed helices can be distinguished in (b) (n-pentane units are light). Imaging conditions: in isopropanol. Size of area: $15*15$ nm^2. Reproduced from reference 10.

Note that the filtering uses 20 pairs of reflections, up to a resolution limit of 0.19 nm^{-1}.

To our knowledge, this study therefore provides the first AFM images in which a helical hand can be unequivocally assigned to *every* stem in the exposed crystal face. It would deserve being pursued a little further. Indeed, electron diffraction evidence suggests that the alternation of helix chirality may not be perfect: such samples would therefore be suitable to test the ability of AFM to detect local (but of course more elusive) "point" defects based on mistakes in the stem chirality alternation.

Contact faces of the β phase of iPP: an AFM illustration of frustration in polymer structures.

The last investigation deals again with isotactic polypropylene, but in its metastable β phase (12). βiPP was discovered in 1959 (13), but its structure solved only in 1994 (14, 15). It is highly original in polymer crystallography in that it is frustrated (15). As seen in Figure 5, it is based on the "standard" three fold helix of iPP, but three isochiral helices pack in a trigonal cell ($a = b = 1.105$ nm, $c = 0.65$ nm, P3$_1$ or P3$_2$ symmetry, depending on the helix chirality). The original feature lies in the different azimuthal settings of the helices: one of them differs significantly from the two other ones. If this azimuthal setting is defined by the orientation of one of the methyl side chains, the structure shown in Figure 5 may be described as North-South-South (NSS), but frustrated structures with NWW (or the equivalent NEE) orientations have been determined (16).

Nucleating agents for βiPP are known and have been patented. A very efficient one is dicyclohexylterephthalate (DCHT) (17) which can be obtained in the form of large single crystals highly suitable for epitaxial crystallization (Figure 6a). The AFM examination of the βiPP contact face with DCHT had several objectives: (a) to establish the frustrated nature of the crystal structure (b) to determine whether the NSS or NWW packing schemes exist at the interface (c) in case of NSS scheme, to discriminate between two structurally different (110) planes which are characterized by different methyl group densities, i.e. chains exposing 1, 2 and 2 methyl groups or 1, 1 and 2 methyl groups, respectively (Figure 5). This problem parallels of course that of the αiPP contact faces structure.

Electron diffraction patterns (Figure 6b) confirm that the contact face is the densely packed (110) plane. Since this plane contains three differently oriented helices, AFM images are expected to, and do indeed reveal more prominent rows of single methyl groups which indicate underlying helices with these methyl groups "sticking out" of the contact plane (Figure 7). These rows are 1.9 nm apart: AFM images therefore reveal the three-chain periodicity, i.e. display a real space "signature" which is characteristic of frustration.

Further analysis of the images is possible, but remains at present more tentative. In particular, the AFM pattern of βiPP is less regular than that of αiPP. This suggests that NSS and NWW patterns may coexist in the contact face, reflecting either the crystal structure as formed initially, or as transformed as a result of surface reconstruction upon dissolution of the DCHT substrate. In any case, the frustrated character of the structure is apparent in the contact face in spite of the fact that two out of the usual six neighbor helices of every exposed stem are missing. This result is in line with the two above studies, which do not indicate any significant disturbance of the contact faces due to AFM tip probing.

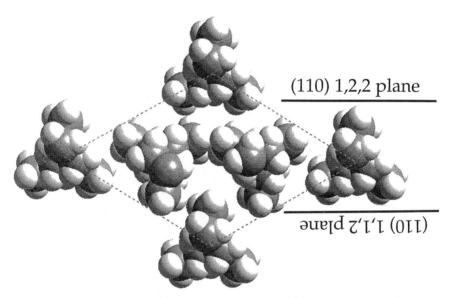

Figure 5. Frustrated crystal structure of the β phase of isotactic polypropylene. The trigonal unit-cell contains three isochiral helices, which differ by their azimuthal settings. The helix settings represented here are North-South-South (cf. text). In the (110) plane, frustration creates a three-helix periodicity of 1.9 nm, which further has a different profile (and therefore AFM "signature") if opposite sides of the (110) plane are contact faces.

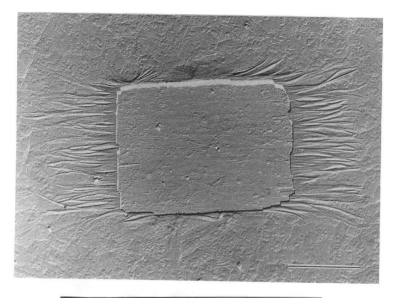

a

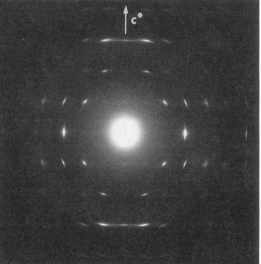

b

Figure 6. (a) Morphology of a thin film of βiPP epitaxially crystallized on a single crystal of DCHT, after removal of the latter organic substrate. The imprint left by the crystal of DCHT is used to locate areas suitable for AFM examination. The present film has been shadowed with Pt/C, and is observed in transmission electron microscopy. Scale bar: 3 μm. (b) Electron diffraction pattern of a film as in (a), in proper relative orientation (chain axis vertical). The diffraction pattern reveals that the contact plane is (110) (zone axis: [120]) and confirms the frustrated nature of the structure (note the characteristic weaker intensity of the meridional 003 reflection compared to 103).

Figure 7. Unfiltered (a) and Fourier-filtered (b) image of the (110) contact face of βiPP on DCHT. The helix axis is tilted at 2 o'clock. Note rows of methyl groups (light) 0.65 nm apart along this direction, and the ≈ 1.9 nm three helices periodicity in a direction transverse to it (i.e. at ≈ 11 o'clock). This periodicity is more prominent in the lower part of the image (with a profile of type 1, 2, 2, cf. Figure 5). Imaging conditions: in isopropanol. Size of imaged area: 5.5*5.5 nm².

66

In combination, the present results suggest therefore that the AFM images do represent genuine features of the epitaxially crystallized contact faces.

Conclusion.

The three studies presented here were partly aimed at evaluating the potential of AFM for crystalline, relatively "soft" polymers. As a result, AFM investigations are the last step in a chain which combines sample preparation and characterization techniques. Suitable samples with geometrically flat and well characterized exposed crystal faces are obtained via epitaxial crystallization. Electron diffraction analysis provides essential information on the crystal phase and the contact plane. Actually, it helps anticipate possible methyl group patterns (for the polypropylenes considered here), i.e. it provides a reference pattern for the subsequent AFM investigation.

AFM examination with methyl-group resolution has made it possible to obtain structural information that is out of reach to other techniques. Investigation of the spp contact surface fulfills (admittedly in a highly favorable case) an old dream: to identify the helical hand of individual stems in crystalline polymers. Recognition of the nature of contact faces in αiPP and γiPP, and of possible surface reconstruction (or structural polymorphism) in βiPP, provides original insights into submolecular details of the crystal structure and contact faces and illustrates powerfully the potential of AFM in the structural investigation of crystalline polymers.

References.

1. Reneker, D. H.; Chun, I., *this issue*
2. Lotz, B.; Wittmann, J. C. *J. Polym. Sci., Polym. Phys. Ed.* **1985**, *23*, 205
3. Rodriguez-Arnold, J.; Zhengzheng, B.; Cheng, S. Z. D. *J. M. S. - Rev. Macromol. Chem. Phys.* **1995**, *C35*, 117
4. Brückner, S.; Meille, S. V.; Petraccone, V.; Pirozzi, B. *Prog. Polymer Sci.* **1991**, *16*, 361
5. Wittmann, J. C.; Lotz, B. *Prog. Polym. Sci.* **1990**, *15*, 909
6. (a) Stocker, W.; Magonov, S.; Cantow, H.-J.; Wittmann, J. C.; Lotz, B. *Macromolecules* **1993**, *26*, 5915 ; *ibid.* **1994**, *27*, 6690 (correction); (b) Stocker, W.; Graff, S.; Lang, J.; Wittmann, J. C.; Lotz, B. *Macromolecules* **1994**, *27*, 6677
7. Khoury, F. *J. Res. Nat. Bur. Std.* **1966**, *A 70*, 29
8. Lotz, B.; Wittmann, J. C. *J. Polym. Sci., Polym. Phys. Ed.* **1986**, *24*, 1541
9. Meille, S. V., Brückner, S.; Porzio, W. *Macromolecules* **1990,** *23*, 4114
10. Stocker, W.; Schumacher, M.; Graff, S.; Lang, J.; Wittmann, J. C.; Lovinger, A. J.; Lotz, B. *Macromolecules* **1994**, *27*, 6498
11. Lovinger, A. J.; Lotz, B.; Davis, D.; Padden, F. J. Jr. *Macromolecules* **1993**, *26*, 3494
12. Stocker, W.; Schumacher, M.; Graff, S.; Wittmann, J. C; Lotz, B., in preparation
13. Keith, H. D.; Padden, F. J. Jr.; Walter, N. M.; Wyckoff, H. W. *J. Appl. Phys.* **1959**, *30*, 1485
14. Meille, S. V.; Ferro, D. R.; Brückner, S.; Lovinger, A. J.; Padden, F. J. Jr. *Macromolecules* **1994**, *27*, 2615
15. Lotz, B.; Kopp, S.; Dorset, D. *C. R. Acad. Sci. Paris* **1994**, *319*, 187
16. Cartier, L.; Lotz, B. *J. Mol. Biol.,* submitted
17. New Japan Chemical Co, Ltd, European Patent EP 93101000.3, and Japanese Patents JP 34088/92, JP 135892/92, JP 283689/92, JP 324807/92

Chapter 4

Morphology, Chain Packing, and Conformation in Uniaxially Oriented Polymers Studied by Scanning Force Microscopy

G. J. Vansco[1], H. Schönherr[1], and D. Snétivy[2]

[1] Faculty of Chemical Technology, University of Twente, P.O. 217, NL-7500 AE Enschede, Netherlands
[2]Institut Straumann AG, CH-4437 Waldenburg, Switzerland

SFM images of the microfibrillar morphology and the molecular architecture of polymers with uniaxial texture are presented. The samples were obtained by drawing, uniaxial compression, gel spinning, electro-spinning, wet spinning, topotactic polymerization and friction deposition. Examples include results on linear poly-ethylene (HDPE), polypropylene (PP), polyoxymethylene (POM), polyamides, poly(ethylene oxide) (PEO), and polytetrafluoro-ethylene (PTFE).

Images of nanofibers of PEO obtained by electro-spinning exhibit fibers with a diameter of 14 nm, and with a chain axis orientation in the spinning direction. Friction deposition of PTFE and HDPE on glass resulted in a transfer of polymer from the slider to the glass substrate. SFM images of the worn surface of these polymers show a microfibrillar morphology with fiber axes and polymer chains oriented in the sliding direction. The low value of the dynamic friction coefficient for PTFE and HDPE is explained by the formation of a highly oriented layer at the sliding interface. Molecular resolution on samples used in friction deposition was achieved using chemically modified SFM tips which were coated by a monolayer of $HS-(CH_2)_2-NH-CO-(CF_2)_7-CF_3$.

Scanning force microscopy for uniaxially oriented polymers

In the years following the appearance of the first paper on atomic force microscopy (AFM) in 1986 (*1*), there has been an exponential growth in the number of publications reporting on AFM studies of nano- and sub-nanometer structures of materials (*2*). This trend can also be seen for scanning force microscopy (SFM) studies of polymers. The first SFM investigations of polymer micro- and nanostructures mostly reproduced known features of polymer morphology and nanostructure. The early research established the credibility of SFM in structural studies of polymers. However, many experimental difficulties and problems with the adequate interpretation of SFM scans were encountered at the beginning, and papers appeared which reported artefacts as real features. This now happens less often, and one can say that SFM has reached a certain maturity. The number of articles reporting on truly original results which could not have been obtained using another method, is continuing to grow.

The unique ability of AFM to study the morphology of polymers in the 1-100 nm size range was realized immediately upon the introduction of the first commercial instruments. The possibility of visualizing polymer chains caused great excitement. A number of research groups built or purchased AFM's. The competition to see polymers from a true molecular perspective began with the first report on an AFM observation of a two-dimensional polymeric monolayer of n-(2-aminoethyl)-10,12-tricosadiynamide (3) with molecular level details. In this work, the Hansma group used a home-built device to visualize rows of hydrocarbon segments that appeared to have a spacing of ≈ 0.5 nm.

In this chapter we present and discuss some of the major results of AFM studies of polymers which possess uniaxial orientation texture. These materials all appear to have a microfibrillar morphology in the sub-micron scale. The axis of the microfibrils is normally parallel to the uniaxial orientation direction, and the chain direction is usually predominantly parallel with the microfibrillar axis. Uniaxially oriented polymers are very well suited for molecular level SFM observations because the direction of the chain axis with respect to the macroscopic orientation is often a priori known. Thus, one ambiguity, i.e. the chain direction, is eliminated compared to imaging of unknown crystal facets. In addition, uniaxially oriented polymer systems have often been the focus of AFM studies because it was quickly recognized that molecular resolution could be achieved using these samples. AFM studies unveiled delicate details of the microfibrillar morphology in the 1-50 nm range, in many instances for the first time. The microfibrillar crystal habit can be linked to the more or less extended form of the low-energy macroconformation of polymers (4). Such microfibrillar crystals can form, in various processes, as will be discussed below.

Routes to polymeric materials with uniaxial texture

High performance, uniaxially oriented polymeric materials can be obtained via different routes (5). Drawing of melt or gel processed polymers (deformation processes) and spinning (or casting) of liquid crystalline polymers constitute the two most frequently used technologies. Fibrillar habits with extended-chain structures can also form as a result of successive polymerization and crystallization (see chapter 6 in ref. (4)), in stirred polymer solutions (shish-kebob morphology, see ref. (6)), and in topotactic polymerization in the solid state (7). The samples with uniaxial texture described in this paper were also obtained by melt-drawing thin films (*8*) and by electro-spinning (*9*). Due to the potential of electro-spinning to obtain fibrillar structures with diameters in the nanometer range, a separate paragraph will be devoted to electro-spun polymer fibers later in this chapter. Uniaxially oriented films and fibers (of e.g. polytetrafluoroethylene (PTFE) and polyethylene) also form during materials transfer in friction deposition processes (*10*). Understanding the friction deposition process itself is important for explaining the low values of kinetic friction coefficient of these two polymers. In addition, friction deposited layers have a remarkable orienting capacity when used as substrate, e.g. in applications of liquid-crystalline materials. Therefore a separate section will be devoted to this problem showing some original, previously unpublished results.

Scanning force microscopy measurements

All SFM experiments reported here as our own work were performed using NanoScope devices made by Digital Instruments. For contact mode AFM both a NanoScope II and a NanoScope III were utilized, while friction images in lateral force microscopy (LFM) and tapping mode images were obtained by NanoScope III devices. If not mentioned otherwise, triangular Si_3N_4 cantilevers (NanoProbe) with nominal force constants 0.38 Nm^{-1} were used for contact mode imaging. Scans in tapping mode were performed by Si cantilevers (NanoSensor). Imaging was done in air at room temperature in most cases, but for some samples (where specified) a liquid cell filled with an appropriate solvent was used.

Polyethylene (PE)

The first AFM images of an oriented polymer with molecular resolution showed PE chains as row-like features at the surface of cold-extruded samples (*11*) (see Figure 1). It was assumed that the rows corresponded to stretched polymer chains, primarily because the expected chain direction matched the direction of the orientation of the features which were imaged. The distances observed in the images helped to prove that this assumption was valid. For example, the periodicity along the chains was found to be 0.25 nm, which corresponds to the repeat unit in the all-trans confor-mation. The zig-zags of the all-trans PE chains visualized in this work correspond to resolved methylene groups.

AFM images of uniaxially stretched ultrahigh-molecular-weight PE (UHMWPE) prepared from gels (*12*) unveiled molecular resolution on the surface of the microfibers (shown in Figure 2 with a typical microfiber diameter in the order of 100 nm). A series of AFM scans with molecular resolution produced two types of pho-tomicrographs: one set having an average chain-chain distance of 0.73 nm, and the other set with an average distance of 0.49 nm . These intermolecular distances, both of which can be observed on the surface of the same microfibril, were identified as the a (0.7417 nm) and b (0.4945 nm) repeat lengths of the orthorhombic crystal unit cell of PE.

70

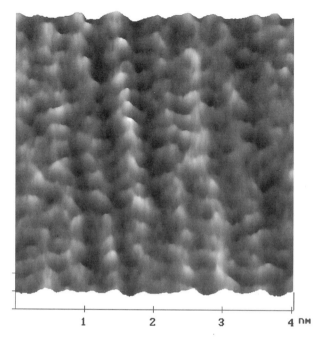

1   2   3   4 ⁿᴹ

Figure 1. AFM surface plot image capturing chain packing at the surface of cold extruded polyethylene. (Courtesy of Dr. S.N. Magonov, Digital Instruments.) Scan size: 4 nm x 4 nm.

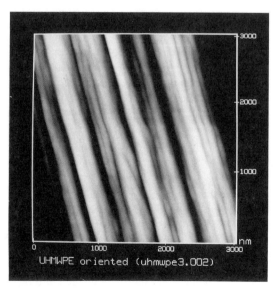

Figure 2. AFM micrograph (height) showing fibers and twisted microfibrils in uniaxially oriented, gel-spun, ultrahigh molecular-weight polyethylene. Scan size: 3 μm x 3μm. Height scale: black - zero, white - 300 nm.

Magonov and coworkers (*13*) reported on results concerning the various levels of the structural hierarchy of the microfibrillar morphology of polyethylene. AFM imaging under water unveiled a longitudinal nanofibrillar layer with a fibrillar diameter of 5-7 nm at the top of the microfibrils, similar to those shown in Figure 2. In addition, at the surface of gel-cast, oriented UHMWPE fibers, a weakly bound overlayer of transverse nanofibrils was observed. These nanofibrils were found to have widths of 2.5-3.0 nm, and were spaced periodically at distances of 5-6 nm. Commercial gel-spun PE fibers (Spectra 900) were found to have a surface which was covered by transversely overgrown ribbons. The mechanism of formation of these transverse structures has not yet been fully explained.

Contact-mode AFM images of microfibrils of uniaxially drawn, gel-cast UHMWPE exhibited periodic contrast variations along the fibers with a repeat distance of ca. 25 nm (*13(e)*). This was attributed to elastic deformations caused by the tip, because the contrast variations disappeared when the contact force was reduced. The authors assumed that darker areas in the contrast corresponded to relatively soft, amorphous regions within the fiber, while brighter regions were described as crystallites which do not become deformed by the pressure of the AFM tip. The so-called long period for the sample studied was found to be 20-25 nm.

AFM was used to study the molecular order of highly textured, quasi-single crystalline HDPE obtained by plain-strain compression in a channel die (*14*). The results showed a remarkable level of long-range coherence in the chain direction with a few regions exhibiting molecular-level imperfections. Surprisingly, no evidence supporting the existence of readily discernible phase-separated amorphous layers was found at high compression ratios. The imperfections observed by AFM included isolated molecular kinks, flip-over of polymer chains, and the gradual, coherent twisting of molecules. These structural features were consistent with the results of SAXS experiments. The existence of well-oriented material with long-range coherence in the chain direction was found to be compatible with an initial random-coil conformation of the polymer chains threading through the lamellae of the initial spherulitic morphology. This result seems to be in conflict with theories assuming any substantial frequency of chain-folding in the lamellae of the unoriented polymer. In the study described in ref. (*14*), several other techniques, as well as AFM, were used to gather experimental evidence which supported the conclusions. In Figure 3 a contact-mode AFM scan obtained by a chemically modified (perfluorinated) tip is shown, exhibiting molecular resolution of uniaxially oriented HDPE obtained in plain-strain compression.

Polypropylene (PP)

The first molecular-level AFM image of isotactic PP (i-PP) was obtained using a specimen which had been epitaxially crystallized on benzoic acid (*15*). This image showed methyl group resolution. The pattern of the methyl groups was found to be consistent with the crystal structure within the (010) facet. Uniaxially oriented i-PP was the subject of a number of early AFM studies performed by our group (16,17(a)). AFM images of cleaved samples unveiled microfibrils with a diameter of 100-150 nm. At the molecular level, chain orientation and chain packing imperfections (twisting and paracrystalline-like disorder) were visualized at the (110) crystal facet. By resolving individual methyl groups, we were able to identify the handiness of the helices in the crystalline i-PP matrix. The packing of the methyl groups, visualized by AFM, was compared with computer-generated structures. The agreement between the expected and the experimentally determined parameters of the crystal structure was within experimental accuracy (ca. +/- 5 %). It is worth mentioning here that the expected chain direction during imaging of uniaxially oriented samples as earlier mentioned, is a priori known. Thus, features parallel to the orientation direction, can usually be identified as polymer chains. On the other hand, images of rod-like features in uniaxially drawn systems do not necessarily correspond to chains. Knowledge of expected values of chain-chain packing is helpful to identify the features. For example, in the image shown in Figure 4, the periodic rows observed cannot be macromolecules since their direction makes an angle of 39° with the fiber (and thus with the expected backbone) direction. If the fiber direction had not been known 'a priori', one might have considered these periodic structures to be macromolecules. In Figure 5 an autocovariance pattern of vertically aligned PP chains is depicted. In this image the PP chains are clearly resolved. In addition there is a structure visible along the chains. The elements which correspond to this structure (identified as methyl groups) can be connected by two straight lines: one making an angle of 40° with the chain direction, and another one at 60° inclination with respect to the chain at the opposite side of the macromolecular backbone. This pattern allowed us to identify the helical hand of individual PP chains in the crystal structure (16(b)).

Polyoxymethylene

In our first communication on polyoxymethylene (POM) (18), we reported on AFM results obtained on mechanically oriented samples. Our images showed features of the microfibrillar morphology (microfibrillar diameter ≈100-150 nm) and molecular resolution of the POM helices packed in the well-known hexagonal crystal structure. Individual turns of the helix, and the five turns that make up the repeat unit of the 9/5 helix (19) were visualized. In a subsequent work, we focused on imaging extended-chain crystals of POM prepared in solid state polymerization of trioxane (17). AFM images in the nanometer range were obtained on these samples. A typical example (raw data and the corresponding autocovariance image) is shown in Figure 6 (scan size: 8 x 8 nm), displaying a periodic structure in two essentially perpendicular directions on the Ångstrom scale. Spots with bright tones correspond to corrugations "sticking out" of the sample surface. Dark areas, on the other hand, represent lower features. Objects parallel to the diagonal that connects the top left and the bottom right corners were identified as images of the polymer chains. This assignment is based on values for the packing distances and on the observation that the direction of

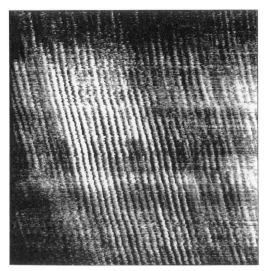

Figure 3. Unfiltered friction force nanograph (15 nm x 15 nm) of highly oriented HDPE obtained by uniaxial compression in a channel die. Imaging was performed with a perfluorinated tip in air, showing close-packed polyethylene chains in the (bc) crystal facet of the orthorhombic unit cell.

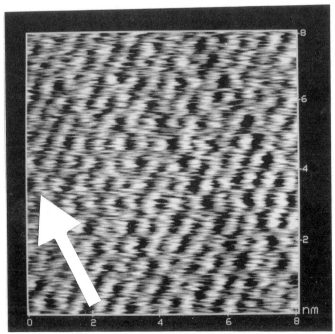

Figure 4. AFM height image of uniaxially oriented, isotactic polypropylene (scan size: 8 nm x 8 nm). The chain direction is shown. The row-like features at 1 o'clock correspond to unresolved images of methyl groups on neighboring chains.

Figure 5. Autovariance pattern of a 10 nm x 10 nm nanograph obtained on the surface of a uniaxially oriented microfibril of isotactic polypropylene. The polypropylene chains, as well as the methyl groups attached to the polymer backbone, are clearly resolved. Twisting of polymer chains, and a paracrystalline disorder can also be seen. The lines show the rows formed by the outermost methyl groups attached to neighboring chains. The unprocessed image is published in ref. (*16(b)*). Image reprinted from Polymer, ref. (*16(b)*), Copyright 1994, with kind permission from Elsevier Science Ltd.

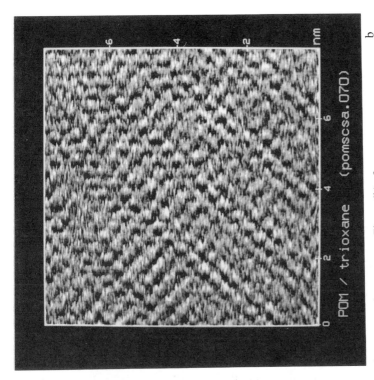

Figure 6. AFM nanograph (raw data: Figure 6a, autocovariance pattern: Figure 6b) of a microfibril of a POM crystal obtained in solid-state topotactic polymerization of trioxane. The polymer chains are parallel to the diagonal connecting the top left and the bottom right corners. Scan size: 8 nm x 8 nm. (Reproduced with permission from reference 17b. Copyright 1994 the Royal Society of Chemistry.)

the features identified as chains coincides with the crystallographic c (or fibre) direction, which was set at 45° with respect to the horizontal scan direction prior to imaging. The length of the c edge of the hexagonal unit cell, based on the AFM image, was found to be c=1.72 nm (expected value from WAXD is c=1.739 nm). The value of the packing distance in the chain-perpendicular direction was a=0.45 nm (expected value from WAXD is a = 0.447 nm).

A simulated molecular layer of POM chains packed in the ac facet helped us to understand the contrast variations in the AFM nanograph. It is known that the POM chain forms 5 turns within one crystallographic c unit. During AFM imaging, the tip will likely follow the contour of the electron density if the contact force is kept at the lowest possible value. Thus, the brighter spots in Figure 6 correspond to the exposed methylene groups of the POM chain, and the individual turns of the helix backbone are hidden behind the methylene units. In Figure 6 there are fuzzy regions that separate three bright spots. These "fuzzy" sections can be identified as parts of the chain where the methylene groups are not directly exposed to the scanned surface, but are turned away to the unexposed side of the macromolecule, and thus are not sticking out of the surface (*17,18*).

Polyamides

It is known that the structure of nylon fibers can be described by a two-phase model, assuming that the fiber consists of alternating, linked, amorphous and crystalline regions (*20*). The long period (the sum of the length of the crystalline and amorphous regions) is about 10 nm for nylon-6. It has already been mentioned in the section on gel-spun UHMWPE that due to the presence of alternating amorphous and crystalline regions, elastically recoverable periodic contrast variations may occur, caused by the tip pressure. In Figure 7 a contact-mode AFM image of a nylon-6 fiber is shown. The area depicted was scanned for ca. 45 minutes. During imaging, the microfibrils which were originally smooth and continuous were broken up into fragments which were pushed out of the fibrillar axis direction. As a result, the image shows not only periodic contrast variations, but also harder, crystalline regions which the scanning tip pushed out to the left and right relative to the fibrillar axis direction. As the pattern formed does not show any relaxation with time, the deformation caused by the tip is a plastic type.

Molecular-level imaging in oriented structures has established the credibility of AFM in polymer research, because it has reproduced many of the structural characteristics (unit cell dimensions, conformation, chain packing) known from measurements made by other well-established techniques. In addition, in a number of cases, AFM has been proven to be a successful technique for determining the local structure in polymorphic systems in which various crystal structures can coexist. For these materials, conventional X-ray diffraction results are often inconclusive due to line broadening and the superposition of diffraction lines. In addition, the determination of the spatial position of e.g. phenylene groups in aramids based on X-ray diffraction

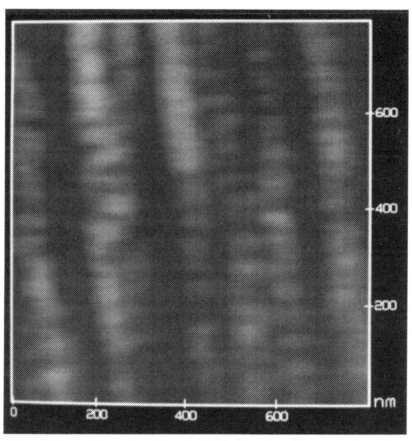

Figure 7. Plastic deformation of the microfibrils of a nylon-6 fiber scanned over 45 minutes. AFM height image, scan size: 800 nm x 800 nm.

pattern is an iterative process which makes a conclusive, accurate structural determination difficult. AFM can probe the microstructure locally, examining in the order of a hundred or so atoms in direct space, and thus it allows one to obtain a unique insight into local atomic arrangements at surfaces of robust, crystalline samples. This has been demonstrated in a number of our papers on the microstructure of aramids.

The macromolecules in extended-chain crystals of as-spun and annealed fibers of poly(p-phenylene terephthalamide) (PPTA) and the microfibrillar morphology were visualized by AFM (21). Nm-scale periodicities observed in the nanographs of the annealed fiber are in agreement with those based on WAXS data reported in the literature. For the as-spun fiber, AFM confirms the presence of at least two polymorphs. Among these, a new lattice structure was observed which has eluded deconvolution from experimental X-ray data. This new lattice exhibits chain conformations and hydrogen bonding structure which is distinct from that seen in either of the two previously reported crystal modifications of PPTA. Interpretation of the AFM images in this case is greatly facilitated by the independent data from molecular simulations. The new crystal facet observed (see Figure 8) is fully consistent with a third crystal modification previously suggested by simulation (22). In addition, WAXS and AFM were used to study the crystallographic structure of poly-(2-fluorophenylene-2-fluoroterephthalamide) fibers (23) (see Figure 9). The techniques used yielded complementary information about the unit cell characteristics. The unit cell dimensions found by WAXS and AFM were again identical to within the experimental error. AFM yielded complementary information about the conformation of the fluorine-substituted phenylene ring. The nanographs were consistent with successive phenylene rings along the chain being rotated in the same direction about the chain axis with respect to the plain of the H-bonds.

Fibers of PPTA with diameters ranging from 40 nm to a few hundred nm were obtained by electro-spinning. AFM was used to estimate the fiber diameter and the surface roughness. Features resembling kink bands were also observed in curved or bent fibers (24).

Poly(tetrafluoroethylene) and polyethylene friction deposition

PTFE and HDPE are polymers which are well known for having exceptional frictional properties such as very low values of dynamic coefficient of friction (25). For PTFE this was attributed for a long time to weak adhesion between the fluoropolymer and the substrate. However, Pooley and Tabor (26) showed that the low friction coefficient value of these polymers can be attributed to the formation of a well-oriented layer at the sliding interface of the polymer. This work demonstrated that transfer of material during sliding friction occurs for semicrystalline polymers that consist of macromolecules with a smooth chain profile. This transfer takes place from the polymeric slider to the surface of a smooth glass substrate on which sliding takes place. In the work of Tabor friction experiments were performed using fresh specimens of various thermoplastics that had not been previously slid on any surface.

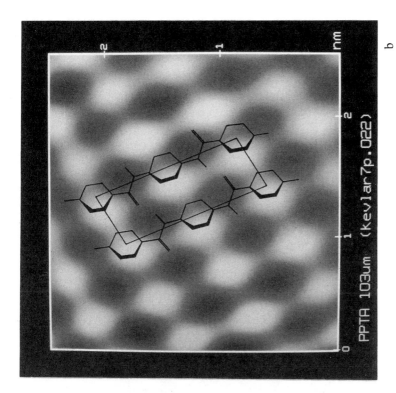

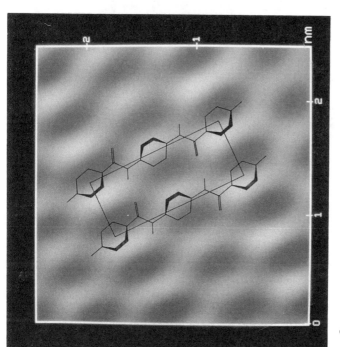

Figure 8. Fourier reconstruction of a typical image of annealed PPTA showing in the overlay plot the (bc) facet of the crystal unit cell of modification I of PPTA (Figure 8a). In Figure 8b the (bc) facet is shown for the modification III, discovered by AFM. Scan size: 2.5 nm x 2.5 nm. Reproduced with permission from ref. (21(b)).

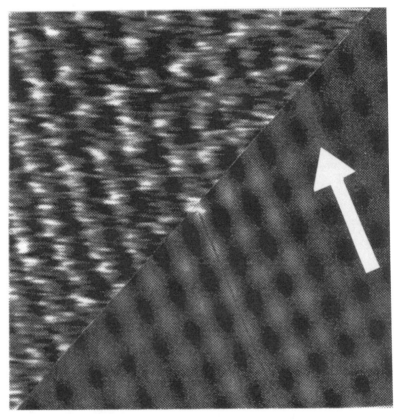

Figure 9. AFM image of a cleaved fiber of poly-(2-fluorophenylene-2-fluoroterephthalamide). Scan size: 6 nm x 6 nm. Top left half: raw data; bottom right half: autocovariance image. Chain direction is shown.

When such a PTFE or a HDPE slider was moved over the smooth glass substrate, and as soon as sliding began, the initially high value of the static friction coefficient dropped to a much lower value of a kinetic friction coefficient. Subsequent optical and electron microscopy experiments showed that during sliding friction of PTFE and HDPE polymer material is transferred from the moving slider to the glass substrate. During the initial traversal of the slider, close to the "stick" region of the static friction, lumps, slabs and streaky fibers are deposited on the surface of the support (glass). In further sliding material is drawn from the slider and gets deposited on the glass in the form of a thin film with a thickness that corresponds to the diameter of only a few macromolecules. Tabor's observations suggested that the high static friction coefficient and lumpy material transfer are related to the high interfacial adhesion of PTFE to glass and to shearing within the bulk of the polymer. For bulk material transfer to occur, a low frictional shear strength and strong adhesion forces are necessary. As a result of the shearing of the bulk, the authors postulated that chain alignment at the sliding surface of the polymer will occur. When sliding takes place in the direction of the chain orientation, some material is drawn from the slider, and the friction coefficient remains low because the sliding takes place along a smooth surface of aligned polymer chains. The morphology of the transferred PTFE layer was the subject of several studies (*26,27,28*).

Interest in friction deposited layers has been revitalized by the work of Smith, Wittmann and coworkers, which showed that the friction deposited layers have a remarkable orienting capacity in epitaxial crystallization, and can serve as orienting substrates for semicrystalline polymers and liquid-crystalline materials (*10,29*). The work of both Tabor and of Smith showed that there is little or no difference in the friction transfer behavior for polymer specimens with different thermal histories and morphologies. These studies indicated that molecular entanglements do not play a significant role in the transfer process. Thus, the fundamental mechanism of the friction deposition process can not be considered as a simple tensile deformation at the interface of the slider. Due to the important role of this interface, it is somewhat surprising that there have not been many morphological investigations of the worn surface of sliders. In this chapter we present, we believe for the first time, experimental evidence obtained from SFM measurements which shows that the surfaces of PTFE and HDPE sliders after sliding exhibit microfibrillar morphology and fibrillar crystals with macromolecular orientation in the sliding direction.

Uniaxially oriented, friction-deposited films and tapes of PTFE have been subjects of AFM studies (*27,28*). Observations of individual PTFE helices have been reported with a chain-chain distance of 0.58 nm, and indications for a 13/6 helical conformation were observed. We used a combination of lateral force microscopy (LFM) and AFM to study the deposited PTFE layers (*28*). During LFM/AFM scanning, PTFE molecules were picked up by the tip, which clearly reduced the frictional force differences measured between glass and PTFE. This observation is in accordance with the model of Tabor which assumes a strong interfacial tension between PTFE and glass during sliding friction. LFM experiments performed on the

PTFE films at scan directions between ca. 40-90° with respect to the polymer main chain orientation showed a "stick-slip" type frictional motion at the molecular level. This phenomenon allowed us to obtain LFM images of a synthetic polymer with molecular resolution. Chain-chain packing distances obtained by LFM and contact-mode AFM were identical to within the experimental error, and had a value of 0.578 nm at 25-30 °C imaging temperatures. Dual-mode contact AFM/LFM imaging was also performed by scanning in the chain direction. Here LFM nanographs showed no measurable "stick-slip" phenomenon, i.e. basically a featureless, flat image was obtained. The contact mode AFM images, however, exhibited clear molecular resolution with the expected chain-chain periodicity. The disappearance of the "stick" component in scans performed in the chain direction is a result of the smooth surface of PTFE on the molecular scale.

Studies of the worn surface of polymeric sliders were performed on PTFE and HDPE. The PTFE blocks used in this research were machined from Teflon rods which were obtained from Warehoused Plastics Sales Inc., Toronto, Canada. The static coefficient of friction on steel for this PTFE had a value of 0.02 as specified by the distributor. The HDPE sample was machined from an injection molded specimen. The material was a polymer received from BASF AG, with a trade label 6031 HX. Friction deposition was performed on thoroughly cleaned microscope slides. The friction experiments were done on the glass substrate using a Mettler FP82HT microscope hot stage for heating the glass. PTFE was deposited at 220°C, while the transfer of HDPE was performed at 130°C. Specimens were hand pressed and slid once over the surface of the substrate.

SFM imaging was performed by using Si_3N_4 non-functionalized and functionalized NanoProbe tips with a spring constant of 0.12 Nm^{-1}. Functionalization was done by using chemical modification by the established method (30) of forming a self-assembled monolayer (SAM) of the thiol $HS-(CH_2)_2-NH-CO-(CF_2)_7-CF_3$ on gold coated Si_3N_4 tips. Details of the tip functionalization procedure are described in a preliminary study (31(a)) and will be published elsewhere (31(b)). With the exception of the image in Figure 10, all scans were obtained using perfluorinated tips.

In Figure 10 a 4 μm x 4 μm micrograph of PTFE obtained in a contact mode AFM scan is displayed. The image shows the surface topology (height) of the work face of the PTFE slider. A thin layer of microfibrils covering a granular structure is visible. Prior to deposition, the sample surface shows only the granular morphology. Thus the overlayer of fibrils was formed during the friction transfer process. The fibrils have diameters in the order of 30-50 nm, and are often are arranged in bundles with a diameter of 100 nm to several hundred nm. The axes of the microfibrils are oriented in the sliding direction. High resolution AFM images, showing molecular resolution, were obtained by "zooming in" at the surface of the microfibrils. A typical AFM contact mode image (friction), showing PTFE molecules, is displayed in Figure 11. The average chain-chain packing distance, observed in nearly hundred images in both

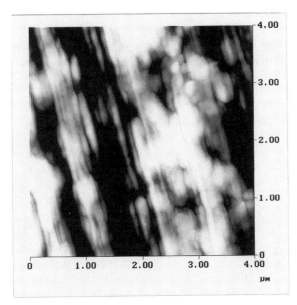

Figure 10. AFM image (height) of the worn surface of a PTFE slider used in a friction deposition experiment. Scan size: 4 μm x 4 μm. The microfibrillar morphology which formed during the deposition process, superimposed on the original granular structure, is clearly visible. The direction of the fibrillar axis is parallel to the sliding direction.

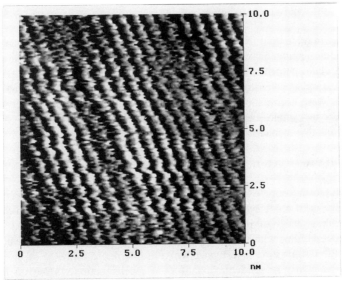

Figure 11. An AFM nanograph (friction image, 10 nm x 10 nm) showing molecular resolution obtained on a PTFE microfibril captured in Figure 10. The chain axis orientation is parallel to the sliding direction. The image was obtained to a perfluorinated tip.

air and ethanol, had a value of 0.556 nm (with a statistical standard deviation of 0.025 nm). This is in excellent agreement with the chain-chain separation distance of 0.555 nm expected from X-ray diffraction data for the phase IV hexagonal structure of PTFE (*32*).

AFM experiments performed on worn (previously slid) polyethylene sliders showed a similar picture. The morphology of the worn slider surface consisted of oriented microfibrils; however no granular structure was observed underneath this layer in any experiments performed until now. Molecular resolution obtained on these fibrils unveiled both the ac and the bc facets on the orthorhombic unit cell of PE (see Figure 12a for the ac, and Figure 12b for the bc facets, respectively). Figure 13 shows histograms for the molecular repeat distances obtained by evaluating a large number of AFM images. The larger chain-chain repeat length was found 0.744 ($\pm$ 0.035) nm, while the smaller packing distance had a value of 0.494 ($\pm$ 0.035) nm. These results are in excellent agreement with a = 0.741 nm and b = 0.494 nm repeat lengths expected from X-ray diffraction measurements (*33*). Thus we conclude that during friction deposition at the worn surface of HDPE sliders, fibrillar crystals form with a fiber axis and a polymer chain orientation in the sliding direction. In addition, statistically, both the ac and the bc facets are exposed, similarly to our earlier results obtained on highly stretched ultrahigh molar mass PE (*12*). A possible explanation for the formation of the fibrillar crystals is that the sliding material in contact with the substrate melts locally under shear and under load, and then recrystallizes under stress, forming extended-chain structures.

Finally, in relation to friction deposition of PTFE, we would like to mention, that in a preliminary study, chain-chain distances and the linear thermal expansion coefficient in the chain-perpendicular direction were estimated as a function of temperature, using samples consisting of friction deposited PTFE layers. In this work, LFM images were used which exhibited molecular resolution (*34*). Further studies are in progress to explain the deviations of thermal expansion coefficient from the bulk value and to expand the temperature range of the AFM observations.

Melt-drawn structures

In order to investigate the surface of uniaxially oriented specimens, Petermann et al. (*8*) developed a device for preparing thin polymer films drawn from the melt. This device was used to produce poly(butene-1) (PB-1), polyethylene, and isotactic polystyrene films with a high degree of uniaxial orientation. Both bare and tin-evapo-rated PB-1 films were investigated by AFM in order to study the surface topography, and effects such as surface relaxations and reconstructions. STM was performed on PB-1 films deposited on HOPG (*35*). The STM images showed flakes of PB-1 with a thickness much larger than the normal tunneling distance between a metallic sample surface and the tip. A superposition of the HOPG substrate and a superlattice, consistent with the helical structure of PB-1 macromolecules, was observed on molecularly resolved images. It is important to note that similar conclusions were

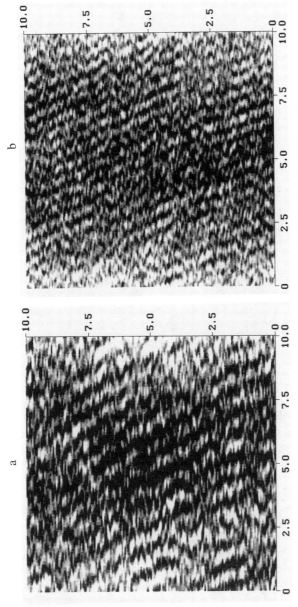

Figure 12. Crystal facets of the orthorhombic unit cell of polyethylene (height image) observed at the worn surface of previously slid polymer. Figure 12a - (ac) crystal facet; Figure 12b: (bc) crystal facet. The images were obtained with a perfluorinated tip. Scan size: 10 nm x 10 nm.

drawn using polyethylene thin films in another STM study (*36*). SFM images of PB-1 films unveiled a close-packed, bundled, needle crystal morphology of the specimens, which protruded out of the surface by 0.5-4 nm (*37*). This morphology is in agreement with the structure found in TEM investigations. Molecular resolution has also been achieved and individual macromolecular helices were visualized with a 0.7 (+/- 0.1) nm pitch height (the expected value from X-ray experiments is 0.65 nm). The structure observed between the stretched backbones might originate from interlocking ethyl side groups (see Figure 14). The difference between the observed intermolecular packing distances in the AFM scans and the expected bulk structural parameters in planes parallel to the surface may result from different molecular packing at the surfaces and in the bulk. It is worthwhile mentioning that this is the first report that gives experimental evidence for surface relaxation effects in oriented polymer systems consisting of macromolecules with side chains. It was only possible to successfully image tin-decorated surfaces with tapping mode AFM, because the metal particles were swept away in contact mode experiments (*37*). AFM images of ultrathin oriented polyethylene films drawn from the melt exhibited a shish-kebab morphology (*38*). In these images, the lamellar crystals also protruded out of the film surface (see Figure 15). Molecular scale images showed extended PE chains with a chain-chain distance of 0.56 +/- 0.1 nm. AFM images of melt-drawn ultrathin films of isotactic polystyrene (*39*) showed close-packed, highly oriented shish crystals which protruded out of the sample surface. Local deviations from the uniaxial orientation were occasionally observed. The diameter of the crystals was larger than the value obtained by TEM. This was explained by assuming that there is a transitional zone in the crystalline core in which no long-range lateral order exists.

Electro-spinning

In electrostatic spinning (*40*) unusually thin fibers are made using an electrostatic field with a high field strength. During this spinning process, electrostatic forces acting at the free surface of a droplet of polymer solution or melt (protruding out of a capillary) overcome the surface tension. The result is that an electrically charged jet can be ejected. Electro-spinning studies have recently received special attention in the group of Reneker (*41*) which has obtained structures with nanometer fibrillar dimension. In a recent paper, we reported on an AFM study of nanofibers and visualization of the chain packing within the fibers for electro-spun polyethylene oxide (PEO) (*42*). In this chapter we discuss further SFM results on electro-spun PEO fibers.

We used PEO with an average molar mass of 2×10^6 g/mol from Aldrich (without further purification) to prepare 1 wt % aqueous solutions. The samples were gently stirred for 12 h at 40 °C in order to obtain homogeneous solutions. Prior to electro-spinning thin fibers, the PEO solution was placed in a pipet. We used a stainless steel pin immersed in the solution as an electrode. A flat metal plate was used as a grounded counter electrode. We connected the stainless steel pin to a high-voltage supply which can generate dc voltages up to 30 kV.

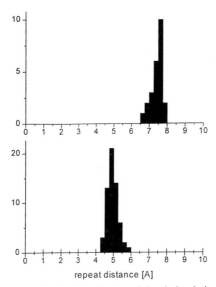

Figure 13. Histogram showing the distribution of the chain-chain distances in the orthorhombic unit cell of polyethylene observed at the worn surface of HDPE sliders in images similar to Figure 12. Top histogram: (ac) facet, bottom histogram: (bc) facet.

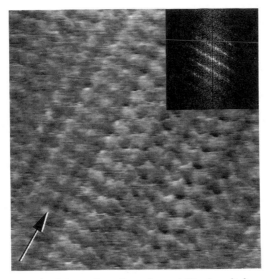

Figure 14. SFM image of the surface of a melt-crystallized, melt-drawn film of poly(butene-1). The inset shows the Fourier transform of the pattern captured in the image. The arrow indicates the direction of the molecular orientation. Courtesy of Dr. M. Miles, reproduced with permission from ref. (*37*).

When we applied a voltage between the solution and the counter electrode, the pendant drop at the orifice of the pipet was deformed into a conical shape. At a threshold value of the surface charge density of the drop, the surface of the drop became unstable, and a jet was ejected from the cone. At a certain voltage, the charge density on the jet reached a critical value where the jet splayed into several finer jets. Other details of our electro-spinning experiment can be found in ref. (*42*).

In Figure 16 a tapping mode image of a split PEO fiber is shown. Without correcting for the tip convolution artefact, the fiber diameter was found to vary between 16 and 20 nm, and in some cases reached as little as 14 nm. To our knowledge, this is the thinnest fiber obtained by an artificial spinning technique that has ever been reported. The use of tapping mode imaging was essential, as during contact mode AFM scans, as expected, the fibers were carried off by the tip.

If we examine a surface area of approximately 10 nm x 10 nm of an electro-spun fiber (contact mode AFM, fiber diameter ≈ 250 nm) with the AFM, the images reveal two sets of evenly spaced rows which form an angle of approximately +50° and -50° with the direction of the fiber axis. If the AFM tip scans parallel or perpendicular to the fiber axis, both sets of rows are simultaneously visible (see Figure 17). If the tip samples the surface topology at an angle of ± 45° to the fiber direction, only one of the sets of rows is visible - the one which is perpendicular to the scan direction.

In order to assign a crystallographic facet, various planes of the PEO structure were drawn by computer (*42*). The symmetry and periodicities observed in the images we obtained with AFM are consistent with the surface topology at the (100) or the (010) facet. The upper turns of the PEO helices, which are exposed at the scanned surface, form ridges which are 0.78 nm apart. The projection of the ridges is inclined by 53° to the crystallographic c direction. The lower turns of the helices also form ridges with a spacing of 0.78 nm. The projection of these ridges forms an angle of -53° with the crystallographic c direction. The surface of the (010) facet of PEO results in a similar pattern. If we look at the ridges in AFM images, such as Figure 17, we see that both the fiber direction and the chain direction bisect the angles between the rows which make an angle of ca. 105°. Based on these observations and the close match between feature distances observed in AFM scans and packing distances in the PEO crystal structure, we identified the row-like features as images of top and bottom turns of helical PEO molecules in the top layer of the (100) or the (010) facet. Within the estimated accuracy of the instrument (± 0.04 nm and ± 4°), it is not possible to distinguish between the two facets.

Some Pitfalls

Correctly identifying the features imaged has traditionally been difficult in AFM. Some features that can be visualized on graphite surfaces by STM (e.g. step edges, dislodged flakes, etc.) may resemble images of "macromolecules" (*43*). Helical

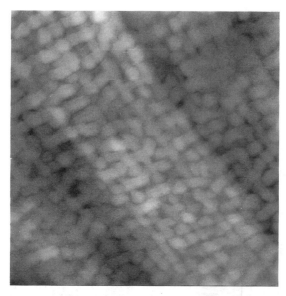

Figure 15. SFM image of the surface of a melt-drawn polyethylene film (scan size: 1 μm x 1 μm) showing a shish-kebob morphology. Note that the crystals protrude a few nanometers out of the substrate surface. Courtesy of Dr. M. Miles. Image reprinted from Polymer, ref. (*38*), Copyright 1994, with kind permission from Elsevier Science Ltd.

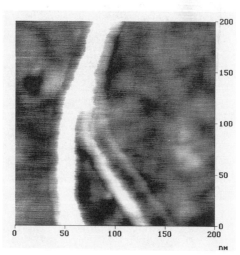

Figure 16. A tapping mode image of a PEO nanofiber (scan size: 200 nm x 200 nm). The smallest fiber diameter observed in this image is 14 nm.

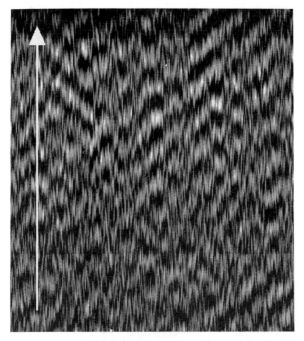

Figure 17. SFM contact mode nanograph (scan size: 11 nm x 11 nm) of the surface of an electro-spun PEO fiber. The white arrow indicates the fiber direction. The features correspond to the helix turns captured in the (100) facet of the monoclinic crystal modification of PEO. The chain orientation is parallel with the fiber (and thus spinning) direction. Image reproduced with permission from ref. (*42*).

features which looked like polymer chains were observed (using STM) on freshly cleaved HOPG, upon which no deposition of polymers had occurred (*43*). Essentially identical nanographs were earlier identified as helical and superhelical structures of oriented polymers (*44*). As demonstrated in the polypropylene section of this review, the observation of row-like features on AFM images with molecular resolution does not always mean that these features correspond to oriented macromolecules. Structural features, protruding out of the imaged plane, might yield contours that resemble polymer chains. A priori knowledge of the backbone orientation direction and a combination of AFM with X-ray, molecular simulation studies and other microscopy techniques help to eliminate these assignment problems. In contact-mode AFM, the tip exerts pressure on the sample surface and deforms it. In addition, due to the convolution of the tip shape with sharp sample features, these features appear to be rounded, and the dimensions are sometimes overestimated. One should be aware of possible imaging artifacts and how they can influence AFM images. Recent improvements in the quality of commercial tips have, however, helped to reduce these problems. It is not the purpose of this review to report on imaging artifacts, so the reader is referred to the literature (*1(c)*), (*45*).

Finally, one other issue needs to be mentioned. In all the examples discussed in this review, molecular resolution was achieved on regularly structured surfaces that exhibit a certain degree of two-dimensional translational symmetry. Resolution of individual atoms or small groups of atoms in non-periodic polymeric structures has not yet been achieved in ambient imaging conditions - apart from some biological molecules with large dimensions. Individual atoms can be "seen" routinely in STM experiments performed in vacuum, and it is hoped that technical progress in SFM will allow us to image (yet unseen) individual, atomic-level details in polymeric materials.

Acknowledgment

In this paper, in addition to the literature review, some representative results obtained in our group in Toronto, Canada (until May, 1995) and in Enschede, the Netherlands (since May 1995) have been presented. Many people contributed to this research during the past six years. Special thanks are due to (in alphabetical order) Prof. A.S. Argon, Dr. S. Förster, Prof. J.E. Guillet, Dr. B.H. Glomm, Dr. C.R. Jaeger, Ms. H. Leist, Mr. G. Liu, Prof. G.C. Rutledge, Dr. D. Trifonova and Mr. H. Yang for their contributions, and for the many stimulating discussions. The financial support given by the Natural Sciences and Engineering Research Council of Canada, the Ontario Center for Materials Research, the University of Toronto, and the University of Twente in the Netherlands, is greatly appreciated. Last but not least, the great help of Ms. Sandra de Jonge and Ms. Anne Klemperer with the editing of this manuscript, is acknowledged.

92

References

1. (a) Binnig, G.; Quate, C.F. and Gerber, Ch. *Phys. Rev. Lett.* **1986**, *56*, pp. 930.
   (b) The various imaging modes together with representative results obtained in polymer science are described by M.J. Miles, in *Characterization of solid polymers*, Spells, S.J., Ed., Chapman & Hall: New York, 1994, pp. 17-55.
   (c) For a recent comprehensive review see Magonov, S.N.; Whangbo, M.H., *Surface Analysis with STM and AFM*, VCH: Weinheim, 1996.
2. Hamers, R.J. *J. Phys. Chem.* **1996**, *100*, pp. 13103.
3. Marti, O.; Ribi, H.O.; Drake, B.; Albrecht, T.R.; Quate, C.F.; Hansma, P.K. *Science* **1988**, *239*, pp. 50.
4. Wunderlich, B. *Macromolecular Physics*, Academic Press: New York, 1976.
5. Ward, I.M. *Adv. Polym. Sci.* **1985**, *70*, pp. 1.
6. Pennings, A.J.; Zwijnenburg, A.; Lageveen, R., *Kolloid Z.Z. Polym.* **1973**, *251*, pp. 500.
7. For a review see Schulz, M.J., Polymer Materials Science, Prentice Hall: Englewood Cliffs, 1974, pp. 113-121 and Chapter 6.4.3. in Vol. 2 of ref. (*4*).
8. Peterman, J.; Gohil, R.M. *J. Mater. Sci.* **1979**, *14*, pp. 2260.
9. Baumgarten, P.K., *J. Coll. Interface Sci.* **1971**, *36*, pp. 71; Larrondo, L.; Manley, R., *J. Polym. Sci. Polym. Phys. Ed.* **1981**, *19*, pp. 933.
10. Motamedi, F.; Ihn, K.J.; Fenwick, D.; Wittmann, J.C.; Smith, P. *J. Polym. Sci. Polym. Phys.* **1994**, *32*, pp. 453.
11. Magonov, S.N.; Qvarnström, K.; Elings, V.; Cantow, H.J. *Polym. Bull.* **1991**, *25*, pp. 689.
12. Snétivy, D.; Yang, H.; Vancso, G.J. *J. Mater. Chem.* **1992**, *2*, pp. 891.
13. (a) Magonov, S.N.; Sheiko, S.S.; Deblieck, R.A.C.; Möller, M. *Macromolecules* **1993**, *26*, pp. 1380; (b) Sheiko, S.S.; Möller, M.; Cantow, H.J.; Magonov, S.N.; *Polym. Bull.* **1993**, *31*, pp. 693; (c) Wawkuschewski, A.; Cantow, H.J.; Magonov, S.N.; Sheiko, S.S.; Möller, M. *Polym. Bull.* **1993**, *31*, pp. 699;
    (d) Wawkuschewski, A.; Cantow, H.J.; Magonov, S.N. *Polym. Bull.* **1994**, *32*, pp. 235; (e) Wawkuschewski, A.; Cantow, H.J.; Magonov, S.N. *Adv. Mater.* **1994**, *6*, pp. 476.
14. Schönherr, H.; Vancso, G.J.; Argon, A.S. *Polymer* **1995**, *36*, pp. 2115.
15. Lotz, B.; Wittmann, J.C.; Stocker, W.; Magonov, S.N.; Cantow, H.J. *Polym. Bull.* **1991**, *26*, pp. 209.
16. (a) Snétivy, D.; Guillet, J.E.; Vancso, G.J. *Polymer* **1993**, *34*, pp. 429; (b) Snétivy, D.; Vancso, G.J. *Polymer* **1994**, *35*, pp. 461.
17. (a) Snétivy, D.; Vancso, G.J. *Coll. & Surf. A.* **1994**, *87*, pp. 257; (b) Snétivy, D.; Yang, H.; Glomm, B.; Vancso, G.J. *J. Mater. Chem.* **1994**, *4*, pp. 55.
18. Snétivy, D. and Vancso, G.J. *Macromolecules* **1992**, *25*, pp. 3320.
19. Sauter, E.Z. *Phys. Chem.* **1933**, *21B*, pp. 186; Uchida, T. and Tadokoro, H. *J. Polym. Sci. A2* **1967**, *5*, pp. 63.
20. Lim, J.G.; Gupta, B.S.; George, W. *Prog. Polym. Sci.* **1989**, *14*, pp. 763.
21. (a) Snétivy, D.; Rutledge, G.C.; Vancso, G.J. *ACS Polym. Prepr.* **1992**, *33(1)*, pp.786; (b) Snétivy, D.; Vancso, G.J. and Rutledge, G.C. *Macromolecules* **1992**, *25*, pp. 7037; (c) Rutledge, G.C.; Snétivy, D. and Vancso, G.J., In *Atomic Force*

Microscopy/Scanning Tunneling Microscopy, Cohen, S.H. et al., Ed., Plenum Press: New York, 1994, pp. 251.

22. Rutledge, G.C.; Suter, U.W.; Papaspyrides, C.D. *Macromolecules* **1991**, *24*, pp. 1934.

23. Glomm, B.H.; Grob, M.C.; Neuenschwander, P.; Suter, U.W.; Snétivy, D.; Vancso, G.J. *Polymer* **1994**, *35*, pp. 878.

24. Srinivasan, G.; Reneker, D.H. *Polym. Int.* **1995**, *36*, pp.195.

25. Cherry, B.W. *Polymer Surfaces*, Cambridge University Press: Cambridge 1981.

26. Pooley, C.M.; Tabor, D. *Proc. Roy. Soc. London A* **1972**, *329*, pp. 251.

27. (a) Magonov, S.N.; Kempf, S.; Kimmig, M.; Cantow, H.J. *Polym. Bull.* **1991**, *26*, pp. 715; (b) Hansma, H.; Hansma, P.K.; Ihn, K.J.; Motamedi, F.; Smith, P. *J. Mat. Sci.* **1993**, *28*, pp. 1372; (c) Hansma, H.; Motamedi, F.; Smith, P.; Hansma, P.; Wittman, J.C. *Polymer* **1992**, *33*, pp. 647.

28. Vancso, G.J.; Förster, S.; Leist, H. *Macromolecules* **1996**, *29*, pp. 2158.

29. Wittmann, J.C.; Smith, P. *Nature* **1991**, *352*, pp. 414.

30. (a) Frisbie, C.D.; Rozsnyai, L.F.; Noy, A.; Wrighton, M.S.; Lieber, C.M. *Science* **1994**, *265*, pp. 2071; (b) Noy, A.; Frisbie, C.D.; Rozsnyai, L.F.; Wrighton, M.S.; Lieber, C.M. *J. Am. Chem. Soc.* **1995**, *117*, pp. 7943.

31. (a) Schönherr, H.; Vancso, G.J. *ACS Polym. Prepr.* **1996**, *37(2)*, pp. 612; (b) Schönherr, H.; Vancso, G.J., *Langmuir*, submitted.

32. *Polymer Handbook, 3rd ed.*; Bandrup, J.; Immergut, E.H., Eds., Wiley: New York. 1989, p. V 37.

33. Alexander, L.E., *X-ray Diffraction Methods in Polymer Science*, Wiley: New York, 1969, p. 478.

34. Förster, S.; Liu, G.; Vancso, G.J. *Polym. Bull.* **1996**, *36*, pp. 471.

35. (a) Fuchs, H.; Eng, L.M.; Sander, R.; Petermann, J.; Jandt, K.D.; Hoffmann, T. *Polym. Bull.* **1991**, *26*, pp. 95; (b) Miles, M.J.; Jandt, K.D.; McMaster, T.J.; Williamson, R.L. *Coll. & Surf. A.* **1994**, *87*, pp. 235.

36. Jandt, K.D.; Bukh, M.; Petermann, J.; Eng, L.M.; Fuchs, H. *Polym. Bull.* **1991**, *27*, pp.101.

37. Jandt, K.D.; McMaster, T.J.; Miles, M.J.; Petermann, J. *Macromolecules* **1993**, *26*, pp. 6552.

38. Jandt, K.D.; Bukh, M.; Miles, M.J.; Petermann, J. *Polymer* **1994**, *35*, pp. 2458.

39. Jandt, K.D.; Eng, L.M.; Petermann, J.; Fuchs, H. *Polymer* **1992**, *33*, pp. 5331.

40. Formhals, A. US Patent, 1,975,504 **1934**.

41. Reneker, D.H.; Chun, I. *Nanotechnology* **1996**, *7*, pp. 216.

42. Jaeger, R.; Schönherr, H.; Vancso, G.J. *Macromolecules* **1996**, *29*, pp. 7634.

43. Clemmer, C.R.; Beebe, T.P. *Science* **1991**, *251*, pp. 640; Legget, G.J.; Davies, M.C.; Jackson, D.E.; Roberts, C.J.; Tendler, S.J.B. *TRIP* **1993**, *1*, pp. 115.

44. Yang, R.; Yang, X.R.; Evans, D.F.; Hendrickson, W.A.; Baker, J. *J. Phys. Chem.* **1990**, *94*, pp. 6123.

45. Goh, M.C. In *Advances in Chemical Physics*, Prigogine, I.; Rice, S.A., Eds., Vol. XCI, John Wiley: NewYork, 1995, pp.1.

Chapter 5

The Formation of Fibrillar Structures from Biopolymers

M. C. Goh, M. F. Paige, P. Markiewicz, I. Yadegari, and M. Edirisinghe

Department of Chemistry, University of Toronto, Toronto, Ontario M5S 3H6, Canada

Applications of the atomic force microscope (AFM) to investigations of fibril formation in biopolymers are presented. The process of self-assembly from monomers to fibrils in collagen was monitored by light scattering and the constituents of the assembly mixture were examined by AFM. Assembly intermediates were identified, and a simple model for the assembly was inferred. Considerations for AFM imaging in these types of problems are discussed.

Polymers in solution can aggregate into larger structures. The parameters that control the occurrence of aggregation are a complex interplay of interparticle forces and solvent conditions. In typical polymer solutions, aggregation results in a tangled mess with a low degree of order. Aggregation of colloidal particles, polymer or otherwise, by a diffusion-limited process leads to ramefied, fractal-like structures [1]. In numerous applications of polymer-based materials, fibres of the polymers are essential. However, polymers typically do not usually self-assemble into such form, and fibres are typically produced by mechanical means, such as extrusion and pulling.

We are interested in investigating the control of self-assembly to form materials with desirable mesoscopic structures such as fibrils. While most biopolymers behave in a similar fashion as polymers and colloids with regard to aggregation, and can in fact be treated in a similar manner [2], there are a number of biopolymers that self-assemble into well-defined and well-controlled structures, which typically are associated with interesting functions. Our approach is thus to study the assembly of such naturally-occuring materials to investigate mechanisms, with the hope of being able to derive a minimal model for the formation of fibrils in general, and the eventual goal of being able to control the structures formed by adjusting the conditions of assembly. While biopolymers generally have very complex arrangement of atoms, and hence have numerous complex interactions, most of these are used for determining the "folded" configuration of the active protein. Our working hypothesis is that the interparticle aggregation, while involving many more atoms, require only a handful of interactions, and can thus be reasonably simple and amenable to investigations.

In this paper we explore the capability of the atomic force microscope (AFM) to study the mechanisms of assembly in collagen, a fibril-forming, structural protein. We shall present results on *in vitro* studies using the AFM for imaging assembly intermediates and light scattering for monitoring the extent of the assembly. A few observations on the AFM imaging of these materials will also be discussed.

The AFM in Filament Studies

Our research group has previously reported results on the use of the AFM for structure determination of a number of biological filaments. In two cases we had investigated, intermediate filaments [3] and paired helical filaments (PHF) [4-6], *in vitro* studies were not possible; these filaments were extracted from tissues. Nevertheless, the structural information obtained by the AFM proved useful in inferring assembly mechanisms [4, 7].

In the case of PHF, the AFM images indicated that instead of the generally-accepted model of a pair of intertwined filaments, the structure more closely resembles that of a twisted ribbon, with underlying ultrastructure [4-6]. This model contravenes the general idea that PHF assembles in stages, by first producing filaments which subsequently intertwines. The AFM results led to the hypothesis that PHF form by a nucleation-elongation mechanism, much like that of a crystallization process, with growth only in one direction [4,7].

The mechanism of formation of intermediate filaments has been derived from a collection of biochemical data and electron microscope images. The prevailing view, summarized in [8], indicates that assembly proceeds in a modular fashion, wherein protein monomers form heterodimers, which then aggregate into tetramers. The tetramers assemble into protofilaments. Two protofilaments form a protofibril, four of which form one intermediate filament. With the AFM, these intermediates were identified [9]. In addition, the high vertical resolution of the AFM enabled measurement of quasi-periodicity along the protofilament and protofibril axes, which could be correlated with the known biochemical information about the sub-units [7].

Imaging: Some Problems and Solutions

AFM imaging of aggregate formation has its peculiarities, and we employ some techniques to assist us in our studies, which will be discussed in the following. First is the commonly recognized tip artifact that is unavoidable in all mesoscale imaging: Due to the non-negligible tip size, the tip geometry is always part of the generated AFM image (see for example, ref. 9). This convolution effect can be of major significance in structure determination as well as in the reliability and reproducibilty of images, which are particularly important in studies of non-periodic, irregularly-shaped samples such as biopolymer aggregates. It is to address this problem that we had previously developed a numerical algorithm for removal of the tip geometry from the AFM image [10,11], which resulted in a more accurate representation of the sample being examined. The main assumption of this algorithm is that only hard body interactions exist between tip and sample, and that the tip and sample cannot occupy the same space. The possible existence of electrostatic and adhesive forces of interaction are extra degrees of complication that is not easily addressed, and could limit the applicability of the algorithm in generating better images. However, perhaps a more important use of the same algorithm is for *in situ* tip inspection. This can be done by codepositing known standards, such as latex spheres or gold colloids, with the sample of interest. Alternatively, a preparation of the standard on a different slide could be imaged, and the tip inspected prior to use on the actual sample. The geometry of the known standard can be removed from the AFM image, leaving a picture of the tip, from which can be seen irregularities that could influence the images.

In determination of complex structures, it may be useful to perform a simulation of the convolution of the tip with a model system, which can be performed with a similar algorithm as above. The simulation can serve as a guide in identifying structures, as well as in exploring the limits of AFM imaging [12]. An example of the utility of this approach is its application to the AFM images of PHF to see the distinction between a pair of helical filaments and a twisted ribbon with ultrastructure [6], and to examine its chirality [13].

Aside from tip effects, imaging non-periodic samples that are cast from a solvent means that there may be a problem with regards to locating the object of interest within the field of the substrate, taking into account that the AFM scanner can explore only ~ 100 μm squares at each scan. Sample preparation must be such that there is enough deposited material that can thus be located with relative ease, but not too much so that they are on top of each other. The appropriate concentrations, the spreading of the solvent on the substrate and the drying process are thus important considerations. With regards to locating an object, we have utilized an address system that can be useful when another technique such as fluorescence microscopy can locate the desired region to within ~ 100 μm. This approach [14] relies on the use of a translucent substrate, such as thin mica or glass, under which can be mounted a fine grid. Indexed locator grids typically used in electron microscopy are ideal for this procedure, although other types of grids, including computer-generated lines, can also be used. The desired address can be found by fluorescence microscopy, for example, prior to mounting the sample on the AFM, and the same address located by a CCD camera, which is a typical attachment to most AFMs. This same technique is useful in re-locating an object that has previously been using the by AFM [14], a situation which may arise for several reasons, such as: (1) to inspect the same sample after it has been removed from the stage, (2) to provide the option of sample rotation, which could be useful in structure elucidation, (3) to study the same sample or the exact area of the substrate, after it has been treated outside the AFM stage, as for example, in annealing studies.

Collagen Assembly

In this paper, we report on the assembly process of fibril forming (Type 1) collagen [8]. Collagen is the main structural protein in animals, which is found outside the cell. The monomer unit consists of three polypeptides intertwined in a triple helix, forming a rod-like structure that is ~1 nm in diameter and ~280 nm in length. The rod-like monomer is rather stiff due to the triple helical configuration. *In vivo*, the collagen monomer is initially prevented from assembly because it has long polypeptide chains attached to the triple helical rod. This "procollagen" is excreted from the cell, and the chains are cleaved enzymatically in the extracellular matrix to initiate assembly. The fibrils are cross-linked after the assembly.

The collagen monomer is acid soluble and its assembly *in vitro* can be initiated by increasing the pH and the temperature of the solution [15]. It is important to note that no covalent linakges are supposed to be formed in this process. (There may be problems at times due to imperfect purification of biological starting materials, which could result in the presence of some crosslinking enzymes in certain preparations).

Materials and Methods. The monomer size is well within easy reach of the AFM. Thus, the various stages of assembly can be monitored, from the monomer to the mature fibril. Type I calf skin collagen (Sigma Chemicals) was utilized in all these studies. The assembly was carried out in a temperature-controlled bath, as described in [16], and light scattered at 90° was measured. At certain time intervals, an aliquot was withdrawn and diluted with buffer to halt the assembly. A drop of this solution was then deposited on freshly cleaved mica and air-dried prior to imaging. Imaging was performed with a Nanoscope III using contact mode, with nominal cantilever force constants of ~0.5 N/m.

Results

Light Scattering. Measurements of scattered and transmitted light are common ways of monitoring aggregation processes, and have been utilized in past studies of collagen. The intensity of the scattered light as a function of elapsed time typically has a sigmoidal shape (Figure 1), with three characteristic regions: the lag phase, where the light intensity barely changes, the growth phase where there is a rapid rise in intensity, and the plateau, where the intensity has levelled off. There are a number of studies that address the aggregation process by examination of these light scattering curves [17,18]; however, detailed interpretation is complicated by the shapes and polydispersity of the particles involved. One of the goals of our studies is to utilize the AFM to provide information about the types of aggregate geometry that are present in the assembly mixture, and to compare these with the assumptions made in the light scattering studies. We use the light scattering measurement as a guide that enables us to identify the extent of assembly that has taken place.

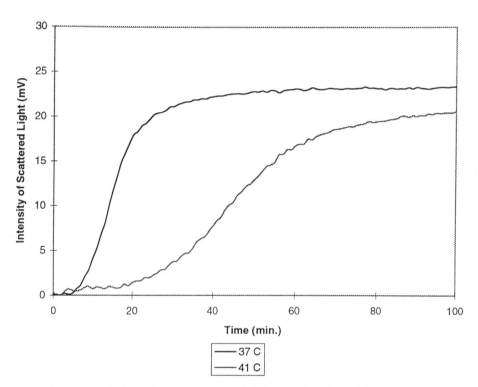

Figure 1. The intensity of the scattered light as a function of time elapsed during assembly for two different temperatures.

The kinetics of assembly, as indicated with the shift in the growth phase and the appearance of the plateau in the light scattering curve, increases drastically with temperature. However, there is a limited temperature range that can be explored since the material appears to denature at ~45 °C, while no change in intensity was observed at 25 °C for 5 days. These studies were done at relatively low concentrations: a typical starting concentration was ~0.8 mg/ml. At a slightly higher concentration of 1.1 mg/ml, non-sigmoidal light scattering curves were observed, which could imply the occurence of a different type of kinetics. However, the complexity in interpretation of the scattering intensity precludes any conclusion at present. The AFM results are qualitatively similar to those at lower concentrations, and further work is needed.

Atomic Force Microscopy. Figure 2 shows a summary of the different structures that have been observed along different points of the assembly at 34 °C [16]. The initial assembly mixture contains mostly monomers, with a few dimers and trimers (Fig. 2a). After a few minutes, and even before there is noticeable change in the scattering intensity, one can find larger aggregates albeit at very low frequency of occurence. Short microfibrils (Fig. 2b), which are ~2 nm in diameter and ~2 μm in length become very abundant early on, and remain the most abundant species until the end. The next stage shows the presence of long microfibrils (Fig. 2c), which are ~4-10 nm in diameter and 10's of μm in length. These can be distinguished easily from the others by their very flexible nature, forming masses of coils everywhere on the substrate. Finally, there are the mature fibrils (Fig. 2d), which are ~10-20 nm in diameter and ~ 10's of μm in length. The mature fibrils exhibit the banding pattern with a ~67 nm repeat, which is characteristic of collagen fibrils. Based on their uncoiled and relatively straight conformations, they are noticeably less flexible than the long microfibrils. These four are the only structural patterns that we have observed. As the assembly progresses, the relative abundance of each type changes. The assembly is halted after the light scattering intensity has reached a plateau, at which point numerous mature fibrils can be observed. However, there are still substantial amounts of long and short microfibrils. We do not find monomers at this point, although this may be attributed simply to the higher degree of difficulty in identifying them.

The same types of assembly intermediates were observed at other temperatures, from 30 °C up to 40 °C, although the relative abundances may be different; we have not quantified the images so far.

These results indicate that there are distinct identifiable intermediates during the assembly of collagen. However, these intermediates are not all of uniform dimensions. Rather, we had grouped together several size scales as belonging to one type of intermediate, because we can differentiate it from another based on the type of pattern that is shown. The presence of distinct intermediates argues against a simple nucleation and growth process for these early stages of firbillogenesis, since there are clear stages through which the assembly proceed. On the other hand, in a strictly modular assembly, the intermediates, which are considered stable, are expected to be well-defined in size. The assembly mechanism in this system appears more complex

a

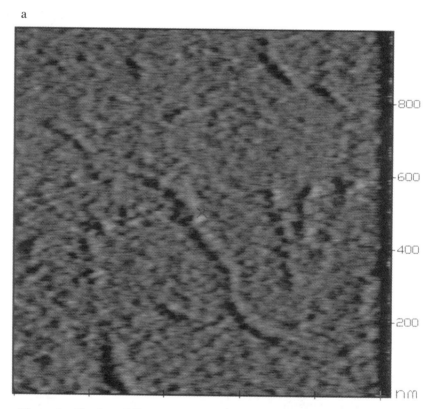

Figure 2. The four different types of assembly intermediates observed: (a) monomers and short oligomers; (b) short microfibrils, forming a network-like pattern; (c) long, flexible microfibrils; (d) mature fibril - note banding.

Continued on next page.

b

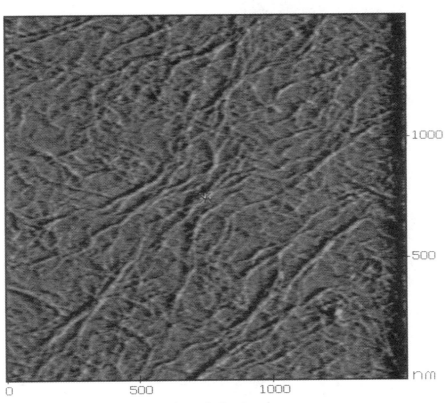

Figure 2. *Continued.*

Continued on next page.

102

c

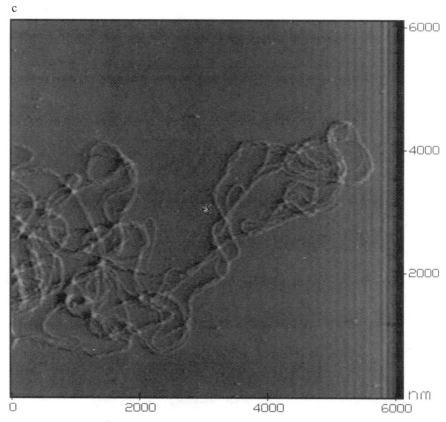

Figure 2. *Continued.*

Continued on next page.

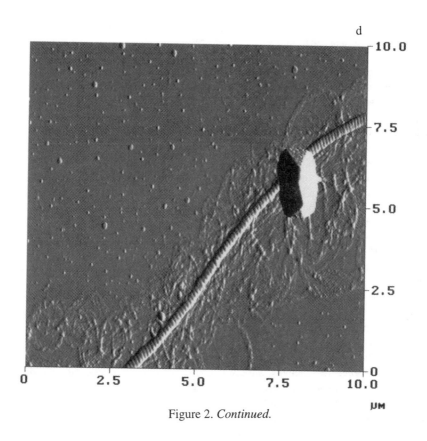

d

Figure 2. *Continued.*

than that in previously studied filaments, which are discussed in [8]: intermediate filaments assemble in a modular fashion, actin and microtubles, which are also important components of the cell infrastructure, form by growth (polymerization) in one dimension. In these other cases, the filament lengthens in one dimension, but the diameter is conserved. What makes collagen particularly interesting its the capability for increasing in both length and diameter, implying the existence of both axial and lateral growth; the control of one over the other offers the possibility of creating other structures, or filaments with different aspect ratios and elastic properties. Yet in spite of the complex mechanism, the assembly itself is experimentally simpler than those of other filaments, and conducive to controlled changes in conditions, since no other material is needed than the monomers.

The simple model we have for for collagen assembly is summarized in Figure 3. While the monomer contains a huge number of atoms and iteractions, the folded triple-helical collagen molecule retains its structure under most conditions, including those of our studies. Thus, we can use the folded protein as our starting sub-unit, ignore all the interactions that served to fold it, and consider only the handful of interpartical interactions that will control the aggregation.

In the simplest terms, we can think of the collagen monomer as a rod, with different axial and lateral interactions, ε_A and ε_L, respectively, such that $\varepsilon_A \gg \varepsilon_L \sim kT$. This would imply that the intial growth will only be along the axial direction. However, beyond a certain length, n^*, the interaction $n^*\varepsilon_L$ is large enough for lateral aggregation to begin, producing the short microfibrils. Meanwhile, axial growth continues forming more n^*-mers, short microfibrils, at the same time that some short microfibrils lengthen to become long microfibrils. These long microfibrils are very flexible mainly due to their diameter-to-length ratio. But this flexibility posses a challenge on the next stage of assembly to the mature fibril: Overcoming the entropy of the flexible chain will be non-trivial. We postulate a mechanism that involves the lateral addition of the short microfibrils to relatively straight regions of the long, flexible microfibrils. This is not unreasonable since the short microfibrils are much less than the persistence length of the flexible fibrils. The local increase in diameter will produce a higher stiffness, causing the microfibril to unfold slightly, which in turn favours more lateral addition of shorter aggregates. Because there are no covalent bonds formed during the assembly, annealing of the components may occur in order to produce a more compact arrangement within the mature fibril.

At present, we have investigated samples wherein we can see seemingly interacting fibrils. We note that (i) short microfibrils are the dominant species at all times of the assembly except the very beginning, and (ii) mature fibrils are always found surrounded by a sea of short microfibrils, which are in agreement with the proposed mechanism. Images such as that in Figure 4 show hints of intertwining that perhaps would lead to fibril growth. However, because these studies are all performed on air-dried samples, we cannot completely eliminate the possibility that the apparent interactions arose during the drying process, although we have found numerous examples of such. We are currently investigating the possibility of following the assembly *in situ*, while changing parameters of temperature, concentration and ionic strength of the solution.

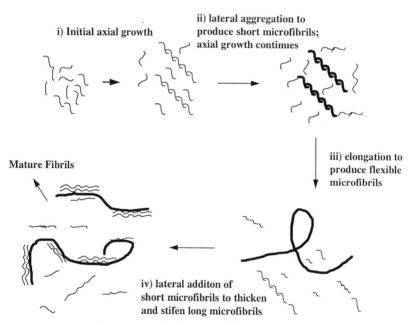

Figure 3.   Model for collagen assembly.

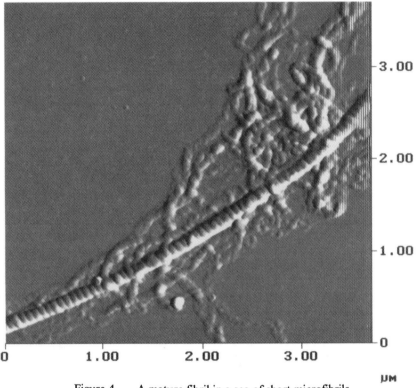

Figure 4.   A mature fibril in a sea of short microfibrils.

Summary

We have shown the utility of the AFM for studies of self-assembly in biopolymers of collagen from monomers to fibrils. The process appears more complex than in other filament-forming biopolymers, with fibril growth in both length and diameter. Nevertheless, there were only four distinct patterns of intermediates observed at different temperatures. Based on our AFM images, we have postulated a simple mechanism of assembly that has elements of both growth and hierarchical processes. At a slightly higher concentration, a non-sigmoidal behaviour of the light scattering intensity was found. While this could imply a different type of kinetics, it is difficult to produce a detailed interpreation of the scattering intensity for such a complex system. The AFM images did not show a difference in the assembly intermediates; however, quantitave data is still needed.

ACKNOWLEDGMENTS

We acknowledge the support of NSERC Canada and the Ontario Laser and Lightwave Research Centre.

Literature Cited

1. see, for example: Meakin, P. in *The Fractal Approach to Heterogenous Chemistry*; D. Avnir, Ed.; J. Wiley: Chichester, 1989.
2. De Young, L.; Fink, A. L.; Dill, K. A. *Acc. Chem. Res.* 1993, *26*, 614.
3. Pollanen, M. S.; Markiewicz, P.; Weyer, L.; Goh, M. C.; Bergeron, C. *Am. J. Pathol.* 1994, *145, 1140.*
4. Pollanen, M. S.; Markiewicz, P. ; Goh, M. C.; Bergeron, C. *Am. J. Pathol.* 1994, *144, 869.*
5. Pollanen, M. S.; Markiewicz, P. ; Goh, M. C.; Bergeron, C. *Colloids and Surfaces. A: Physicochemical and Engineering Aspects*, 1994, *87, 213.*
6. Pollanen, M. S.; Markiewicz, P. ; Goh, M. C. *J Neuropathol Exp Neurol*, 1997, *1, 79.*
7. Pollanen, M.S.; Goh, M. C. to be submitted.
8. Lodish, H. *et al. Molecular Cell Biology, 3rd ed.*; Scientific American Books: New York, 1995.
9. Schwarz, U. D.; Haefke, H.; Reimann, P.; Guntherodt, H. *J. Microscopy*, 1993, *173, 183.*
10. Markiewicz, P. ; Goh, M. C. *Langmuir* 1994, *10, 5.*
11. Markiewicz, P. ; Goh, M. C. *Rev. Sci. Instrum.*, 1995, *66, 3186.*
12. Markiewicz, P. ; Goh, M. C., *J. Vac. Sci. Technol. B* 1995, *13, 1115.*
13. Markiewicz, P. ; Pollanen, M. S.; Goh, M. C., submitted.
14. Markiewicz, P. ; Goh, M. C., submitted.

15. Holmes, D. F.; Capaldi, M. J.; Chapman, J. A. *Int. J. Biol. Macromol.* 1986, *8, 161.*
16. Gale, M. A.; Pollanen, M. S.; Markiewicz, P ; Goh, M. C. *Biophys. J.* 1995, *68, 2124.*
17. Fletcher, G. C. *Biopolymers* 1976, *15, 2201.*
18. SIlver, F. H.; Langley, K. H.; Trelstad, R. L. *Biopolymers*, 1979, *18, 2523.*

POLYMER MORPHOLOGY AND STRUCTURE: CRYSTALS TO FIBERS

Chapter 6

Deformation Induced Changes in Surface Properties of Polymers Investigated by Scanning Force Microscopy

S. Hild, A. Rosa, and O. Marti

Department of Experimental Physics, University of Ulm, D-89609 Ulm, Germany

In this study the possibility of combining commercial Scanning Force Microscopes (SFM) with stretching devices for the investigation of microscopic surface changes during stepwise elongation is investigated. Different types of stretching devices have been developed either for Scanning Platform-SFM or for Stand Alone-SFM. Their suitability for the investigation of deformation induced surface changes is demonstrated. An uniaxially oriented polypropylene film is stretched vertically to its extrusion direction. The reorientation of its microfibrillar structure is investigated and correlated to macroscopic structural changes determined by taking a force-elongation curve. Microtome cuts of natural rubber filled with 15 PHR carbon black are stretched. Changes in topography, local stiffness and adhesive force are simultaneously reported by using a new imaging method called Pulsed Force Mode (PFM).

Mechanical properties of polymers, e. g. Youngs-modulus or tensile strength, are correlated to their structure. Generally, the structure of a polymer is given by its particular structural arrangement (1, 2). In crystallizable polymers, like isotactic polypropylene, the structural arrangement is given by the arrangement of crystalline and amorphous region. The morphology of crystalline polymers is influenced by the crystallization conditions. During crystallization from the unstressed melt spherolithic structures will grow. The crystallization, which occurs in an external flow field, e. g. during the extrusion process, creates highly oriented structures. Depending on the extrusion conditions lamellar rows, shish-kebab or microfibrillar structures are generated (3, 4, 5). During uniaxial deformation, the differences in mechanical behavior can be observed by taking force-elongation curves (Figure 1). These variations are caused by specific changes in the original microscopic structure during the deforma-

110

tion process. Spherulites are transformed in fibrillar structures (*3*) or fibrillar structures can reorientate (*6*).

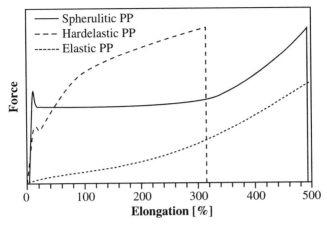

Figure 1: Schematic diagram of force-elongation curves for different morphologies of polypropylene.

In amorphous especially elastic polymers, like natural rubber, the mechanical properties are mainly influenced by the density of tie-points (*7, 9*). These tie-points can either be caused by molecular chain entanglements (*8*) or originated by chemical cross-linking (*9*), e. g. interlace rubber with sulphur. Another often used technique to modify mechanical properties of rubber is adding carbon black (*10, 11*). The unique mechanical properties of filled rubber such as increasing mechanical strength if the amount of filler increases are well known on macroscopic scale (Figure 2), but their microscopic origins however still not have been understood in detail.

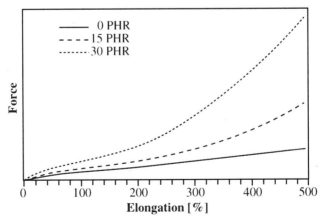

Figure 2: Schematic diagram of force-elongation curves of natural rubber filled with increasing content of carbon black.

On macroscopic scale, measuring force-elongation-curves during uniaxial deformation, e. g. stretching, is a conventional technique to determine mechanical properties, like Youngs-modulus or tensile strength. To investigate their microscopic origins stretching experiments are combined with other analyzing techniques. Optical microscopy (*12*), IR-spectroscopy (*13*) or X-ray diffraction (*5,12*) are successfully applied to investigate deformation induced changes in molecular structure, crystallinity or to observe orientation processes. These techniques are mainly suitably to investigate bulk changes. To study changes in surface structure predominately scanning electron microscopy (SEM) (*14, 15*) is used. For surface analysis by SEM polymers need metallization, because of this it is difficult to investigate more than one deformation state per sample. Scanning force microscopy (SFM) enables to investigate polymer surfaces with respect to their topography and surface properties like local stiffness, adhesion or friction from the μm-range down to the nm-scale without special pretreatment (*16, 17*). Therefore the combination of a SFM with a stretching device should provide a new method to investigate surface changes of polymers during uniaxial deformation (*18, 19*).

Whereas for the conventional analyzing methods like X-ray or SEM different commercial stretching devices are available, up to now no devices are developed for SFM. Few years ago, only SFMs have been available where the sample is scanned (Scanning-Platform-SFM or SP-SFM). These limits size and weight of sample. Because of this, only miniaturized stretching devices can be used. During the last years a new type of SFM, so called Stand-Alone-SFM (SA-SFM), has been developed, where the scanning unit is moved. These microscope can be placed on top of the sample, therefore the limitation in sample dimensions is cancelled. Therefore SA-SFM will be more suitable for the combination with a stretching device than SP-SFM.

In this study we will introduce different types of stretching devices which can either be combined with SP-SFM or SA-SFM. Their suitability for the investigation of deformation induced surface changes of polymers will be tested using two samples: A melt extruded Polypropylene film (OPP) is stepwise stretched and deformation induced changes in surface morphology are documented by SFM. Natural rubber filled with carbon black is investigated to document changes in surface topography and surface properties like local stiffness and adhesion during the stretching.

Instrumentation

In SP-SFM, the size and weight of the stretching device is limited, e. g. in Nanoscope III (*Digital Inc., Santa Barbara, CA, USA*) the maximum floor space of a stretching device can be 17 mm x 10 mm and the maximum height 10 mm. Because of this, a miniaturized stretching device is constructed, which fits in this SFM. To reduce weight the main parts of the stretching apparatus -two clamping jaws to fix the sample (Figure 3, 1a and 1b)- are made out of aluminium. One clamping jaw is fixed. The second, mounted on two sliding bars (Figure 3, 2), can be moved by turning a fine pitch thread screw (Figure 3, 3). Pressed the screw towards the fixed part of the apparatus the disengaged part moves. To mount the stretching device into the SFM, the fixed clamping jaw (Figure 3, 1b) is glued onto a magnetic holder. Because of the small size no force sensor and even no stepper motor can be implemented. Using a SA-SFM, this setup enables to measure under water.

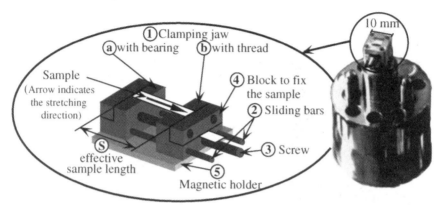

Figure 3: Schematic diagram of a miniature stretching device for conventional scanning force microscopes.

The sample, which must have a length of about 20 mm, can be mounted in the device as shown in Figure 3. Two blocks (Figure 3, 4) will fix the sample. An minimum effective sample length (Figure 3, S) of 3.5 mm for stretching is reached, when both clamping jaws are close together. To avoid slipping of the foils during stretching emery paper is glued on this blocks. Because of the SFM dimensions, the maximum elongation which is possible with this setup is 4 mm. By turning the screw elongation steps of 0.1 mm distance can be reached. To determine the stretching ratio the distance between the two clamping jaws is measured by a sliding gauge. The maximum force which can be applied to the sample before slipping takes place is determined to be 5 N. It has been shown, that polymer foils with a thickness major than 20 µm directly can be imaged. For elastic materials and or thin films, e. g. thin cuts, an additional support in the area of scanning is necessary to avoid vibrations.

Combining a stretching device with a SA-SFM enables to enlarge their dimensions. In Figure 4 a stretching device developed for a commercial SA-SFM (*CSEM Inc., Neuchâtel, Switzerland*) is shown. Here, stretching is done by two shifting tables driven by computer controlled stepper motors (Figure 4, 1). This balanced assembly is used to secure the symmetric deformation of the sample. Although very slow stretching rates of 0.01 mm/s can be used, no on-line imaging during stretching is possible. This is mainly due to vibrations of the stepper motors. Because of this, for all SFM investigations stepwise stretching and in-situ taking images is necessary.

To determine the applied force during the measurement, a force sensor with 5 N range is integrated (Figure 4, 2). This low maximum force is installed for the study of microtome cut sample. To investigate foils or compact samples the force sensor has to be changed. Force values are read out by computer. This enables either to measure force-elongation curves or to control the force during taking SFM images.

To control the force is especially important for the investigation of elastic materials, where relaxation processes can occur when the stretching process is interrupted.

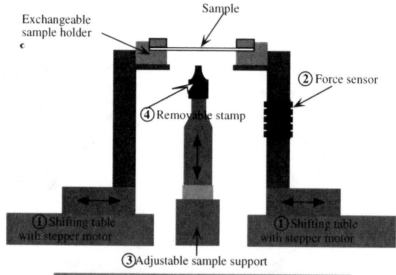

Figure 4: Stretching device for Stand Alone Scanning Force Microscopes. Deformation parameters are controlled by computer. Integrated force sensor allows direct monitoring of force.

In this setup the minimum size of the sample is limited by the dimensions of the cantilever holder (15 mm) of the used SFM. With respect to this size, a maximum elongation of 80 mm is possible. To avoid any sample vibrations during imaging a support (Figure 4, 3) with removable stamp (Figure 4, 4) is mounted below the sample. The height of the support can be adjusted by a micrometer screw. Exchanging the small stamp by heating or cooling plates allows additional control of sample temperature.

For imaging, the SFM is placed on top of three fixed supports (Figure 4, 5). The position of the microscope can be slightly changed. The supports are arranged beside and in the middle that way, that the cantilever position is aligned in the center of the sample. The main scan direction is vertical to deformation direction. Although this setup can easily be modified, the large dimensions of the stretching device and the fixed position of the SFM limit the suitability for other commercial SA-SFMs. Therefore, a smaller movable setup has been developed (Figure 5).

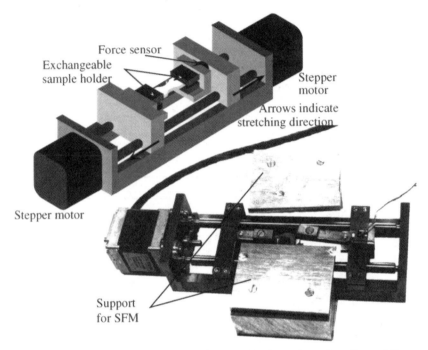

Figure 5: Stretching device suitable for Stand Alone Scanning Force Microscopes. The combination with an inverse microscope allows the position control of sample and cantilever.

In this setup, a SA-SFM can be placed on top of the apparatus by removable supports. If this support is adjusted on x-y-sliding tables, the position of the SFM can be changed during the experiment. This can be necessary if macroscopic changes, e. g. necking, take place outside the region where the cantilever is placed. This compact setup also can be placed on top of an inverse optical microscope. On one hand this allows position control of the cantilever on the other hand this enables the simultaneous analysis of optical properties like birefringence.

116

Experimental

To investigate deformation induced morphological changes a semicrystalline material is used. As sample material isotactic polypropylene has been chosen, because its deformation behavior is well studied (*3, 4, 6*). Two commercial melt drawn polypropylene films (*Hoechst, Factory Kalle, Wiesbaden, Germany*) are investigated: An uniaxially drawn film (OPP) with a thickness of 50 μm and a biaxially oriented film (BOPP), produced by two-step-stretching process from OPP, with a thickness of 25 μm. The crystallinity, established by Differential Calorimetry, is 67 % for OPP and 71 % BOPP. Both samples show a glass transition at about 12 C.

For the stretching experiments, the OPP film is cut into a rectangle with 20 mm length and 1 mm width. The narrow side is parallel to extrusion direction. After mounting the sample in the miniature stretching device (Figure 3) the effective sample length is 3.5 mm. A force-elongation-curve (Figure 6) for a similar sized sample have measured on a separate stretching device with integrated force sensor. The points of elongation where SFM images have been taken are marked by letters A - D.

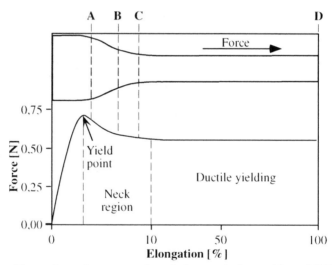

Figure 6: Force-elongation-curve taken during vertical stretching of OPP. Macroscopic changes of the foil are shown. Points A-D mark the elongation ratios, where SFM images are taken.

To study the surface in the unstrained and strained state, the miniature stretching device (Figure 3) is combined with a commercial SA-SFM (*Explorer, Topometrix, Santa Clara, CA, USA*). These combination allows in-situ imaging of a stepwise elongated film under water. Imaging under water is necessary for this sample to diminish electrostatic interactions between tip and polymer surface, which can falsify the determined sample height. The SFM measurements have been performed by operating in the static contact mode. As force sensors microfabricated silicon-nitride (Si_3N_4) tips with a spring constant of 0.05 N/m have been used.

Natural rubber (polyisoprene, PI) is used exemplary as elastic material. The investigated PI has an admixture of 1.8 PHR DCP for chemical crosslinking and 15 PHR carbon black N660 (*Degussa AG, Hürth, Germany*). The average diameter of the carbon black particles is determined to be 67 nm (*20*). For stretching experiments, these samples are cut using a low-temperature microtome to a thickness of 10 μm. The unstrained dimensions of the thin cuts are 4.0 x 1.0 mm². Because this is too small for mounting in the stretching device, the slice is glued onto a metallic holder similar to those used for SEM experiments.

The rubber thin-cuts are stepwise stretched to a maximum elongation of 450 %. After each interruption of stretching the sample is relaxed until force is constant, which is about 10 minutes, before taking SFM measurement. The force of the relaxed sample has been recorded. The resulting stress-strain curve is shown in Figure 7. This curve is similar to a curve taken with low deformation rate of about 0.1 mm/sec.

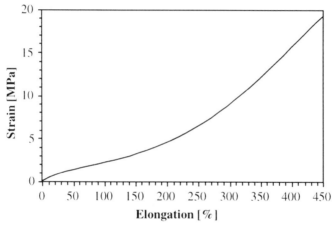

Figure 7: Stress-strain-curve of rubber sample filled with 15 PHR carbon black cut by low-temperature microtome.

SFM images are taken by a commercial SA-SFM (*CSEM inc., Neuchâtel, Switzerland*). Because of the softness and stickiness of the rubber surface, imaging in contact mode is difficult. Using Tapping mode (*Digital Inc., Santa Barbara, CA, USA*) enables to image natural rubber, but no significant difference between filler particles and rubber can be distinguished (*21*). Due to the high modulation frequency of the cantilever of about 200 kHz rubber appears hard in Tapping mode. Recently, we have introduced the Pulsed Force Mode (PFM) for the investigation of soft materials without disruptive lateral forces (*22*). In this mode the cantilever is modulated with frequencies in the range from 500 Hz up to 2 kHz. Because of the lower modulation frequency rubber appears softer than filler particles. Beside this, the PFM enables the additional imaging of surface properties. Topography, local stiffness and adhesive forces are simultaneously mapped. A detailed description of this mode is given in

118

(Rosa, A. *Measurm. Sci. Techn., submitted*). SFM images of PI thin cuts in un-strained state and strained up to 400 % elongation have been taken using the PFM at a modulation frequency of 1.6 KHz and a modulation amplitude of about 90 nm. Si_3N_4-Cantilevers with a spring constant of 0.9 N/m are applied.

Results and discussion

Morphological changes of semicrystalline films

In Figure 8 the topography of the unstrained OPP film is shown. In unstrained state the force micrograph is dominated by a fibrillar structure, which is highly oriented parallel to extrusion direction. The lateral extension of these fibrils is larger than the chosen scan range of 2.5 μm. The surface corrugation perpendicular to the fibril or-ientation is about 25 nm. The average diameter of the fibrils determined by cross-section (Figure 8, A) varies between 20 nm and 50 nm. Imaging the surface with a scan range of 150 μm shows a similar fibrillar structure like Figure 8, but here the diameters of the fibrils are about 200 nm. Because of these observations we assume, that the fibrillar structure detected on 150 μm scan range consists of bundles of smaller, singular fibrils. At the smaller scan range of 2.5 μm these singular fibrils are imaged. The structure of a singular fibril can either be microfibrillar or shish-kebab like (*23*). Microfibrils consist of highly oriented polymer chains with a small con-tent of crystalline lamellae. Shish-kebabs are assembled of a nucleus of extended chains with lamellar rows, which grow perpendicular from the extended chains into the melt.

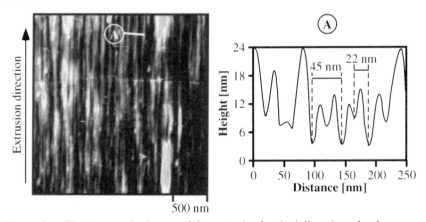

Figure 8: The topography image of the unstrained uniaxially oriented polypropy-lene (OPP) has a maximum height of 25 nm.

To determine the structure of the singular fibrils illuminated maps are used (Figure 9). In this diagram the fibrils seem to have a weak corrugation vertically to the extrusion direction. Because of these corrugation we assume that their structure is shish-kebab like. Microfibrils have to have a more or less plain surface. To simpli-fy a singular fibril further will be called "fibril" without differentiation. Figure 9 also shows, that the fibrils are not perfectly aligned one parallel to each other. They are more or less twisted. Although the fibrillar structure can clearly be distinguished, the surface structure appears slightly distorted. We propose to explain this by an amorphous layer on top of the fibrils.

Upon straining the foil perpendicularly to the extrusion direction (ED) a typical force-elongation-curve for semicrystalline polymers with Hookean region, yield point, necking and ductile yielding can be detected (Figure 6). On macroscopic scale during stretching the neck-in of the film and a zone of ductile yielding can be observed (Figure 6). The points of elongation where SFM images have been taken are marked by letters A-D in the force-elongation curve (Figure 6, A-D). Corresponding SFM images are shown in Figure 10, 11, 13 and 14.

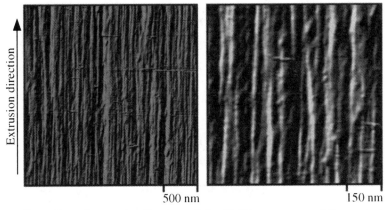

500 nm   150 nm

Figure 9: For a better visualization of the fibrillar structure of the unstrained uniaxially oriented polypropylene (OPP) illuminated maps can be used.

Up to 3 % deformation the film shows Hookean elasticity with a linear increase in force. Then the yield point with maximum force of 0.75 N appears in force-elongation curve. either on macroscopic or on microscopic scale changes in the sample shape are visible. Nevertheless on microscopic scale, the interfibrillar areas should be slightly stretched. But these changes in interfibrillar distance are too small to be detected by SFM, because of the tip radius of about 10 nm. Stretching the film closely beyond the yield point, the formation of a neck occurs. Simultaneously the detected force decreases. Now, changes in surface structure can be observed (Figure 10).

The fibrils start to reorientate parallel to the applied force. This reorientation process seems to occur stepwise at different surface layers. Whereas on the layers more far away from the surface the parallel alignment still can be seen, on the first layer near to the surface, the parallel alignment of the fibrils is dissolved into a less ordered one. Additionally, at this point of elongation a lot of fibrillar entanglements can be seen. We assume that these fibrillar entanglements will appear at points where former twisting of the fibrils can be observed. Because of this "knots" the failure in between the highly oriented fibrils is prevented during the vertical elongation. Without these strong tie-points, when the fibrillar structure is only connected by amorphous material a film fails if only small load is applied (*19*).

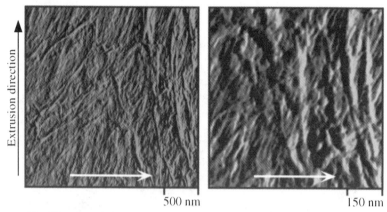

Figure 10: Illuminated SFM images taken after stretching beyond yieldpoint. Arrow indicates the stretching direction. Reorientation of fibrillar structure starts.

At further deformation up to 7 % elongation, on macroscopic scale a neck is improved. In the neck region (Figure 6, B) no preferred orientation of the fibrils can be imaged. The whole structure is transformed into a net-like or woven structure (Figure 11).

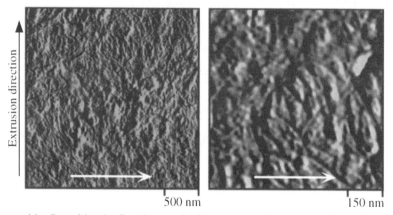

Figure 11: Stretching in direction marked by arrow until macroscopic necking occurs. Illuminated SFM micrograph shows no preferred orientation.

Based on X-ray and SEM measurements such woven structures have been proposed by several authors (*24, 25, 26, 27*) to format during the two-step biaxial orientation of polypropylene. In Figure 12 the force micrograph of a biaxially oriented polypropylene film (BOPP) prepared by two step stretching process (Figure 12) is shown.

Again, the layer-like structure of the fibrillar network can be seen. Comparing the illuminated image with the structure of the stretched OPP shown in Figure 11 similar, woven structure can seen.

a) Topography   b) Illuminated image

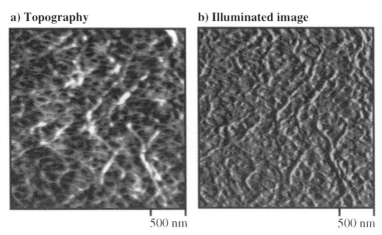

500 nm   500 nm

Figure 12: SFM image of biaxially oriented polypropylene (BOPP). The topography shows maximum height of 40 nm (a). For better comparison to OPP images the illuminated map is shown (b).

After a small increase of elongation up to 10 % the net-like structure is transformed into a deformation state, where the fibrils are again preferentially oriented (Figure 13). Now, the preferential orientation is tilted perpendicular to ED. At this strain state also different surface layers can be distinguished.

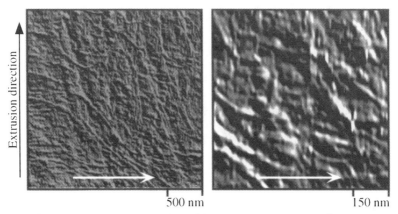

500 nm   150 nm

Figure 13: Stretching near to ductile yielding (10 % in direction of the arrow), reorientation of fibrillar structure parallel to external stress becomes visible in illuminated SFM images.

122

Stretching the film to elongation rates higher than 10 % causes ductile yielding of the sample. The detected force becomes constant. SFM micrographs of 100 % elongated OPP reveal that the fibrillar structures is aligned parallel to the stretching direction (Fig 14). The thickness of the microfibrils reduced about 20 %. Whereas in Figure 9 the surface seems to be slightly blurred, now the singular fibrils clearly can be distinguished. The amorphous surface layer seems to disappear. An explanation for this effect can be, that the amorphous chains are converted into fibrillar ones.

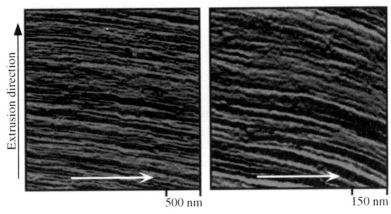

500 nm   150 nm

Figure 14: Illuminated SFM images taken at 100 % elongation when ductile yielding takes place. The fibrillar structure is reoriented parallel to stretching direction, which is indicated by arrow.

Based on the SFM images and literature date we propose a schematic model for orientation of the fibrillar structure during stretching perpendicular to extrusion direction, which is shown in figure 15.

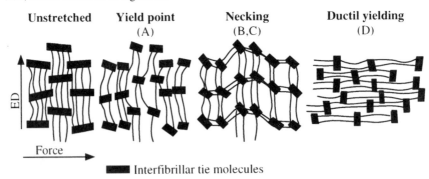

Figure 15: Schematic diagram of fibrillar reorientation. Interfibrillar tie molecules prevent failure between adjacent fibrils.

When elastic behavior of the film can be observed the interfibrillar areas are slightly stretched (Figure 15, Yieldpoint). During macroscopic necking, the fibrillar structure starts to reorientate parallel to the applied force. This reorientation process first happens near to the surface. By way of an intermediate state where no preferred orientation of the fibrils can be observed (Figure 15, Necking) the fibrillar structure is tilting. Further elongation results again in a fibrillar structure where the fibrils are highly oriented parallel to the applied force. This reorientation process is preferred of failure of the film if strong intermolecular tie-points exist. In the investigated film these tie-points seem to be the twisted fibrils, which prevent the film of interfibrillar failure.

Changes in surface properties

The topography of both unstrained and strained natural rubber filled with 15 PHR carbon black is shown in Figure 16. In unstrained state two regions of height can be distinguished. A low, dark one, which, we assume to be the polymeric matrix. In this dark area higher particles (light) are visible. The maximum height of this features is about 120 nm. Because of their lateral size of about 70 nm, these should be filler particles. After stretching up to 400 % deformation, filler particles are orientated parallel to the stretching direction. Whereas in unstrained state the polymeric matrix appears homogeneous, after stretching a slight line-like structure is visible. These can be due to the orientation of molecular chains parallel to the extrusion direction and reflects the inhomogeneous deformation of the rubber (28).

Unstretched rubber (15 PHR) **Stretched rubber** (15 PHR)

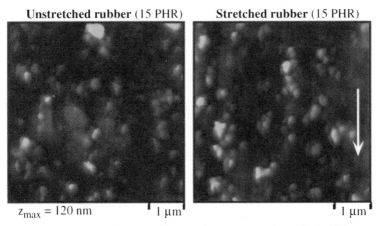

z_{max} = 120 nm   1 μm   1 μm

Figure 16: Topography of rubber filled with 15 PHR carbon black. Filler particle appears higher. Stretching up to 400 % causes an orientation parallel to stretching direction (marked by arrow).

Changes in surface properties like local stiffness or adhesive force can be used to get more information about the microscopic structure and changes during stretching. First, local stiffness is imaged (Figure 17). In the unstrained state areas of different compliance can be visualized which can be correlated to topography. The main part of the surface appears soft (dark), this correspondents to the area, we assumed to be the polymeric matrix. Because of the chemical structure carbon black particles should have lower compliance than the rubber matrix. The regions with higher topography are harder. This confirmed the assumption that here the filler particles are located.

Whereas in unstrained image the filler clusters show random orientation, they get orientated after deformation to 400 %. The local stiffness of the matrix is reduced. This can be due to the orientation of molecular chains. Like in topography, linear structures parallel to the strain direction appear in the polymeric matrix. But in the local stiffness image these lines are more distinctive than in the topography image. It seemed, that the line-like structures of the matrix connect adjacent filler clusters. Therefore, deformation of the matrix between filler clusters is larger than in the surrounding areas. The rubber is unhomogeneously deformed.

Unstretched rubber (15 PHR)   **Stretched rubber** (15 PHR)

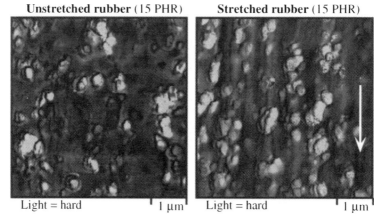

Light = hard   1 µm   Light = hard   1 µm

Figure 17: SFM micrograph of local stiffness of rubber filled with 15 PHR carbon black. Filler particles are stiffer (lighter) than the polymer matrix (dark). After stretching up to 400 % (arrow indicates stretching direction) the local stiffness increases and the line like structure becomes more visible.

A more careful analysis of the filler particles shows differences in their compliance. A possible explanation for this observation based on the fact, that the local stiffness signal also contains information of layers below the surface layer (29). Therefore it is difficult in the local stiffness images to distinguish clearly between a surface-near-volume up to a depth of about 100 nm and the very surface itself. It is known that in filled rubber the carbon black is coated by thin layer of polymer, so called "bound rubber" (30). The differences in the local stiffness are a hint that some of the filler particles should be coated by such a polymeric layer.

A possibility to discriminate between uncoated and coated filler clusters is to investigate the chemical interaction between surface and tip (31, 32, 33) by measuring the adhesive force. Adhesive force is sensitive only for the surface. PFM enables to map adhesive forces simultaneously with topography and local stiffness (Figure 18). In the unstrained sample the polymer matrix shows higher adhesion (lighter) than the filler particles. In the areas where filler particles are presumed two values of adhesive force can be distinguished, which shall be due to uncoated and coated carbon black. Uncoated filler appears black, which means the adhesive force is low. Coated filler shows higher (gray) adhesion, but the adhesive force is lower than in the surrounding matrix. To explain the difference in adhesive force between the polymeric matrix and the bound rubber layer a simple model is used: We propose that the adhesive force is correlated to the formation of adhesive contact between tip and polymer surface. In the unstretched natural rubber the polymer chains are non ordered like in a spaghetti ball and a lot of free chain ends are available to form an adhesive contacts to the tip. In bound rubber reactive sides of the chains are bound to the carbon black, therefore the number of free chain ends and the probability of forming adhesive contacts, is reduced. The determined adhesive force decreases.

Unstretched rubber (15 PHR) **Stretched rubber** (15 PHR)

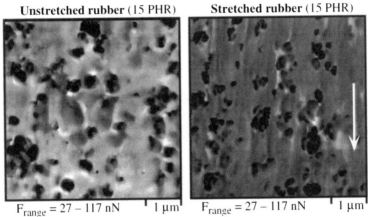

$F_{range} = 27 - 117$ nN 1 µm $F_{range} = 27 - 117$ nN 1 µm

Figure 18: SFM adhesion micrograph of rubber filled with 15 PHR carbon black. In unstretched state, the polymeric matrix shows high (light), the filler particle low adhesion (dark). After stretching up to 400 %, the adhesion of matrix decreases.

After stretching, the adhesion of the filler clusters is unchanged. Comparing the SFM adhesion micrographs for the unstrained and strained sample, a decrease in adhesive force of the rubber matrix can be observed that way, that it is difficult to discriminate between polymeric matrix and coated filler. For a more detailed investigation of the adhesion forces histogram technique is used. Here the amount of pixels belongs to the same intensity, which means the same adhesive force, is determined. In Figure 19 histograms of adhesion images taken at different stain ratios are shown. Two peaks with different intensity are visible: The small one at an adhesive force of about 10 nN does not change the position with increasing elongation. The higher one shifted with increasing deformation towards from a maximum value of 80 nN to lower adhesion 40 nN.

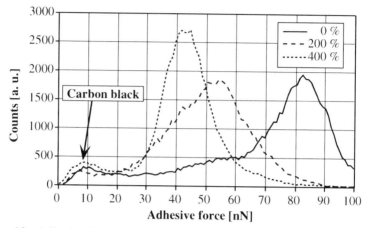

Figure 19: Adhesion distribution of rubber filled with 15 PHR determined at different strain states shows two peaks: The filler particles have an adhesive force at about 10 nN, adhesive force of rubber decrease with respect to deformation ratio.

Based on the SFM images the small peak at low adhesive force can be allocated to the carbon black particles. The peak with higher intensity belongs to the rubber. The reduced difference in adhesive force between coated filler and stretched polymeric matrix, confirmed the model of structure dependent adhesive force: During stretching the polymer chains will be oriented. For stretched chains the possibility of adhesive bond formation is reduced. This results in a decrease of measured adhesive force. This model neglected the influence of contact area or contact time. Therefore quantitative analysis of adhesive values can not be given here.

Conclusions

This study shows possibilities for the investigation of microscopic surface changes during stepwise elongation by SFM. For isotactic polypropylene, the reorientation of microfibrils can be visualized and correlated to macroscopic changes detected from the force-elongation- curves. A model for the microscopic surface changes is given. From SFM images it seems that the reorientation processes first starts at the surface. Bulk reorientation appears at higher elongations. For a detailed analysis the experiment has to carried out again with a stretching device where definite small steps of elongation can be done and the force can be controlled during the whole investigation. This will be possible with the stretching device shown in Figure 5.

Using a new mode, called PFM, enable to investigated deformation induced changes in topography, local stiffness and adhesion of filled rubber simultaneously. The inhomogeneous deformation of the polymeric matrix can be shown. Besides this, filler particles consisting of carbon black are determined. Comparing local stiffness and adhesive force images allows to distinguish the very surface of coated and uncoated fillers. Histograms of adhesive force enables the direct observation of the influence of deformation in the adhesive force. Depending on this a simple model is proposed, which describes the changes in adhesive force by microscopic changes in rubber structure. For a more quantitative description of these phenomenon it is necessary to take into account changes in local stiffness, which results in changes of contact area, and the influence of contact time.

Acknowledgements

This work was supported by the Sonderforschungsbereich 239. We gratefully acknowledges the Friedrich Ebert Foundation, Bonn, Germany and the DFG for their fellowships. We are thankful to E. Weilandt, B. Zink, R. Brunner, B. Heise, M. Pietralla and H. G. Kilian for fruitful discussions. Sample materials were prepared by M. Kienzle. Control electronics were implemented by G. Volswinkler.

References

1. Kreibich, U.T.; Lohse, F.; Schmid, R.; Wegner, G. in Polymere Werkstoffe, Bd. 1.; Batzer, H.; Thieme Verlag, New York, NY, 1985, 588 - 714

2. Holliday, L. in Structure and properties of oriented Polymers; Ward, I. M.; Materials Science Series; Applied Science Publishers: London, UK; **1975**

3. Samuels, R. J. Journ. Polym. Sci. Phys. **1970**, 4, 701

4. Peterlin, A. in Structure and properties of oriented Polymers; Ward, I. M.; Materials Science Series; Applied Science Publishers: London, UK; **1975**, 36-56

5. Noether, H. D.; Whitney, W. Kolloid-Z. u. Z. Polym. **1973**, 251, 991

6. Okijima, S. N.; Tanaka, I. H. Polym. Let. **1971**, 9, 797

7. Treloar, L. R. G. The physics of Rubber Elasticity; Claredon Press: Oxford, UK; **1958**

8. Mehnert, K.; Paschke, E. in Polyolefine; Schley, R.; Schwarz, A.; Kunststoff-Handbuch, 4; Carl-Hanser-Verlag, München, D,

9. Wood, L. A.; Bullman, G.W. Journ. Polym. Sci. Phys. **1972**, 10, 43

10. Hess, W. M.; Herd, R. In Carbon black; Donnet, J-B., Bansal, R. C., Wang, M-J.; Marcel Dekker, Inc. New York, NY, **1993**, 108 - 113

11. Kilian, H.G.; Strauss, M.; Hamm, W. Rub. Chem. Techn. **1994**, 67, 1

12. Stein, R. S.; Wilkes, G.L. in Structure and properties of oriented Polymers; Ward, I. M.; Materials Science Series; Applied Science Publishers: London, UK; **1975**, 57-149

13. Read, B. E. in Structure and properties of oriented Polymers; Ward, I. M.; Materials Science Series; Applied Science Publishers: London, UK; **1975**, 150-186

14. Miles, M.; Petermann, J.; Gleiter, H. Journ. Macrom. Sci. **1976**, B4, 523

15. Michler, G. H. Kunststoff-Mikromechanik, Hanser Verlags: München, D, **1992**

16. Yang, A. C.-M.; Terris, B. D.; Kunz, M. Macromolecules **1991**, 24, 680

17. Overney, R. M.; Lüthi, R.; Haefke, H.; Frommer, J.; Meyer, E.; Güntherrodt, H.-J.; Hild, S.; Fuhrmann, J. Appl. Surf. Sci. **1993**, 64, 197

18. Hild, S.; Gutmannsbauer, W.; Lüthi, R.; Fuhrmann, J.; Güntherrodt, H.-J. Helv. Phys.Act. **1995**, 256, 759

19. Hild, S.; Gutmannsbauer, W.; Lüthi, R.; Fuhrmann, J.; Güntherrodt, H.-J. Journ. Polym. Sci. Phys. **1996**, 34, 1953

20. Stumpf, M. Thesis, Ulm, **1992**

21. Rosa, A. Thesis, Ulm, **1997**

22. Marti, O. in Micro/Natribology and its Application; Bhushan, B.; Nato ASI Series, E, Appl. Sci . Vol. 330; Kluwer Academic Publisgers: Dordrecht, NL; **1996**, 17-34

23. Wunderlich, B. Macromolecular Physics, Academic Press: New York, NY, **1976**, Vol. 2

24. Adams, G. Polym. Sci. Techn. **1973**, 1, 169

25. Uejo, H.; Hoshino, K. Journ. Appl. Polym. Sci. **1970**, 14, 317

26. Khoury, F. J. Res. Nat. Bur. Std. **1966**, 70A, 383

27. Padden, F. J.; Keith, H. D. Journ. Appl. Phys. **1966**, 19, 4013

28. Vilgis, T. A. Macrom. Theory Simul. **1994**, 3, 217

29. Overney, R. M.; Leta, D. P.; Pictroski, C. F.; Rafailovich, M.; Liu, Y.; Quin, J.; Sokolov, J.; Eisenberg, A.; Overney, G. Phys. Rev. Lett. **1966**, 19, 4013

30. Donnet, J. B. ; Vidal, A. Adv. in Polym. Sci. **1986**, 76, 104

31. Frisbie, C. D.; Rozsnyai, L. W.; Noy, A.; Wrington, M. S.; Lieber, C. M. Science **1994**, 265, 2071.

32. Nakagawa, T.; Ogawa, K; Kurumizawa, T.; Ozaki, S. Jpn. J. Appl. Phys. **1993**, 32, 294.

33. S. Akari, D. Horn, Keller, H. Adv. Materials **1995**, 7, 594,

Chapter 7

Scanning Probe Microscopy: A Useful Tool for the Analysis of Carbon Black-Filled Polymer Blends

Ph. Leclère[1], R. Lazzaroni[1], R. Lazzaroni[1], F. Gubbels[2], M. De Vos[3],
R. Deltour[3], R. Jérôme[2], and J.L. Bredas[1]

[1]Service de Chimie des Matériaux Nouveaux, Centre de Recherche en Electronique
et Photonique Molécules, Université de Mons-Hainaut, B-7000 Mons, Belgium
[2]Centre d'Etude et de Recherche sur les Macromolécules (CERM), Institut de
Chimie, Université de Liège, B-4000 Sart-Tilman, Belgium
[3]Laboratoire de Physique des Solides, Université Libre de Bruxelles,
B-1050 Bruxelles, Belgium

Conducting polymer composites, that consist of a conducting filler
randomly distributed throughout insulating polymers or polymeric
materials, deserve interest for different applications such as sensors and
electromagnetic radiation shielding. The electrical resistivity of Carbon
Black (CB)-filled multiphase polymer blends depends on the CB
localization. This study has emphasized that Lateral Force Microscopy is
a powerful tool to investigate the morphology of CB-filled polymer
blends in relation to the blend composition and CB loading. As a rule,
CB can be selectively localized in one phase or at the interface of a
binary polymer blend, depending on the thermodynamics of the ternary
system and the kinetics of the dispersion process. The selective
localization of CB at the interface of polyblends with a co-continuous
two-phase morphology is a unique situation that allows the CB
percolation threshold to be decreased down to 0.5 wt%.

Electrical conductivity can be imparted to insulating polymers by loading with a
conductive filler (1,2). Processing, electrical properties and theoretical analysis of
these materials have been reported for the last four decades in the scientific
literature (3). The basic research strongly developed in the 70's, namely because of
the progress in the understanding of physics and physical chemistry of dispersed
systems and randomly heterogeneous materials.

The choice of the filler/polymer pair is crucial for the production of electrically conducting polymer composites since the electrical properties depend on the filler-filler and filler-polymer interactions. If the filler self-interactions dominate, then the filler particles tend to stick together and to form aggregates rather than chains of particles in contact. Conversely, an exceedingly good adhesion of the polymer to the filler results in an insulating layer around the particles and prevents the formation of current-conducting chains. Another major problem in the production of such systems is the filler content, which must be kept as low as possible in order to make the processing conditions, the mechanical properties, and the final cost as attractive as possible. For this purpose, a general strategy has been proposed that relies upon a combination of the structural characteristics of *composites* (polymer/filler combinations) and *polymer alloys* (polymer/polymer combinations)(4-9). Indeed, percolation of conductive filler particles in one phase or at the interface of a co-continuous binary polyblend can lead to a double percolation (one for the polymer phase and one for the filler in this phase or at the interface) and accordingly provide the polymeric material with electrical conductivity at a very low filler content.

Extrinsic conductive polymers are not only attractive as valuable models for the experimental study of disordered matter and composites, but also for potential applications. For instance, they can be use as anti-static and electromagnetic shielding materials to protect electronic equipment from incoming electromagnetic interference or to prevent radiation emission in the outer space. When pressure is applied to conductive filled polymers, the difference in the moduli of the matrix and the filler can markedly change the particle-to-particle contacts and result in resistivity modifications that may appear as either a positive or negative piezoresistance depending on the filler grade and contact, and the nature of the polymer matrix (10). These piezoresistive materials can be used as pressure sensors or switches. When a conductive filled polymer is heated above room temperature, it can undergo a huge resistance increase, known as the positive temperature coefficient (PTC) effect which may be used in electrical safety devices and self-regulated heaters.

A better knowledge of the actual structure of the dispersed particles, in relation to the material processing and the final macroscopic properties is still required, and proves to be a difficult problem since length scales from ten nanometers (typical diameter of filler particles) to a few microns (possibly millimeters) are involved. Tools such as electronic microscopies and scanning probe microscopies (Scanning Tunneling Microscopy and Atomic Force Microscopy) can be of major help. For instance, scanning probe potentiometries (11,12) have been used to analyze the electrostatic forces on the surface, however with a low lateral resolution. Recently, Electric Force Microscopy (EFM) has been proposed (13) as a new type of scanning probe microscopy, which is able to measure electric field gradients near the surface using a sharp conductive tip.

Experimental section

High-density polyethylene (PE) (Solvay Eltex B3925: M_n = 8,500, M_w = 265,000, density 0.96, melt index < 0.1), polystyrene (PS) (BASF Polystyrol 158K: M_n = 100,000, M_w = 280,000, density 1.05, melt index 0.39) and two grades of carbon black (CB) particles (Degussa Printex XE-2 (XE), and Cabot Black Pearls BP-1000 (BP), which is less purely graphitic than XE) were mixed in an internal mixer (Brabender Rheomixer) at 200°C. The initial mixing rate was 25 rpm for two minutes and then increased up to 64 rpm for 10 minutes. Polymer blends were compression molded at 200°C for 10 minutes as sheets suited to electrical resistance measurements. Phase morphology was usually investigated immediately after melt mixing and rapid cooling of the samples in liquid nitrogen. Melt viscosity and melt elasticity were measured with a dynamic mechanical analyzer (Rheometrics DRA-700) at 200°C. Electrical measurements were performed with the four-probe technique (in order to avoid resistance at the sample/electrode contacts). Atomic Force Microscopy (height and friction) images were recorded with a Digital Instruments Nanoscope III microscope, operated in contact mode under ambient conditions with a 100 μm triangular cantilever (spring constant of 0.58 Nm^{-1}). All the AFM images were recorded with the maximum number of pixels in each direction (*i.e.*, 512 pixels) and from several regions of each sample. The scanning time was *ca.* 5 minutes. The images in this paper were not filtered and thus shown as captured.

Electrical and morphological characterization of CB-filled homopolymers

Figure 1 shows how the resistivity of PE or PS depends on CB nature and content. As a rule, at low CB content, the electrical resistivity is close to that of the polymer matrix, *i.e.* 10^{11} to 10^{16} Ω.cm. When the CB content is increased, the resistivity sharply decreases by several decades over a narrow concentration range typical of the percolation threshold (14-18). It then decreases more slowly towards the limiting resistivity of the compressed filler (*ca.* 10^4 to 1 Ω.cm).

According to the percolation theory, there exists a connectivity threshold below which the conducting particles are only partially connected and above which in any sample of large enough dimensions, full connection occurs. The resistivity of the composites obeys a power law of the form $\rho = (p - p_C)^{-t}$ near the transition, where ρ is the bulk resistivity of the composite, p is the concentration of the conductive component, p_C is the percolation threshold concentration, and t is a universal exponent. The currently accepted theoretical value for the universal conductivity critical exponent in three-dimensional lattices is t = 2.0 (19).

Although the percolation theory predicts a percolation threshold of *ca.* 20 vol% (20), the experimental observations lie in a large range of 5-30 vol% (21-23). The values of p_C are listed in Table I for PE and PS homopolymers filled with either XE or BP particles. In the case of XE, p_C is 5% and 8% in PE and PS, respectively. Higher p_C values are observed for BP, *i.e.* 12% and 25%. In agreement with previously published data (24), it appears that substitution of monophasic polystyrene by a two-phase semicrystalline polyethylene results in a decrease of the percolation threshold (from 8% down to 5% for XE and from 25% to 12% for BP). This observation is consistent with the selective localization of CB particles in the amorphous phase of PE. Mixing of PE with a non-miscible polymer in which CB has no tendency to accumulate is an alternative approach for decreasing the CB percolation threshold, that has been worked out successfully (24, 25).

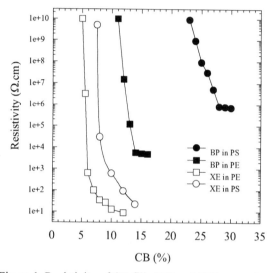

Figure 1. Resistivity of CB filled PS and PE homopolymers.

Table I - Percolation threshold and critical exponent observed for various CB and polymers (32).

System	Ω_0 ($\Omega.cm$)	p_C (%)	t
XE in PE homopolymer	3.8×10^{-3}	5.0	2.2
XE in PS homopolymer	10×10^{-3}	8.0	2.2
BP in PE homopolymer	1	12.0	1.9
BP in PS homopolymer	10^2	25.0	1.9

It is clear from Figure 1 that the CB percolation threshold in one homopolymer changes with the grade of CB particles. This behavior is not predicted by the conventional percolation theory that considers ideal non interacting spherical particles. Actually, CB consists of clusters of primary particle aggregates, whose fate depends on the interactions with the dispersion medium and the mechanical dispersion forces. It is therefore not surprising that the percolation threshold (1) depends on the grade of CB, the polymer matrix, and the main processing parameters.

Characterization of CB-filled polymer blends

CB-filled polymer blends can be characterized in direct space at the nanometer scale by conventional Atomic Force Microscopy and Lateral Force Microscopy (or Friction Force Microscopy) as well (26). Figures 2 and 3 compare the height image and the friction mode image of a blend of 45% PE, 55 % PS, and 1% XE. It is straightforward that LFM is more efficient and useful than contact height image. The three constitutive components (PE, PS and CB) are clearly discriminated by LFM as the result of a difference in their friction coefficients μ. It appears from the collected images that PE has a higher μ value, which provides it with the lighter color on the gray-scale. Since there is almost no interaction between the tip and CB these particles must be black. The μ value is intermediate for PS, which is therefore dark gray on the image.

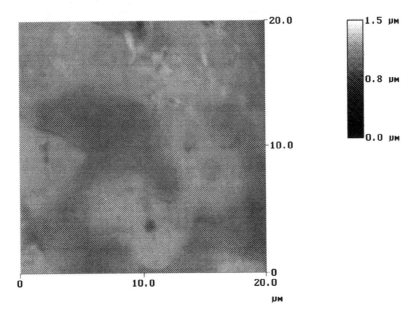

Figure 2. Contact mode AFM height image of the PE/PS/XE (45/55/1) blend.

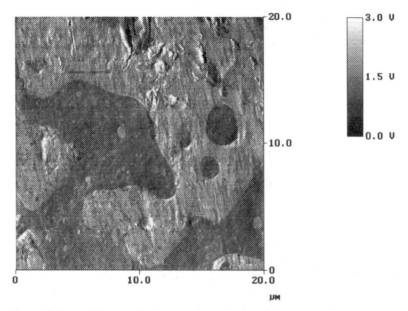

Figure 3. Lateral Force Microscopy image of the PE/PS/XE (45/55/1) blend.

Figure 4 is a more detailed LFM image of the PE/PS/XE (45/55/1) blend where the black spots are assigned to CB particles, which are selectively dispersed in the PE phase. Taking the blend composition and p_C for XE in PE into account, a p_C value of 2.25 % should be predicted, which is in a pretty good agreement with the experimental value (2.9 % in Table II). In the case of BP-type CB, the CB particles are localized in the PS phase, and the percolation threshold is then $ca.$ 10.9 %, in a reasonable agreement with the theoretical value (13.75 %).

So changing the CB grade may deeply change the CB distribution between the polymer phases. Indeed, BP has a marked preference for PS, whereas XE strongly prefers PE. A better understanding of the nature of the interactions between CB and PE or PS has emerged from theoretical calculations based on quantum-chemical approaches (27, 28) (post-Hartree-Fock ab $initio$ technique including electron correlation via second-order perturbation theory) and empirical molecular mechanics techniques (27, 29). It is important to note that all these methods lead to very similar conclusions. The presence of functional groups (such as -COOH) on the surface (as it would be the case for BP particles) leads to a higher binding energy with PS than with PE as result of the interactions with the -COOH groups of BP and the phenyl rings of the PS chains. Moreover, the binding energy density on a clean surface (such as XE surface) is higher for PE than for PS.

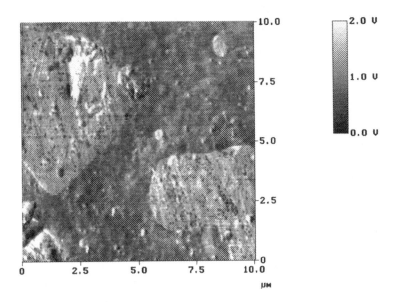

Figure 4. Detailed LFM image of the PE/PS/XE (45/55/1) blend. PE is brighter and PS darker; black spots in the PE phase are the CB aggregates.

Table II - Percolation threshold and critical exponent in relation to CB localization in PE/PS blends (32).

Morphology	Ω_0 $(\Omega.cm)$	p_C (%)	t
XE in PE phase in PE/PS	5.0×10^{-3}	2.9	2.0
BP in PS phase in PE/PS	2	10.9	2.0
XE at PE/PS interface (Mech.)	0.14	0.39	1.5
XE at PE/PS interface (Kin.)	0.60	0.60	1.3
CB at PE/PS interface (Thermo.)	1.40	0.46	1.3

Selective localization of CB in PE/PS blends

The percolation threshold is thus expected to decrease when the volume fraction of the continuous phase in which CB is selectively localized is decreased (24, 25). The double percolation has been discussed elsewhere (25, 30, 31).

The value of p_C can however be further decreased by the selective localization of the CB particles at the interface of a co-continuous PE/PS blend. This very specific localization can be achieved as result of either a kinetic or a thermodynamic control.

Kinetic localization of CB at the PE/PS interface. Since the CB grade is a key parameter for the preferential, if not selective localization of CB particles in one phase of a polyblend, Gubbels (32) has proposed to mix first the CB particles with the less interacting phase (*e.g.* XE with PS, and BP with PE). When the second polymer is added, the CB particles tend to migrate from the phase in which they have originally been dispersed to the second phase. By stopping the phase mixing at regular time intervals and recording a LFM image, the optimum time t_C for the CB particles to reach the PE/PS interface can be determined. Figure 5 shows the LFM image for this optimum mixing time t_C in case of the 45% PE, 55% PS and 1% BP blend. A CB layer at the PE/PS interface is clearly seen. The thickness of this layer is about 100 nm. For this system, the percolation threshold is as low as 0.6 wt%.

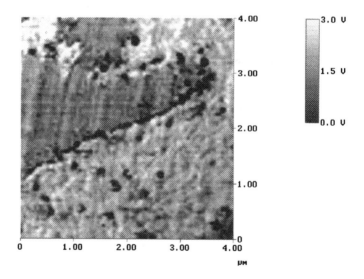

Figure 5. LFM image of the CB localization at the PE/PS interface as result of a kinetic control. (PE/PS/BP = 45/55/1). (t = t_C)

When 1% BP is originally dispersed in the PS phase, in which the percolation threshold is about 25 wt% (Table I), the electrical resistivity is very high. Upon increasing mixing times, the resistivity decreases so as to reach a value of 2.3×10^4 Ω.cm when t = t_C, thus when CB is localized at the PE/PS interface as shown in Figure 5. Figure 6 schematizes observations reported by LFM analysis. There is a complete agreement between the resistivity measurements and the CB particles localization as observed by LFM. At mixing times longer than t_C, resistivity starts going up again, as result of dilution of CB in the PE phase. When using BP instead of XE, the same general behavior is observed with another optimum mixing time. Since the optimization of the mixing time is a time-consuming process, it would be better to drive the CB particles to the interface in a spontaneous way.

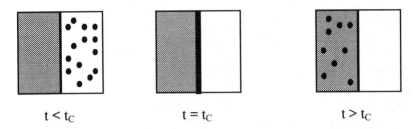

Figure 6. Effect of mixing time on the localization of CB in a PE (white)/PS (gray)/CB (black) polymer blend.

Thermodynamic localization of CB at the PE/PS interface. BP, which is preferentially localized in the PS phase, has a higher pH than XE which tends to accumulate in the PE phase. It thus appears that the oxidation state of the CB surface, or its pH by extension, is a key parameter in the control of the CB/polymer interactions. According to ASTM D 1512-90 standard, pH for XE and BP is 7.0 and 2.5, respectively. Therefore, Gubbels prepared a series of CB particles of increasing pH (32). He observed that the electrical resistivity of a co-continuous PE/PS (45/55) blend loaded with CB went through a minimum when the pH of the CB particles was increased, all the other conditions being the same. This observation indicates that CB particles of an intermediate pH are preferentially localized at the interface. These data will be detailed elsewhere (33) and are schematized in Figure 7. These previous conclusions are now confirmed by the LFM image of the different samples. Indeed, CB particles are localized: (i) in the PS phase when of a low pH, (ii) at the interface when pH of CB has intermediate values. As a representative example, Figure 8 shows that CB particle of an intermediate pH form an *ca.* 120 nm thick layer at the interface, and (iii) essentially in the PE phase for CB of a high pH. This thermodynamic control of the CB localization at the polyblend interface is by far more interesting than the kinetic control.

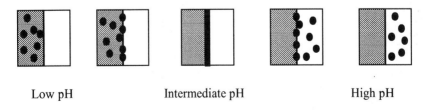

Low pH   Intermediate pH   High pH

Figure 7. CB localization in the PE/PS/CB blend in relation to the acidity of CB.

138

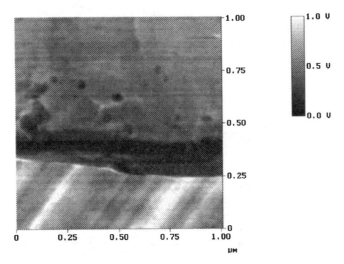

Figure 8. LFM image of a PE/PS/CB (45/55/2) blend in case of CB
particles of intermediate pH.

Conclusions

The experimental results reported in this paper emphasize that blends of two
insulating polymers (PE and PS) of a co-continuous two-phase morphology can be
endowed with electrical conductivity although using very small amounts of
conductive CB particles. The key criteria for the design of such conducting polymer
composites are: (i) percolation of the polymer phases and (ii) percolation of CB at
the polyblend interface. Since the constitutive components of the ternary systems
have different friction coefficients, LFM proves to be a useful technique for the
analysis of the morphology at the nanometer scale.

A double percolation is the basic requirement for electrical conductivity.
Provided that CB is selectively localized at the polymer blend interface, the CB
percolation threshold p_C may be as low as 0.4 wt %, *i.e.,* a striking 0.002 volume
fraction. This strategy is not restricted to the loading of polymer blends with carbon
black fillers; since particles of intrinsically conducting polymers could be used as
well.

Acknowledgements

All the authors are very much indebted to the "Ministère de la Région
Wallonne (DGTRE: Programme mobilisateur ALCOPO)". The research in Mons
was also supported by the Belgian Federal Government Office of Science Policy
(SSTC) "Pôles d'Attraction Interuniversitaires en Chimie Supramoléculaire et
Catalyse", the Belgian National Fund for Scientific Research FNRS/FRFC, an IBM
Academic Joint Study, and the European Commission (Human Capital and Mobility

139

Network: *Functionalized Materials Organized at Supramolecular Level*). The research in Liège was also supported by the SSTC "Pôles d'Attraction Interuniversitaires: Polymères". RL is "chercheur qualifié" by the National Fund for Scientific Research (FNRS - Belgium).

References

(1)   Norman, R.M. *Conductive Rubbers and Plastics*; Elsevier: New-York, NY, 1970.
(2)   *Carbon-Black Composites*; Sichel, E., Eds.; Dekker:New-York, NY, 1982.
(3)   Carmona, F. *Ann de Chim. Fr.* **1988**, 13, 395.
(4)   Geuskens, G.; Gielens, J.L.; Geshef, D.; Deltour, R. *Eur. Polym. J.* **1987**, 23, 993.
(5)   Sumita, M.; Sakata, K.; Asai, S.; Miyasaka, K.; Nakagawa, H. *Polym. Bull.* **1991**, 25, 265.
(6)   Geuskens, G.; De Kezel, E.; Blacher, S.; Brouers, F. *Eur. Polym. J.* **1991**, 27, 12615.
(7)   Asai, S.; Sakata, K.; Sumita, M.; Miyasaka, K. *Polym. J.* **1992**, 24, 415.
(8)   Klason, C.; Kubát, J. *J. Appl. Polym. Sci.* **1975**, 19, 831.
(9)   Michels, M.A.J.; Brokken-Zijp, J.C.M.; Groenewoud, W.M.; Knoester, A. *Phys. A* **1989**, 157, 529.
(10) Carmona, F.; Canet, R.; Delhaes, P. *J. Appl. Phys.* **1987**, 61, 2550.
(11) Martin, Y.; Abraham, D.W.; Wickramasinghe, H.K. *Appl. Phys. Lett.* **1988**, 52, 1103.
(12) Muralt, P.; Pohl, D. *Appl. Phys. Lett.* **1986**, 48, 514.
(13) Viswanathan, R.; Heaney, M.B. *Phys. Rev. B* **1995**, 75, 4433.
(14) Quivy, A.; Deltour, R.; Jansen, A.G.M.; Wyder, P. *Phys. Rev. B* **1989**, 39, 1026.
(15) Sheng, P.; Klafter, J. *Phys. Rev. B* **1983**, 27, 2583.
(16) Abeles, B.; Sheng, P.; Coutts, M.D.; Arie, Y. *Adv. Phys.* **1975**, 24, 407.
(17) Mehbod, M.; Wyder, P.; Deltour, R.; Pierre, C.; Geuskens, G. *Phys. Rev. B* **1987**, 36, 7627.
(18) Kirkpatrick, S. *Rev. Mod. Phys.* **1973**, 45, 574.
(19) Song, Y.; Noh, T.W.; Lee, S.-I.; Gaines, J.R. *Phys. Rev. B* **1986**, 33, 904.
(20) Clerc, J.P.; Giraud, G.; Laugier, J.M.; Luck, J.M. *Adv. Phys.* **1990**, 39, 190.
(21) Heaney, M.B. *Phys. Rev. B* **1995**, 52, 12477
(22) Webman, I.; Jortner J.; Cohen, M.H. *Phys. Rev. B* **1975**, 11, 2885.
(23) Grannan, D.M.; Garland, J.C.; Tanner, D.B. *Phys. Rev. B* **1981**, 46, 375.
(24) Gubbels, F.; Jérôme, R.; Theyssié, Ph.; Vanlathem, E.; Deltour, R.; Calderone, A.; Parente, V.; Brédas, J.L. *Macromolecules* **1994**, 27, 1972.
(25) Gubbels, F.; Blacher S.; Vanlathem, E.; Jérôme, R.; Deltour, R.; Brouers, F.; Teyssié, Ph. *Macromolecules* **1995**, 28, 1559.
(26) Wiesendanger, R. *Scanning Probe Microscopy and Spectroscopy: Methods and Applications,* Cambridge University Press, Cambridge, 1994.
(27) Calderone, A.; Lazzaroni, R., Brédas, J.L., to be published.
(28) Calderone, A., Parente, V.; Brédas, J.L. *Synth. Met.* **1994**, 67, 151.

140

(29) Parente, V.; Lazzaroni, R.; Brédas, J.L. to be published in the Proceedings of International Conference on Polymer-Solid Interfaces, Namur, Aug. 12-16, 1996.

(30) Sumita, M.; Sakata, K.; Hayakawa, Y.; Asai, S.;Miyasaka, K.; Tanemura, M. *Colloid Polym. Sci.* **1992**, 270, 134.

(31) Levon, K.; Margolina, A.; Patashinsky, A.Z. *Macromolecules* **1993**, 26, 4061.

(32) Gubbels, F. Ph. D. thesis, University of Liège, 1995.

(33) Gubbels, F.; Jérôme, R.; Vanlathem, E.; Deltour, R. to be published.

Chapter 8

Effect of Thermal and Solvent Treatments upon Thermoplastic Olefin Surface Morphology and Composition Studied with Surface-Sensitive Techniques

M. P. Everson[1], M. Mikulec[2], and P. Schmitz[3]

[1]Physics Department, Ford Research Laboratory, MD3028/SRL, P.O. Box 2053, Dearborn, MI, 48121–2053
[2]Paint, Materials and Process Engineering Department, Ford Motor Company, ACD-ETF, P.O. Box 6231, Dearborn, MI, 48121
[3]Chemistry Department, Ford Research Laboratory, MD3083/SRL, P.O. Box 2053, Dearborn, MI, 48121–2053

Thermoplastic olefins (TPOs) are now in common use in the automotive industry as paintable fascia materials. An improved understanding of how the as-molded TPO material and pretreatment affect paint adhesion is desirable. Previous experiments using optical microscopy have established a connection between TPO surface morphology during a simulated paint process and final paint adhesion and durability.

We have used Scanning Force Microscopy (SFM) and x-ray photoelectron spectroscopy (XPS) to investigate the morphology and phase composition of as-molded TPO surfaces and surfaces exposed to conditions simulating a painting process. SFM-measured high-friction components at the TPO surface correlate with good paint adhesion in a similar painted part. XPS on a subset of these samples establishes a connection between the high-friction areas and the rubbery copolymeric constituent in TPO. These results imply that paint performance on TPO depends on high rubber density at the TPO surface.

The TPOs currently used in production are blends combining semi-crystalline polymers with an amorphous copolymer. For example, the main components of a representative TPO (TPO #1), are polypropylene and a random ethylene-propylene copolymer. The copolymer is in an elastomeric state in TPO, and will be referred to as rubber throughout this report for the sake of brevity. The polypropylene can be in a range of states from purely polycrystalline to amorphous. TPOs typically also contain other polymers, inorganic fillers, like talc, and various additives, such as thermal stabilizers. This study involved investigation of the behavior of three different TPO materials currently in use in the automotive industry, referred to as TPOs # 1-3.

The surface of injection molded TPO is relatively inert (nonpolar with a low surface energy), so paint adheres poorly to an as-molded part. One of a variety of chemical or physical pretreatments are required to modify the TPO surface to allow successful initial paint adhesion. Treatments can include trichoroethane (TCE vapor degreasing), primer applications, low temperature flame treatments or combinations of the above. For TPO #1, currently in production, an adhesion promoting (AP) primer, suspended in an aromatic solvent blend, is used as part of the paint system to enhance adhesion. The principle constituent of the AP primer, believed to be responsible for adhesion, is a chlorinated polyolefin (CPO), in most cases a chlorinated polypropylene. The exact mechanism of how the AP interacts with the TPO substrate is unclear. However, it is postulated that the CPO entangles with the rubber phase, providing mechanical adhesion across the interface (1,2). Swelling of the rubber phase by solvents during the adhesion promoter/paint bake process is also believed to be important, allowing for penetration and interdiffusion of the AP and TPO. Previous work in this area has utilized a number of analytical techniques to study the influence of coating solvents on TPO morphology and their subsequent influence on paintablilty (3,4).

The development of methods for assessing the surface morphology and composition (distribution of rubber phase) of TPO, and how process-related treatments can influence these properties, is important in generating an understanding of the underlying mechanisms of paint adhesion and durability. Of equal importance is the establishment of methods which can be utilized in the plant, either on-line or off-line, to monitor material composition and surface changes to assure that process and material formulations are optimal. A major motivation for this study was to support the development of an on-line optical quality assurance method for the TPO painting process (3,5). Magnified optical images are a potential inexpensive, nondestructive process control tool for monitoring the TPO painting process. During an earlier optical study (3), correlations were noted between surface appearance of treated TPO parts at 100-800x magnification and final paint adhesion and durability. The study observed the surface morphology of different materials through a number of simulated paint processes involving both thermal and chemical treatments. The surface morphologies at various steps of the process that affected the quality of the final paint adhesion (peel test) and durability (thermal shock test) were determined. Pattern Recognition and Neural net systems are now being evaluated in conjunction with a library of optical images to determine if such a system can adequately monitor the quality of the paint process on the line.

The phenomenological connection between a material's surface morphology resulting from pretreatment and paint adhesion and durability are not fully understood at this time. Initial material factors such as composition, component molecular weight distributions and mold face cooling rates, in addition to the paint process itself, may all be important in obtaining good paint adhesion and durability for TPO. In the special case of scuff resistance a connection has been established between the presence of optically observed "droplets" (discussed in detail later) on the surface after thermal treatment and poor scuff resistance of the painted part (6). Establishing the connection between morphological changes (for example "droplet" formation) and surface composition is of interest. In the present study we have used Scanning Force Microscopy (SFM) and x-ray photoelectron spectroscopy (XPS) to investigate these changes, measuring both morphology and phase composition of the surfaces of TPO materials after a variety of pretreatments. The SFM measures these properties on a sub-micron scale, providing correlations between surface features and their composition. XPS provides surface composition from the top most 100Å , averaged over an ~500μm diameter area. In this study we wish to address how SFM and XPS information help in understanding the structures seen in optical micrographs that either enhance or diminish paint performance.

The overriding issue in this work is whether morphology and surface composition of the TPO after pretreatment provide sufficient correlation with processes occurring in the near-surface region during AP application and painting that ultimately affect adhesion and durability. The surface information obtained by both SFM and XPS are potentially important since no other work has probed the surface region on the scale investigated using these techniques. Additionally, information from the immediate surface may be very important because the surface is the path for ingress of solvent and AP into the near-surface region. Finally, a correlation between surface morphology and near-surface behavior would help to establish the validity of the assumptions made for the optical quality assurance method.

Experimental Procedures

Scanning Force Microscopy - Background. In the SFM (7) a cantilever with an ultra-fine tip at the end is scanned over the sample surface while a constant force is maintained between the tip and the sample. This force is proportional to the cantilever deflection, which is usually measured optically. A surface profile, which can have atomic resolution in favorable cases, is produced from the scan data. If the detection scheme also allows for measurement of the twist or bending of the cantilever, frictional forces can also be measured simultaneously with the topographic information. The resolution of the SFM on compliant samples, such as TPO, is limited to roughly the radius of curvature of the tip. Commercially available tips with radii of curvature of approximately 300Å were used for these experiments.

The SFM data, which consist of a 3D mesh of points describing the surface profile, are frequently shown as a 2D gray-scale image, where the brightness of a point indicates the z height (or friction level) at that location on the surface. Scan data can also be presented as a light-shaded image derived from the measured surface mesh as shown in Figure 3. Such images allow the observer to get an immediate qualitative feel for the surface that has been measured.

An additional capability of the SFM is the measurement of the dependence of the force upon the tip-sample distance, allowing investigation of material hardness or attractive/adhesive forces. This latter mode was not used in these experiments, but it does hold promise for sensing near-surface features using the hardness of the surface as a guide. The SFM data in this report were measured with an Autoprobe LS SFM, manufactured by Park Scientific Instruments, Sunnyvale, Ca.

One caveat that must be noted here is that the SFM is extremely surface-sensitive. Consequently, SFM friction measurements can be rendered useless by too much surface contamination. The friction results for newly treated TPO samples were typically stable for periods as short as a few weeks when stored in a desiccator. Longer shelf-life samples tended to have the friction of rubber areas decreasing to near that of other surface components. This effect may be due to oxidation, surface segregation of some additive or component, or a surface chemistry effect like surface reorientation. Oxidation of one or more surface components is discussed later in the report. One other consequence of surface sensitivity is that there may be disagreement between SFM measurements and real paintability due to rubber near but below the surface, since solvent from the AP and paint clearly can diffuse into the surface to a limited extent. Investigations using freshly made cross-sections should help to resolve problems presented by the difficulties noted above.

X-ray Photoelectron Spectroscopy (XPS). Core level XPS has been used extensively in the study of the surface composition of polymeric materials (8). For the study of hydrocarbon based materials, such as TPO, traditional XPS core level spectra can not discriminate between the different olefinic constituents present and therefore is not useful for determining the component distributions at the surface of TPO materials. However, in addition to the measurement of core level information, XPS instrumentation is also capable of probing the valence band, resulting in information from electrons residing in molecular orbitals, i.e. those directly involved in the bonds between the atoms of molecules (9). XPS valence band spectra can quite easily discriminate between olefinic materials such as polypropylene, polyethylene and ethylene-propylene copolymers and has been shown to be useful in evaluating changes in the surface composition of blends of these components (10).

All valence band spectra used in this study were collected on an M-Probe X-ray photoelectron spectrometer, manufactured by Fisons Surface Science Instruments, Mountain View, CA. The X-ray source provided monochromatic Al-K$_\alpha$ radiation (1486.6 eV) focussed to a line-spot 250 µm wide by 750 µm long. The x-ray power used was 100 W. The analyzer was operated at 50 eV pass energy during the acquisition of all valence band spectra. A low energy (1-3eV) flood gun and a Ni charge neutralization screen were utilized to minimize charging effects. All spectra were collected with the analyzer positioned normal to the sample surface resulting in an average sampling depth of about 100 Å.

TPO Surface Morphologies Measured with Optical Microscopy and SFM

Optical Microscopy Summary. Prior to this work, optical investigations (3) had discerned changes in the TPO surface as a function of pretreatments ranging among thermal, chemical and physical processes (e.g. process temperature, TCE pretreatment, solvent wipe and flame treatment). The optical measurements, as well as those for SFM and XPS reported below, used sections of 4 by 8 in. plaques of the relevant TPO materials, injection molded under supplier-determined optimal conditions. There was a pronounced variation in the as-molded plaques depending upon position on a given item. For instance, the surface morphology near the mold gate was markedly different from that seen in the center of a plaque. Different lots of material tended to have the same features.

The optical study observed a number of important features about the effects of temperature and solvents upon the surface of TPO plaques. Although the results are summarized in other reports (6) a small part of the data will be presented here in order to make this report more self-contained. The study examined a total of approximately 4000 images of magnified views of the TPO #1 surface for 23 different combinations of bake time and temperature. It found that purely thermal treatments can bring to the surface droplets of material as seen in figure 1B. The presence or absence of these droplets on a plaque was a function of bake time and temperature. Increasing either bake time or temperature resulted in increased creation of droplets up to a point, and for further increased bake time or temperature the droplets would grow in size and start to disappear from the plaque. For high temperatures or long times the droplets would disappear from the plaque. This behavior is consistent with a mechanism wherein a constituent of TPO migrates to the surface, forming droplets, but eventually the droplets grow and merge into a continuous film that cannot be resolved by a magnified optical view of the surface. Figure 1 shows an example of this general type of behavior. In as-molded TPO #1 there are no droplets (Figure 1A), after a 40 min. bake at 86°C the droplets appear (Figure 1B). For a bake at 121°C for 50 min. (Figure 1C) the droplets are larger, and some have begun to fuse together. For longer bake times or higher temperatures it becomes difficult to identify any droplets using the optical microscope. Subsequent investigation in collaboration with the supplier, has shown that replacing the current EP rubber with a different formulation can eliminate droplet formation upon baking for TPO #1.

Another set of experiments were undertaken to simulate the effect of solvents upon the surface region of TPO. This experiment used a TPO (TPO #2) material that is very similar to TPO #1. First the plaque was treated with TCE as in the usual TPO #2 paint process. Following this the plaque's surface was treated at paint bake temperature with the solvents present in the AP package (AP without the CPO component). Specifically, after a 121°C 30 min bake treatment a quantity of solvent was placed on the sample with a dropper. The slowly cooling sample was observed in an optical microscope as a function of time, until after the solvent on the plaque evaporated away (a few minutes). Shortly after the solvent exposure, globules appeared on the surface of the TPO. As the solvent evaporated away, the globules retracted into and subsided onto

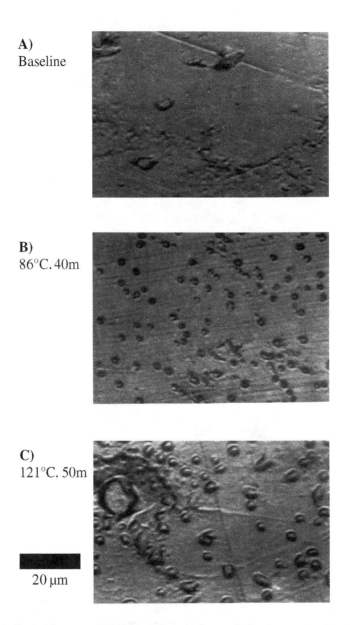

A)
Baseline

B)
86°C, 40m

C)
121°C, 50m

20 μm

Figure 1. Optical images of TPO #1 surfaces before and after heat treatments. (A) As molded. (B) After 86°C bake for 40 min., droplets (round dark features) cover the TPO #1 surface. (C) After a higher temperature bake (121°C) for 50 min. the droplets (dark with light centers) have grown, and some have fused together into oblong shapes. The density and size of droplets can vary over a given plaque.

the surface of the material. This process is shown in Figure 2. Swelling of the rubber phase by solvents is thought to be important in the conventional paint adhesion mechanism for TPO, as described in the introduction.

The surface effects created by both the bake temperature and the solvent presumably also have counterparts in the uppermost strata of the molded material. This region, along with the surface, is critical for good paint adhesion and durability. Providing further explanation of the processes occurring in these two sets of optical experiments is the reason the body of AFM and XPS work reported here was undertaken.

Scanning Force Microscopy. The SFM was used to determine the vertical size of features and roughness of TPO sample surfaces. Figure 3 shows pseudo-3D images of two as-molded TPOs. TPO #1 is shown at top, and TPO #3 is shown in the bottom of the figure. The vertical axis has been enlarged by a factor of 20 relative to the lateral scale to highlight surface features. The difference between typical regions of the two materials is apparent from the figure. The TPO #1 material is relatively smooth aside from a few large scale defects, whereas the TPO #3 material shows both rough and relatively smooth, curved regions. The SFM data examples shown in the figures are characteristic of those samples, but substantial variation from the presented data was observed in several cases.

Table I summarizes the results for a series of measurements conducted for TPO #1, TPO #2 and TPO #3. The data were recorded at several randomly selected areas near the center of each plaque. The sample column shows the various samples and pretreatments used in these measurements. Materials in the as-molded state are marked with "untreated". The TPO #1 samples were subjected to three different purely thermal treatments to observe the effect heat alone has on the surface. In addition the TPO #1, TPO #2 and TPO #3 materials were characterized after being treated with a thermal/solvent process similar to the first stages of that used for painting TPO #2 parts. This process was selected for these experiments because it had 8 years of field experience establishing that it was an effective procedure for painting TPO parts. Since the final step of painting the part would prevent any frictional information from being obtained, only the pre-painting part of the TPO #2 process was simulated. Our laboratory procedure for this "TCE / AP solvent" process consisted of: (A) TCE pretreatment at 70°C, and (B), a 40 min. bake at 90°C with periodic placement of a thin layer of xylene and toluene upon the hot TPO. The purpose of the xylene/toluene mixture was to mimic the effect of the solvents in the AP, without the chlorinated polyolefin resin present. The SFM measurements of the TCE / AP solvent treated samples were performed in ambient air after all solvents had one day or greater to evaporate away. The topography column contains a brief description of the main attributes of the surfaces measured. Approximate RMS roughnesses for typical $80 \times 80 \mu m^2$ areas are shown in the third column of Table I. The roughness values are rounded to the nearest 100Å because only a few measurements from each sample were used, so the data are not statistically relevant to much more precision than shown. Results of thermal shock tests carried out for painted plaques corresponding to the simulated pretreatments, when data were available, are shown in

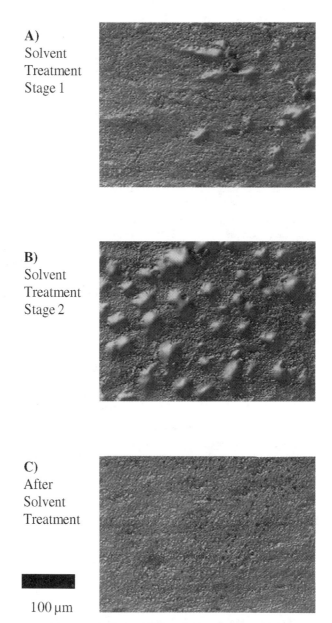

A)
Solvent
Treatment
Stage 1

B)
Solvent
Treatment
Stage 2

C)
After
Solvent
Treatment

100 μm

Figure 2. TCE / AP solvent treatment of TPO #2, as described in the text. (A) A short time after the solvent is placed on the heated plaque, globules appear at the TPO surface. (B) Shortly thereafter the surface is covered with globules. (C) After the solvent evaporates away the globules shrink again, partly retracting below the surface, and partly subsiding onto the original surface.

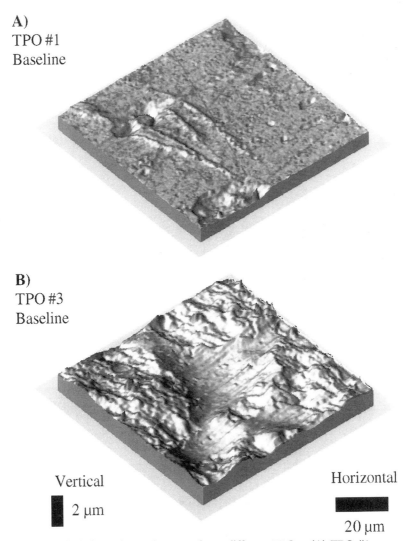

A)
TPO #1
Baseline

B)
TPO #3
Baseline

Vertical

2 µm

Horizontal

20 µm

Figure 3. Light-shaded pseudo-3D images of two different TPOs. (A) TPO #1 as-molded and (B) TPO #3 as-molded. The vertical scale in both images is enhanced by a factor of 20 over the horizontal scale.

Table I. SFM Measurements of Topography and Friction of TPO Samples

Sample	Topography	RMS Rough. (Å)	Thermal Shock	Friction Low / High	High-Friction Area
TPO #1 Baseline* (Untreated)	flat, pores to 1 μm deep	300	Fail**	0.19 / None	0%
TPO #1 Thermal A T=88°C, 50 min.	flat + some small droplets 0.2 μm high	300	Pass/ Fail**	0.20 / None	0%
TPO #1 Thermal B T=88°C, 80 min.	flat + many small droplets 0.2 μm high	300	Not Avail.	0.20 / None	0%
TPO #1 Thermal C T=121°C, 60 min.	flat + fewer larger droplets	300	Pass**	0.24 / 0.57	2-3%
TPO #1 with TCE / AP solvent treat.	rougher, but no very large features	400	Pass	None/ 1.5	100%
TPO #2 Baseline (Untreated)	flat, pores to 1 μm deep	300	Not Avail.	0.25 / None	0%
TPO #2 with TCE / AP solvent treat.	Rough, a few droplets 5μm across	1100	Pass	0.24 / 0.5-0.9	90%
TPO #3 Baseline (Untreated)	Distinct rough and flat regions	2500	Not Avail.	0.18 / None	0%
TPO #3 with TCE / AP solvent treat.	Rough with flat regions	2300	Pass	0.39 / 1.5	10%

* This TPO #1 is typical of lots from before 4/93. When painted to obtain thermal shock data an AP flash at 22°C was used.

** Thermal shock results for these trials used an Adhesion Promoter treatment at the temperature shown in the "Sample" column. The thermal series (#1-3) for the SFM used no AP treatment, observing only the effects of heat treatments upon the surface of TPO #1.

the fourth column. The remaining columns are devoted to friction measurements discussed in the next section.

One feature that the surfaces of the thermally treated TPO #1 samples all have in common is droplets (3) ranging in size from approximately 0.2-3µm across and about 0.05-0.2 µm high (Figure 6A, for example, shows two different types of such "droplets"). There are several possible explanations for these features. One explanation is that they are agglomerations of low-molecular-weight polymer that have diffused to the surface of the material due to differences in surface energy between lower and higher molecular weight species. Another possibility is that they are surface material recrystallized during the thermal treatments. A third possible explanation is that the droplets are composed of a TPO additive that migrates to the plaque surface upon baking. The Thermal Treatment A sample has the droplets present but does not always pass thermal shock. Understanding the method of formation and physical properties of these droplets thus may be important in determining optimal paint process conditions. Additionally, these features are thought to be important in scuff durability of painted TPO parts (6). Further information on the composition of the droplets was obtained with friction measurements and is discussed in the next section.

Surface Composition of TPO

Friction Measurements of TPO Surfaces. The nature of the surface and near-surface region of TPO is the most important factor in paint adhesion and durability. The AP is thought to function through entanglement of the chlorinated polyolefin with the amorphous rubber phases within the TPO surface. Thus the function of the AP is adversely affected by crystalline regions of the uppermost stratum of TPO, in which entanglement would be very difficult. Because of this functionality, the density of amorphous versus crystalline regions near the surface of the TPO is potentially an important element with regards to paint adhesion and durability. Therefore, frictional contrast between amorphous (good AP interaction) and crystalline (bad AP interaction) regions seemed promising as a rough measure of the AP / surface interaction strength.

A local friction probe like the SFM should be able to distinguish between crystalline polymer and a rubbery elastomeric phase. Friction coefficients (μ) for steel running against poly-crystalline polypropylene or polyethylene are reported to be in the range $\mu = 0.1 - 0.3$, whereas for bulk rubbers $\mu \approx 1 - 4$ or greater is typical (11). These values suggest a friction contrast of a factor of 3 or greater between rubber (copolymer and possibly amorphous polypropylene) and polycrystalline polypropylene in TPO. The values measured in the SFM are for a sharp silicon tip, and so might differ significantly from the above values, but some frictional contrast between rubbery and crystalline components would be expected.

As an initial feasibility test for the use of friction contrast measurements on TPO, we measured relative friction of an as-molded sample and one that was thought to have a high density of rubber. The high-rubber sample was a TPO #2 material that had been subjected to the treatment described for the sample in Figure 2. This sample was thought to be the most certain to have rubber at the surface. Topography and friction for the "globule" sample are shown in Figure 4 A) and B) respectively. Similar data for the as-molded TPO #1 sample are presented in Figure 5. Average relative friction coefficients for the high-rubber sample were 3-6 times as large as those for the baseline (as-molded) sample. One notable feature in Figure 4 is the presence of low-friction regions (dark in fig. 4B) that correspond to the highest topographic features in Figure 4A (brighter areas). These might be pushed-up and distorted pieces of the original polycrystalline skin. The friction signal is so large, especially at the lower-center and lower-right of the Figure, that it has saturated the friction detector signal.

These data illustrate that friction contrast should be useful in distinguishing amorphous from crystalline phases at the TPO surface. A few additional cross-checks were performed to validate the measurement technique. First, we verified that washing the surfaces with soap and water resulted in a uniformly low friction coefficient. This is expected since the surfactant layer left behind will modify the friction coefficient, in this case lowering it on areas of previously high friction. The second test involved removal of rubber from the surface of a TPO #1 plaque treated with TCE and toluene solvents. The rubber residing on the surface was removed by brief (30s) ultrasonic agitation in cold toluene. It was reasoned that this treatment would strip amorphous rubber from the surface very rapidly, without significantly affecting polycrystalline components or swelling sub-surface rubber. In this case also we obtained the expected low-friction result.

After proving out the approach, we performed friction measurements upon the sample surfaces shown in Table I to determine what information the SFM friction mode could obtain about the presence and areal densities of rubber. The quoted friction coefficients are averages, different regions can have values varying by about ±20%. For those cases where an even larger spread from the average was seen the range is quoted.

TPO #1 Thermal Series. We measured several types of samples with a range of thermal shock (an adhesion test) performances. The TPO #1 samples were treated in three basic ways: no treatment (As-molded), pure thermal treatment, and simulated AP treatment. These treatments give thermal shock results of fail, sporadic failure, and pass respectively. It is reasonable to conjecture that the variation in thermal shock results might be related to the presence or absence of an amorphous component at the TPO surface. Figures 5 and 6 show both topographic and frictional images of baseline and thermal treated samples respectively. For the simulated AP treatment the TPO #1 results are roughly similar to that for the TPO #2 material shown in Figure 4. (The scales for Figures 4-6 are the same for both the topographic and frictional data, with the exception of Figure 4B.) There is a definite, but subtle, evolution of the topography across the data set that is also noted in Table I. Thermal treatment results in the presence of many small droplets on the surface, but otherwise the topography is unaffected (the deep pores still remain, there just happen to be none present in Figure 6A). In the AP treated sample the surface is roughened on a small scale, and many of the pores are eliminated or reduced in size.

TPO #2 After TCE / Toluene Treatment

A)   Topography   B)   Friction

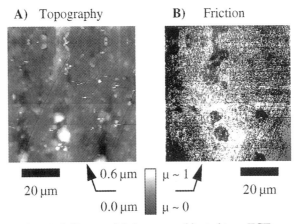

20 µm   0.6 µm   µ ~ 1   20 µm
         0.0 µm   µ ~ 0

Figure 4. Data for TPO #2 material that was subjected to a TCE vapor degrease followed by several xylene and toluene applications at temperature to simulate a painting process: (A) topography and (B) friction, for the same area. Note low-friction regions (dark in B) that correspond to the highest topographic features (brighter areas in A).

TPO #1 As Molded

A)   Topography   B)   Friction

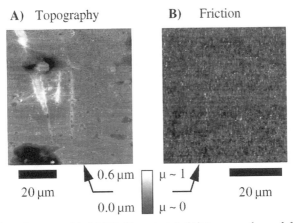

20 µm   0.6 µm   µ ~ 1   20 µm
         0.0 µm   µ ~ 0

Figure 5. Data for as-molded TPO #1 material: (A) topography and (B) friction. This surface is entirely low-friction (about 0.2) aside from non-reproducible noise in the friction signal. Figure (B) is from a flatter piece of surface than (A) to eliminate the geometric effects in the friction signal due to sloping edges of the pores.

TPO #1 After Thermal Treatment at 121°C

A) Topography   **B)** Friction

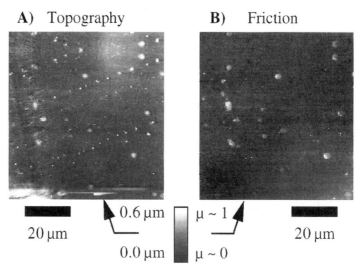

$0.6\,\mu m$ $\mu \sim 1$

20 μm   20 μm

$0.0\,\mu m$ $\mu \sim 0$

Figure 6. Thermally treated TPO #1 (T=121°C, 60 min): (A) topography and (B) friction, for the same area. The surface is flat with both large and small droplets, whereas frictionally only the larger droplets are distinct. The frictional data have been low-pass filtered to make the high-friction regions more clear.

The change in friction of the surface across the series is somewhat more illuminating. For the baseline sample the friction value is low (0.19 average) across the entire surface, the only exception being noise spikes, which are not reproducible. For the thermal samples treated at 88°C there are features which take the form of spherical caps or droplets. The droplets had essentially the same low friction coefficient as baseline. However at the highest temperature, 121°C, interesting features appear in the frictional image. First it should be noted that in the 121°C topographic image 6A there are two different types of "droplets", they have approximately the same height of 0.2 μm, but the wider ones are about 2 μm across whereas the thinner ones are about 0.5 μm across or less. The smaller droplets also tend to lie in lines on the surface. In the frictional images the smaller droplets have about the same friction values as baseline, whereas the wider droplets are seen to have a significantly higher friction value of ≈ 0.7. Finally, in the simulated AP treatment sample the friction is high (≈1.9) across the vast majority of the surface. The areal density of amorphous component at the surface becomes measurable, although small for the 121°C pure thermal sample. This correlates with the first reasonable pass in the thermal shock test. The rubber areal density is, however, large only for the simulated AP treatment. We examine next how the three TPO materials, TPOs #1-3, compare after the simulated AP treatment.

As-molded versus Solvent Treatments. The TPO #2 process (with TCE pretreat) resulted in 100% pass for the thermal shock test for all three materials. SFM friction measurements for all three materials in as-molded condition gave low friction values, in the range 0.2 - 0.35, indicating the surfaces have little or no amorphous component and are probably largely polycrystalline. After TCE / AP treatment all three samples showed some higher-friction (of order μ=1) regions over 10% (TPO #3) to 100% (TPO #1) of the measured surfaces. Clearly the combination of solvents and high temperatures is effective in driving amorphous material to the surface, either through diffusion, or the opening of pores or micro-cracks.

XPS Valence Band Measurements of Heat Treated TPO. XPS valence band spectra have been acquired from TPO #1 materials heated at 121°C for 30 min. As mentioned earlier, XPS valence band measurements can discriminate between polypropylene, polyethylene and ethylene-propylene copolymers and should be useful in evaluating changes in the surface composition of blends of these components. Shown in Figure 7 are valence band spectra for pure polypropylene, polyethylene and an ethylene-propylene copolymer (rubber). As is evident, the C(2s) region of these materials are quite different. For the sake of brevity, a detailed description explaining the reasons for the spectral differences will not be given here but will be presented in a more detailed report and references therein (10). In short, the differences observed result from a different density of electronic states for each polymer due to the differences in the molecular bonding of the basic repeat unit of the carbon back-bone.

The primary components of TPOs #1 and #2 are polypropylene and an ethylene-propylene copolymer (rubber phase). It has been shown that the valence band spectrum resulting from a blend of polyolefins (polypropylene and an ethylene-propylene

156

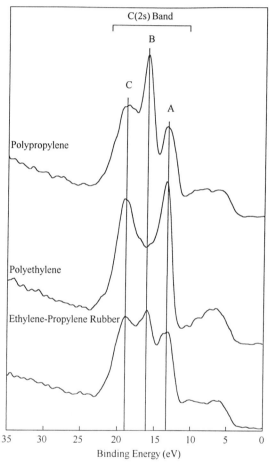

Figure 7. Valence band spectra acquired from the surfaces of plaques prepared using pure polyolefin components: polypropylene, polyethylene, and ethylene-propylene copolymer

copolymer) has a characteristic intensity distribution representative of a proportional combination of the component spectra (10). Therefore, the experimental TPO valence band spectra obtained from TPO surfaces can be used (at least qualitatively) to assess the relative surface composition of TPO following various treatments. Valence band spectra were acquired from untreated (baseline) and heat treated (30 min at 121°C) TPO #1 material. The spectra observed are shown in Figure 8. The baseline spectrum is very similar to the pure polypropylene spectrum shown in Figure 7. These results suggest that the untreated surface is very high in polypropylene. As is evident in Figure 8, the result of heat treating is to shift the C(2s) spectral features in a direction consistent with a higher rubber concentration. This is more clearly displayed in Table II in which peak ratios from the pure components and the TPO #1 spectra are compared.

Table II. Measured peak ratios from C(2s) region of XPS valence band spectra from Figures 7 and 8.

Material	B/A Ratio	B/C Ratio
Polypropylene	1.98	1.62
Ethylene-Propylene Rubber	1.25	1.10
TPO #1, Baseline	1.85	1.57
TPO #1, 30min @ 290°F	1.59	1.35

The use of peak ratios for determining the relative composition of a binary blend of polypropylene and rubber has been previously demonstrated (10). As is evident, the baseline TPO #1 spectrum is very similar to pure polypropylene. The measured B/A and B/C ratios are 1.98 and 1.62 for pure polypropylene and 1.85 and 1.57 for the TPO #1 material. Heat treatment of the TPO #1 material results in B/A and B/C ratios of 1.59 and 1.35, consistent with an increase in the rubber concentration in the near surface region. However, it should be noted that the changes observed would also be consistent with an enhancement of polyethylene (or some other linear hydrocarbon material) with heat treatment which therefore can not be unequivocally ruled out. Also evident in the heat treated spectra of Figure 8 is the appearance of an additional peak at a binding energy of about 25 eV. This feature can be assigned to the O(2s) band. The appearance of this feature indicates that some oxidation of the surface, or possibly some migration of an oxygen containing species has occurred as a result of the heat treatment. To determine whether polypropylene or the rubber was more prone to oxidation and/or component migration under the thermal treatments used, pure polypropylene and rubber were heat treated under the same conditions (30min @ 121°C in air). The resulting valence band spectra are shown in Figure 9. No differences were observed in the C(2s) region of the heat treated polypropylene spectra, and no peak was observed in the O(2s) band following heat treatment. However, it is evident that the rubber did change upon heat treatment, as evidenced by the appearance of the O(2s) peak at 25 eV. Therefore

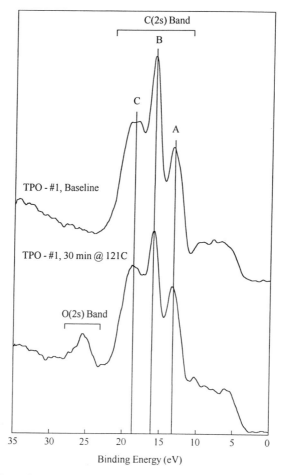

Figure 8. Valence band spectra acquired from TPO #1, as molded (baseline) and following heat treatment at 121°C for 30 minutes.

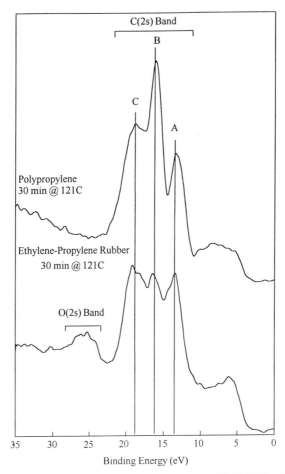

Figure 9. Valence band spectra acquired from heat treated (121°C for 30min. in air) polypropylene and an ethylene-propylene copolymer.

both the changes in the C(2s) feature of 161 with heat treatment and the appearance of oxygen at the surface are consistent with an increase in the rubber concentration (or additive associated with the rubber phase) at the surface as a result of the thermal treatment. It should be noted that additional measurements from heat treated TPO #1 materials have shown a similar trend in increased rubber levels, although to a smaller extent, while the appearance of oxygen species at the surface following heat treatment are not consistently observable. In addition, similar observations are not observed with simple binary blends suggesting that these changes are specific to the TPO #1 material (10).

Discussion

A few comments about the SFM friction technique in general are in order before application of these results to TPO paintability. The origin of the variation in the friction value for rubber on the different surfaces, especially the simulated AP treated TPO #2 material, is unknown. Of course the distribution of globules is uneven in the TPO #2 AP treated case, and we may simply be seeing the most pronounced variation between regions where a globule was present and those where no globules existed. Certainly the thickness of rubber constituents on the surface after treatment can vary, and this could affect the measured friction values. For SFM measurement of a soft material, there is expected to be some penetration of the tip into the material itself. The measurements in these experiments typically used a force of 120nN on a tip with radius of curvature of 200-500Å. Thus the initial contact pressure is high, but rapidly becomes lower as more of the tip area is supported. If a soft film's thickness (the rubber) is less than or near to the distance the tip would normally sink in for that force, a friction value different from either rubber or polycrystalline material may result. Another possibility for the variation is differences in oxidation or other surface reactions for the rubber, which might induce a friction variation across the surface of otherwise similar rubber constituents. Finally, as noted in the introduction, "rubber" could be either copolymer or amorphous polypropylene, or a blend of the two. The variation in friction for the rubber phase may be a combination of all three of the above phenomena.

The results in this report bring some new information to bear upon the question of the nature of TPO paintability. Both the SFM friction results and the XPS valence band data support the conclusion that the surface of as-molded TPO is largely devoid of rubber. SFM measures a continuous low-friction skin on all the as-molded TPO samples studied here. The simplest interpretation of this is that the immediate surface region has little or no amorphous-phase components, and are composed of a polycrystalline film. It is also possible that rubber is present in this skin, but is sufficiently dispersed that it does not have the frictional characteristics associated with bulk rubber. The XPS data are also consistent with there being little rubber within about 100Å of the surface of baseline TPO #1. Presumably the material nearest to the part's surface is rapidly quenched by the mold face, and has different characteristics from the bulk of the TPO material. The primary purpose of chemical/thermal treatments in the painting of TPO is thus to break through or modify this skin so that sufficient amorphous material is present to which the AP can anchor. The thickness of this skin can be addressed through measuring cross-sections in SFM frictional mode. Regions where there is bulk segregation of rubber would be recognized by their higher friction values. The SFM friction data also show that the use of TCE, xylene and toluene in the TCE / AP solvent process results in significant accumulation of amorphous material at the surface of TPO, correlating with good paint durability. Based on limited data, purely thermal processes seem to generate an inadequate surface density of rubber, with the possible exception of the T=121°C treatment. The amount of solvent used in the TCE / AP treatment makes it undesirable as a production process due to plant emissions concerns. The current TPO painting process involves power-washing of the part followed by an AP application and bake at 121°C for 15 min., followed immediately by painting and bake at 83°C for 32 min. Apparently the solvents present in the AP application (AP typically =93% solvents + 7% solids) are sufficient to produce the changes in rubber density at or near the surface that are needed for good adhesion. Changing the TPO composition or molding process so that the rubber density at the surface is higher would significantly help the process' environmental friendliness, since the use of volatile solvents could be reduced. It appears that toluene or other paint solvents might themselves be important in the accumulation of rubber at the TPO surface. If this is true, changing to a water-born process for painting TPO may be difficult.

Conclusions

We have found a correlation between the SFM-determined surface areal density of amorphous material and paint durability of TPO as measured by the thermal shock test. XPS valence band data strongly imply that the amorphous material at the surface after pretreatment is the ethylene-propylene rubber, or some component which segregates from the rubber phase. It may be that the surface areal density of rubber is a critical factor for initial paint adhesion and durability. However, it is also possible that rubber on the surface after pretreatment is just an indication of another more important parameter, such as the diffusion rate for solvent or polymer through the top layers of the TPO. Exploring this distinction is an interesting direction for future experiments. It is clear from this work that during pretreatment and painting it is not sufficient to consider only solvent swelling of TPO and AP diffusion into it, as might have been thought previously. A great deal of diffusion and sometimes mass transport of rubber, apparently through pores or micro-cracks, also occurs during this process.

SFM is a valuable analytical tool for determination of both surface morphology and phase / material composition of the surfaces of technological materials. XPS valence band measurements are useful for distinguishing between chemically different polymeric constituents at the surface of a molded material. These experiments provide us with a contrast mechanism to examine the phases of material present on the TPO surface both as-molded and after a variety of treatments. Combination of these results with those from other analytical techniques, as well as durability tests, should be able to identify the phenomena involved in good TPO paintability and paint durability. The combined use of reflective light microscopy, SFM, and XPS valence band measurements as a window on the surface dynamics related to heat treatment and painting will also be valuable for the evaluation of new plastics with improved characteristics.

Acknowledgments

The authors would like to thank Dr. David Bauer, Dr. Joseph Holubka, Dr. Jeff Helms and Dr. Robert C. Jaklevic for their helpful suggestions regarding this work. The authors would also like to extend special thanks to Dr. Giuseppe Rossi.

Literature Cited

1. Aoki, Y. J. *Polymer Science, Part C* **1968**, *23*, 855.
2. Schmitz, P.J.; Holubka, J.W. *J. Adhesion* **1995**, *48*, 137.
3. Mikulec, M. *Ford PTPD Technical Report* **1994**, "Effect of Manufacturing Process on TPO surface morphology".
4. Ryntz, R.A. *Ford PTPD Technical Report #00008530* **1994**.
5. Mikulec, M.; Lougheed, R.; McCubbrey, K.; Trekel, J. "Development of image analysis algorithms and software for automatic classification of TPO surfaces" private communication 1994.
6. Mikulec, M. *Ford PTPD Internal Report* **1994**, "Reduction in TPO Bumper Scuffing based on surface morphology".
7. Rugar, D.; Hansma, P.K. *Physics Today* **1990**, *43*, 23.
8. Clark, D.T.; Thomas, H.R. *J. Polymer Sci.* **1978**, *16*, 791.
9. Pireaux, J.J.; Riga, J.; Caudano, R.; Verbist, J. In *Photon, Electron and Ion Probes for Polymer Structures and Properties*, Dwight, D.W.; Fabish, T.J.; Thomas, H.R. Ed.;ACS Symposium Series, 1981, *162*, 169.
10. Schmitz, P.J.; Holubka, J.W.; Srinivasan, S. *Surf. Interface Anal.*, in press
11. Bowden, F. P.; Tabor, D. *The Friction and Lubrication of Solids*, Clarendon, Oxford, 1950, 169.

Chapter 9

Scanning Force Microscopy of Conjugated Anisotropic Thin Films Grown in High-Vacuum

Fabio Biscarini

CNR—Instituto di Spettroscopia Molecolare, Via Gobetti 101,
I-40129 Bologna, Italy

Scanning force microscopy has been used to investigate the quantitative aspects of the morphology transition from isotropic grains to anisotropic lamellae in sexithienyl thin films. 2-D islands made of few molecular layers grow up to a critical thickness which depends on the deposition temperature, then the film grows as a rough 3-D surface. A statistical analysis of SFM data yields domain size and shape distributions, diffusional barriers, roughness scaling parameters, and orientational ordering. This framework provides an explanation for the mechanism of growth in terms of competition between diffusion and capture.

Vacuum-grown thin films made of conjugated molecules have attracted a considerable interest as prototype organic semiconductors,[1,2] which can be used as active systems in molecular devices.[3-6] The advantages in using conjugated molecules and oligomers are: i) their chemical structure is well defined, ii) the molecular design tailors the optical and band gaps, and the intermolecular interactions, [7, 8] iii) they can be grown under clean and controlled conditions as conventional semiconductors.[9-11] By controlling the nature of the molecule, the substrate, and the growth conditions it is possible to investigate the interplay among intrinsic electronic structure, intermolecular and epitaxial ordering, anisotropy and dimensionality, charge and energy transport, and collective optoelectronic response.[12] Although different types of systems, such as conjugated polymers, molecular and liquid crystals, Langmuir molecular films possess some of these features, rarely they combine all of them as effectively as the vacuum-grown conjugated thin films.

Even though the interest towards these types of thin films dates back more than one decade, the influence of the film morphology onto the spectroscopical and electrical observables is still an open question. Scanning tunneling (STM) and scanning force microscopies (SFM) play an important role in trying to establish such correlations. STM provides high resolution imaging (13-15) combined with electronic and chemical information,[16-18] while SFM is very sensitive to changes in the film morphology upon variation of experimental parameters (19-21)

In this paper we show how SFM data can be used to investigate quantitatively the growth of thin films made of α-sexithienyl (T6), which is both an excellent model compound and a system with potential applications in multilayer structures (22) and devices.(4-6) T6 is a rodlike conjugated molecule whose electronic and optical properties have been extensively studied (23-26). Its structure has been resolved by X-ray diffraction (27, 28), and in the following the single crystal nomenclature will be adopted.(28) In high-vacuum sublimed thin films, T6 forms spontaneously ordered domains whose structure depends upon the growth conditions.(5, 21, 24, 29, 30) The order yields anisotropic conductivity and large hole mobility (30, 31), and polarized emission (32). These properties appear related to the film morphology, which has been investigated by scanning electron microscopy (30), STM and SFM (21, 32-34) and transmission electron microscopy (34). It is therefore important on one hand to assess the influence of the experimental conditions on the film morphology, on the other hand to extract information on the film properties and to understand the process of growth . These issues will be discussed in the following. First, the dependence of the morphology on the deposition temperature and film thickness is presented. Then, we describe the statistical procedures and the results of the analysis. Finally, in the fourth section we discuss the mechanism of growth emerging from the SFM observations.

Experimental

T6 was synthesized by coupling of the dimer via Grignard reagents catalyzed by Li. (35) T6 films are grown by sublimation on ruby mica in high - vacuum (HV) (base pressure 1×10^{-7} mbar). Reproducible heating ramps of the polycrystalline T6 powder yield deposition rates in the range 0.2 - 1 Å/s. The substrate deposition temperature and the film thickness are the experimental parameters which are systematically varied. The deposition temperature is measured with a thermoresistor in contact with the mica surface, and the film thickness is measured with a quartz oscillator. T6 films are transparent up to 10-20 nm thickness, then become pale yellow in the range 25-50 nm, and orange above 150 nm.

SFM images have been obtained with a commercial instrument (TMX 2010, Topometrix, Santa Clara CA) operated in contact mode at 1-5 Hz scan rates, using standard silicon nitride cantilevers with 0.03 - 0.06 N/m spring constant. Controlled atmosphere (relative humidity <20 %) was useful to avoid removal of material.

Film Morphology

The morphology of T6 thin films is characterized by 2 different type of domains: grains and microcrystalline lamellae. The dominant type of domain is determined by the growth conditions. (21, 30) At a given film thickness, the effect of deposition temperature is to increase the size of domains. This is clearly shown in the sequence in Figure 1 for 300-nm thick films. The domains up to 150°C exhibit round shapes which suggest an isotropic growth. As the temperature increases from 25°C (RT) to 150°C, the contacts between adjacent grains become extended, and a network of large grains joined by convex necks forms. At 200 °C the film morphology changes from grains to lamellae. The transition is accompanied by a change in the film roughness, as it is inferred from an increase of the height range by few times. A similar evolution of the morphology was observed by SEM and a sequence of phase transitions for micron-thick T6 films has been reported by XRD in the same temperature range. (30)
The axes of the lamellae form 30, 60, 90, 120 degree angles. Polarized microscopy shows that the orientation of the large lamellae and the crystallographic directions of mica are correlated, which suggests that mica induces the preferential orientation.
Another relevant growth parameter is the nominal film thickness, which controls the amount of material deposited. In Figure 2 it is shown that for a given

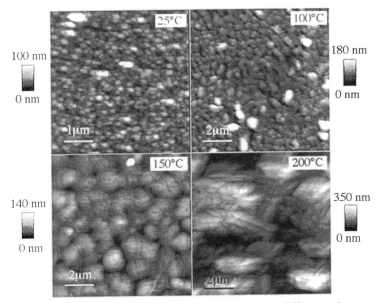

Figure 1. SFM images of 300-nm thick films on mica at different substrate temperatures. (Adapted from ref. 21)

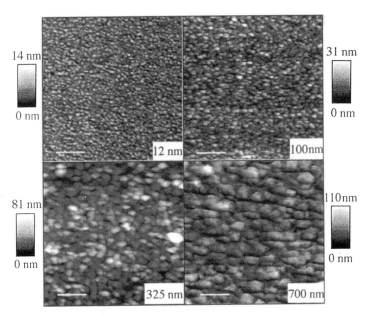

Figure 2. Evolution of the morphology with nominal film thickness at 25°C. (Adapted with permission from reference 39. Copyright 1997 The Royal Society of Chemistry

deposition temperature the domain size and the height range increase with the nominal film thickness. As a consequence, the mean grain size and the spread of the height distributions should increase accordingly.

This monotonic behavior breaks down at the early stages of growth when the film is formed by disjointed islands. The islands in Figure 3a appear much larger than the small grains of the 12-nm-thick film in Figure 2. This implies that the distribution of domain size undergoes a sudden decrease of its mean value as the film gets compact. A similar behavior has been recently associated to a Stranski-Krastanov-type of growth transition (36) in which a layered growth is followed by a random growth above a critical thickness. (37) A sensitive parameter for the onset of 2-D growth is the aspect ratio of the islands. The typical island height with respect to mica (which is identified by the flat grey background in Figure 3a) is 10-20 nm, while the lateral size is on the order of several hundred nm. The corresponding height-to-diameter ratio of the islands is smaller or comparable to 10%. This upper limit is a conservative estimate which takes into account the fact that the apparent lateral size might be larger due to the tip finite size (30-50 nm radius of curvature). Although no extended terraces are resolved, a closer look at the histogram of the heights (Figure 3b) reveals few broad but distinct peaks: the left-most one is the background (set to zero), then a first hump spreads between 3 and 5.5 nm, a second one between 5.5 and 9 nm, and the third one between 9 and 11.3 above the background respectively. The 2-3 nm height difference between the modes is consistent with the interlayer spacing along the a-direction measured by XRD (2.35 nm). (30). This evidence suggests that at low coverage the 2-D islands are made of stacks of discrete layers where the molecules stand mostly upright, and that the growth is a-oriented. Upon film completion at $\approx$12 nm thickness the grains form as in a 3-D growth.

The 2-D growth is even more apparent on films grown at 150°C. The branched islands in Figure 4a are few-μm-wide and the island height with respect to the background is 6-8 nm. Some of the islands in Figure 4a have coalesced and merged into sponge-like domains characterized by a large number of voids. The latter features are stable in time and do not heal, as a consequence of large diffusional barrier along the island walls with respect to the thermal energy. The profile in Figure 4b shows large terraces (marked by arrows) whose height is 1.8 - 2.4 nm and multiple integer. Depositing more material as in the 5-nm thick film (Figure 4c) does not yield a flat continuous film. The height difference between the islands (labelled b) and the bottom of the dark canyons (labelled c) in Figure 4c is still 6-8 nm. Instead, mesas with sharp edges (features a) begin to form on top of the islands. The step-height, which was measured as the mean difference between the height of the mesas and the islands, is 2.3±0.4 nm, in agreement with the 2.2±0.1-nm interlayer spacing from (l00) XRD peaks. (30) The layered growth continues up to 50-nm films (Figure 4d), when a secondary nucleation sprouts off at domain boundaries. The capture rate of these secondary nuclei is considerably larger than that of the terraces, and further deposition of material yields a network of few-micron-long and 100-200-nm-high lamellae. (38, 39)

Quantitative Analysis of Film Growth

T6 films evolve from layered islands and lamellae to grains as the thickness increases, and viceversa as the deposition temperature increases. In particular:
i) the domain size increases with both deposition temperature and film thickness;
ii) the roughness increases in the transition from grains to lamellae;
iii) the anisotropy of the domains increases with the deposition temperature;
iv) the lamellae give rise to orientationally ordered domains.
In the following the first two aspects are discussed, while the evolution of shape and orientation are presented in detail elsewhere. (40)

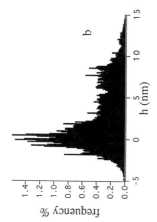

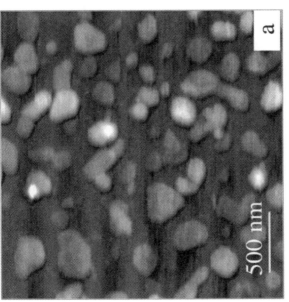

Figure 3. Morphology of 5-nm T6 film on mica at 25°C: a) SFM topography; b) height histogram. The zero corresponds to the background peak.

168

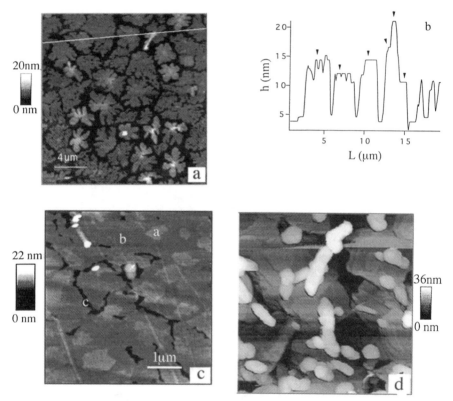

Figure 4. SFM images of T6 films on mica at 150°C and different nominal film thickness: a) 3 nm, b) profile across line in a), c) 5 nm, d) 50 nm.

Statistical Analysis of Domains. Individual domains (grains and lamellae) are selected from several images of different scan size by drawing manually a 1-pixel rough boundary with a black line so that the boundary coincides with the lowest feature in the color scale. Then, thresholding is done within the manually drawn boundary at a slighthly upper color, which reduces the arbitrariety of the operator's judgement in following the optimal profile. NIH-Image software (National Institute of Health, Bethesda, Maryland, USA) has been used for pixel statistics within the selected domains. Robust statistical sets are built out of 400-500 elements.

The distribution of the area A is analyzed in terms of its moments :

$$\left\langle A^m \right\rangle = \frac{1}{N} \sum_{i=1}^{N} A_i^m \tag{1}$$

where the sum runs over the N elements of the set, and $m=1,2,3$ so that mean, variance, and skewness are estimated. The errors on the mean values are estimated as the average absolute deviation $\Delta A \approx \frac{1}{N} \sum_i |A_i - \langle A \rangle|$.

The resulting histograms are shown in Figure 5. The distributions spread out and their mode shifts to higher particle sizes as the substrate temperature increases. The average domain size <A> exhibits an Arrhenius-type behavior vs. the inverse deposition temperature (Figure 6), with an activation energy $E_A = 0.36\pm0.04$ eV. The analysis of the higher moments yields the analytical form for the domain size distribution:

$$P(x;\sigma) = \frac{\sqrt{3}}{2\sigma} x^2 \exp(-x) \tag{2}$$

where $x = \frac{\sqrt{3}A}{\sigma}$, and σ is the standard deviation of the size distribution. The good agreement between experimental and theoretical distributions (Figure 5) proves that the distributions are effectively skewed towards large-size populations. Thus, smaller domains coalesce into larger ones. This process is driven by the minimization of the surface energy at the contact points which takes place by the diffusion of material and its incorporation at the contact interface. Here films are many monolayer thick and diffusion on mica should not be relevant. Thus, the 0.36 eV activation energy of the growth process can be ascribed to the barriers experienced by T6 molecules diffusing on a T6 surface . The diffusional barriers were estimated using a parametrized model of the van der Waals interaction of a single T6 molecule on a T6 *bc* crystal surface. The expected values for translation and reorientation barriers turn out to be 0.27 and 0.50 eV respectively. (*21*)

The evolution of the film morphology with deposition temperature of 50 and 100 nm - thick films follows the same type of activated growth as the 300 - nm series, but the grain - lamella transition occurs at lower deposition temperatures for decreasing film thickness (Figure 6). Thus, it is possible to grow ordered films at lower temperature if the film thickness is kept below a critical value. (38). The transition temperatures in Figure 6 are estimated by analysing the evolution of the anisotropy and orientation of the domains. It turns out that while the anisotropy changes continuously as the lamellae form, the orientational order parameter (which describes the average orientation of the domains) has a discontinuity in the slope. This phenomenology, which resembles a second-order phase transition such as for instance uniaxial to biaxial

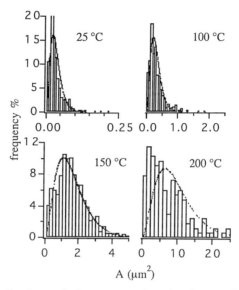

Figure 5. Distributions of the average domain size at different substrate temperatures. Dots represent the distribution density in Equation (2) multiplied by the respective histogram bin size.

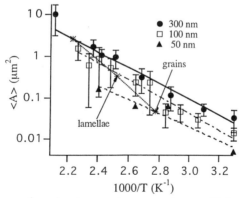

Figure 6. Average domain size vs. inverse substrate T for different film thickness. Error bars represent the mean absolute error, and straight lines the best fit for each film thickness. Dotted line is the locus of the transition from grains to lamellae. Data for 50-nm films are estimated from the correlation length of the topography fluctuations.

ordering (but with the caveat that here the films are grown under non-equilibrium conditions) allows one to estimate the transition temperature as the onset of the change of slope. This method is described in detail elsewhere (40), and can be useful for constructing the "phase diagram" of a molecular thin film from SFM data. (Biscarini, F. ; Marks, R. N. unpublished data)

Roughness Scaling Analysis. The apparent disordered behavior exhibited by the film morphology in Figure 1 and 2 can be statistically analyzed using the framework of the dynamic scaling theory. The theoretical framework rests upon the scaling behavior of the surface roughness σ, which measures the interface width as the mean square fluctuation of the topography. In general, the surface morphology of real systems depends on the scalelength of observation L. A fractal surface instead would be scale-invariant. There is a class of surfaces termed self-affine whose morphology does change upon a change of magnification, but it looks similar after the third dimension (z-axis) is properly rescaled. (41-43) The self-affinity implies that the surface roughness follows simple power-laws with deposition time t, which is proportional to the film thickness when the deposition rate is constant:

$$\sigma(t) \approx t^{\beta} \qquad \text{for } t < t_x \qquad (3)$$

where β is the growth exponent, and scalelength of observation L:

$$\sigma(L) \approx L^{\alpha} \qquad \text{for } t > t_x \text{ and } L < \xi \qquad (4)$$

where α is the roughness exponent. Crossover time t_x and correlation length ξ set the upper limits of the self-affine behavior, and they are related through the dynamic exponent $\zeta = \alpha/\beta$

$$\xi \approx t_x^{1/\zeta} \qquad (5)$$

At scalengths $L > \xi$ the roughness saturates. The exponents α, β, ζ identify the universality class of growth, thus the relevant basic growth mechanism can be inferred independently from the detailed microscopic knowledge of the system. (43, 44)

SPM techniques yield directly the topography on a wide range of spatial lengths and the surface roughness can be calculated from SPM images. Alternatively, other direct space methods (45-47), or the power spectral density (PSD) of the topographical profiles $h(x)$ (48-50):

$$PSD(f) = \frac{1}{L} \left| \int_0^L dx h(x) e^{i2\pi fx} \right|^2 \qquad (6)$$

yield the scaling exponents. Here the mean height $<h>=0$. Equation (5) is calculated on an image with scan length L, x is taken as the fast scan direction which is less affected by 1/f noise, and f is the spatial frequency. The scaling relation in equation (3) translates into the power-law decay of PSD:

$$PSD(f) = K_0 f^{-\gamma} \qquad (7)$$

in a finite range of f, and the roughness scaling factor $\alpha = (\gamma\text{-d})/2$ where the linescan dimension d equals 1 in this case. (*41*)

The typical PSDs of T6 films exhibit self affine behavior in an intermediate range of spatial frequencies (Figure 7). The self-affine range spans one order of magnitude of the spatial frequencies at RT and reaches about 3 orders of magnitude above 200 °C. This is an example of an actual system where a modest self-affinity in the grains blows out in the lamella, showing that lamellae arise from aggregation of smaller lamellae. (*51*)

The PSD in Figure 7 represent a statistical set of images. First the images are corrected for trends by removal of the best-fit plane, then for each image Fast Fourier Transform algorithm (*52*) is applied to each scan line to estimate a 256-frequency point PSD and single-line PSD's are averaged. The representative PSD as in Figure 7 are obtained by averaging the logarithmic spectra of 3-8 images of each different sample zones and cropping together spectra at different scan lengths. 20-50 images yields a sufficient number of "typical" PSD, and on the basis of them we reject images with anomalous topographical features (such as protruding grains or large cavities) or affected by excessive instrumental noise (between 5 and 20 % of the total images). (*21, 38, 51*)

The roughness scaling factor α (Table I) decays with temperature from about 1 at RT down to 0.7 at 200 °, i.e. within the limits predicted by diffusion-limited aggregation ($\alpha = 1$) (*53*) and molecular-beam epitaxy (MBE) growth ($\alpha = 0.66$). (*54*) The latter model describes a mechanism of growth where the relevant mechanism is the adsorption at step edges, and works effectively in systems with strong molecule - surface interactions. The preliminary data on the growth exponent β obtained from the analysis of films at different thickness are to be compared with the theoretical values $\beta=0.25$ for pure diffusion, and $\beta=0.2$ for MBE.

The evolution of α suggests that surface diffusion controls the growth at low temperature, and as the mobility of the molecules on the T6 surface increases at higher temperatures the capture of the molecule at the terrace edges becomes dominant. In the latter regime, the strong intermolecular interactions between parallel T6 molecules favor the lateral growth of the (100) T6 plane. Therefore, the growth of T6 films is determined by the interplay of 2 sequential processes, diffusion and capture, and the nature of the transition from grains to lamellae is the progressive depletion of the diffusion term relatively to the capture. This also implies that the capture process is either not activated or has a larger barrier than that of diffusion. It is remarkable how this analysis, which is performed on a micro- to mesoscopic scalelengths, can provide insights on the subtleties of the mechanism of growth. The conclusions of the dynamic scaling analysis confirm the picture that was qualitatively drawn on the basis of the evolution of the grain size distributions. (*21, 51*)

A final remark concerns the correlation length , i.e. the inverse frequency which separates the plateau from the self-affine branch. From the analysis of the PSD at several temperatures, it is shown that ξ has an Arrhenius behavior and that is proportional to the effective diameter of the domain. Thus, the estimate of ξ is a very convenient method for measuring the average domain size compared to the statistical averaging described in the previous section: it rests on solid physical basis, is straightforward, does not require additional image manipulation, and is independent on arbitrary statistical sampling. (*51*)

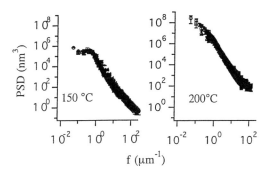

Figure 7. SFM topography power spectral density vs. spatial frequency at 2 different temperatures.

Table I. Roughness Scaling Parameters for T6 Thin Films

T (°C)	ξ (μm)	$\Delta\xi$ (μm)	α [a]	$\Delta\alpha$ [a]	ζ [b]	β [c]
25	0.61	0.03	1.09	0.06	2.7	0.4
100	2.6	0.4	0.87	0.07	–	–
150	2.3	0.3	0.82	0.06	3.6	0.23
200	3.3	0.8	0.70	0.06	–	–

a For 300-nm thick films
b Estimated from the correlation length vs. film thickness.
c Estimated as α/ζ.
SOURCE: Reproduced from reference 38. Copyright 1996 American Chemical Society.

Nature of the Growth Transition

The SFM observations show that the grain-lamella transition is governed by kinetics: when diffusion is dominant the grains form, viceversa lamellae appear when capture is the slowest step. The evolution of the morphology with film thickness shows that such a transition has the character of a Stranski-Krastanov growth. In the following the microscopic nature of the transition is discussed in terms of the specific nature of T6 and mica.

According to XRD (*30*) the films grown at RT onto a dielectric substrate are formed by small ordered domains ($\approx$30 nm coherence length) and at least 2 crystalline phases: the dominant α-phase where the molecules stand normal to the surface (homeotropic arrangement), and the minority β-phase where the molecules lie flat on the surface (parallel arrangement). The usual assumption is that the phase mixture arises from a random nucleation on the substrate due to the low molecular mobility. This would be consistent with a 3-D growth. As it has been shown, this is not the case for T6 on mica. Since a major restructuration is unlikely due to the large diffusional barriers of T6, then the film is made of ordered layers in contact with mica, and isotropic (polycrystalline or amorphous) material on top. As a consequence, the properties sensitive to order, e.g. polarized absorption, (*25*) photoluminescence, and carrier mobility, should be varying across the film thickness. For instance, in a device where the molecules with homeotropic orientation, the in-plane carrier transport would be mainly contributed by the first few ordered layers.

The fundamental question is to understand why the ordered growth is disrupted as more material is deposited. In epitaxially grown films the driving force of the Stranski-Krastanov transition is attributed to the relaxation of the increasing strain due to lattice misfit between the substrate and the overlayer. (*36, 37*) In the case of T6 islands on mica at RT, high-resolution SFM imaging has not yielded reproducible observations of any periodic arrangement. Therefore, there is not yet conclusive evidence, neither from SFM or other techniques, about heteroepitaxial growth with T6 molecules in registry with the mica lattice. Therefore, the explanation of the Stanski-Krastanov transition might lie in the competition between the intermolecular interactions and interactions between molecules and mica surface.

A homeotropic 2-D growth requires several conditions: i) a weak interaction between T6 molecules and mica, (*55*) ii) small diffusional barriers experienced by T6 on mica , and iii) an internuclei distance which is smaller than the product between the time of thermalization of the T6 molecule ($\approx$300 °C) and the parallel component of the velocity. Upon such conditions, if T6 molecules impinging the surface retain a sufficient momentum, they migrate to an island edge and get steadily locked by other T6 molecules. The adsorption energy of a T6 molecule at the terrace edge is $\approx$1.5 eV considerably above the thermal energy.(*21*) The weak attraction towards mica can also explain why single molecules would stand up on mica and start a homeotropic nucleation. Mica usually has a negative charged surface with sheets of O atoms and mobile counterions such as K^+. (*56*) In a T6 molecule the charge distribution is inhomogeneously distributed within the thienyl units, with the negative charge and dipoles polarized on the electronegative S atoms on opposite sides with respect to the mirror plane (molecular symmetry is C_{2h}) (*57, 58*); as a consequence there is not a permanent dipole moment in the ground state. When T6 lies flat on mica, there is repulsion between the negative charges on the O atoms and those on the S atoms on T6. At the same time, the surface dipole (from the mica double layer) -T6 dipole interactions cancel because of the molecular symmetry. There is not a rotation around the long molecular axis which effectively screens the repulsion or makes non-zero the dipole-dipole interaction. A configuration with S atoms on one side would result in an attractive dipolar interaction, but it requires that T6 overcomes a large torsional barrier due to the breaking of the conjugation. Thus, the T6 molecule must tilt upward in order to decrease the electrostatic repulsion and to gain in the surface dipole - induced dipole energy. The tilt angle depends ultimately in the balance between the different polarizability component of T6.

The 3-D growth starts when the mica surface is completely covered by T6 molecules. The induced-dipole field decays with the film thickness, and the van der Waals attractive interactions between the impinging T6 molecule and the film surface becomes dominant. Upon such conditions, the energetically favorable orientation is with the T6 molecule adsorbed flat on the surface (≈ 1 eV binding energy) (21). Since the 0.36 diffusional barrier experienced across the T6 surface is considerably higher than the thermal energy at RT, there will be a fraction of T6 molecules which nucleate flat. As the deposition temperature increases, the impinging T6 molecules diffuse considerably across the T6 surface, until they get captured at a terrace edge. Therefore, at higher deposition temperature, the critical thickness of the 2-D -> 3-D growth transition shifts towards larger values.

Conclusions

In this paper SFM data have been statistically analyzed to yield valuable information on the mechanism of growth in high-vacuum of conjugated molecular thin films. The proposed framework consists of analyzing the domain size distributions, the surface roughness scaling behavior, and shape and orientation order parameters. Estimated values for the T6 diffusional barriers, roughness scaling parameters and orientational order parameters of the domains have been extracted from SFM measurements.

We have presented an example of an actual organic system exhibiting a Stranski - Krastanov transition, and discussed how this transition arises because of the strong van der Waals interactions between T6 molecules. The deposition temperature shifts this transition to larger thickness, and the transition from grains to lamellae occurs when the rate-limiting step swaps from diffusion to capture. We have shown that when coupled to statistical analysis, SFM is a powerful tool not only for assessing the quality of the film, but also to gain insights in the basic growth phenomena.

Acknowledgements

I am grateful to O. Greco, A. Lauria, R. N. Marks, P. Samorì, P. Ostoja, C. Taliani and R. Zamboni, without whom this work would have not been possible. I thank also V. Dediu, C. Matacotta, A. Degli Esposti, P. Pasini, A. Stella, J. Krim, for stimulating discussions. Partial support from the Italian CNR - Comitato 12, and from the EC-TMR project SELOA is also acknowledged.

Literature Cited

1) Pope, M.; Swemberg, C. E.*Electronic Processes in Organic Crystals*; Clarendon: Oxford, United Kingdom, 1982.
2) *Spectroscopy and Excitation Dynamics of Condensed Molecular Systems*; Agranovich, V. M.; Hochstrasser, R. M. Eds.; Modern Problems in Condensed Matter Sciences; North-Holland: Amsterdam, The Netherland, 1983; Vol. 4.
3) Tang, C. W.; VanSlyke, S. A.; Chen, C. H. J. *Appl. Phys.* **1989**, *65*, 3610.
4) Horowitz, G.; Peng, X.; Fichou, D.; Garnier, F. *J. App. Phys.* **1990**, *67*, 528.
5) Ostoja, P.; Guerri, S.; Rossini, S.; Servidori, M.; Taliani, C.; Zamboni, R. *Synth. Met.* **1993**, *54*, 447.
6) Dodabalapur, A.; Torsi, L.; Katz, H. E. *Science* **1995**, *268*, 270.
7) Pichler, K.; Halliday, D.A.; Bradley, D.D.C.; Burn, P.L.; Friend, R.H.; Holmes, A.B. *J. Phys.; Condens. Matter* **1993**, *5*, 247.

176

8) Cornil, J.; dos Santos, D.A.; Beljonne, D.; Brédas, J.L. *J. Phys. Chem.* **1995**, *99*, 5604.
9) Tada, H., Saiki, K; Koma, A.; *Jpn. J. Appl. Phys.* **1991**, *30*, L306.
10) Ludwig, C.; Gompf, B.; Glatz, W.; Petersen, J.; Eisenmerger, W.; Möbus, M.; Zimmermann, U.; Karl, N. *Z. Phys. B - Cond. Matter* **1992**, *86*, 397.
11) Forrest, S. R.; Zhang, Y. *Phys. Rev. B* **1994**, *49*, 11297; Forrest, S. R.; Burrows, P. E.; Haskal, E. I.; So, F. F. *Phys. Rev. B* **1994**, *49*, 11309.
12) Blinov, L. M.; Palto, S. P.; Ruani, G.; Taliani, C.; Tevosov, A. A.; Yudin, S. G.; Zamboni, R. *Chem. Phys. Lett.* **1995**, *232*, 401.
13) Lippel, P. H.; Wilson, R. J.; Miller, M. D.; Wöll, C; Chiang, S.; *Phys. Rev. Lett.* **1989**, *62*, 171.
14) Soukopp, A; Glöckler, K.; Bäuerle, P; Sokolowski, M.; Umbach, E. *Adv. Mater.* **1996** in press.
15) Stabel, A.; Rabe, J. P. *Synth. Met.* **1994**, *67*, 47.
16) Sautet, Ph.; Joachim, C. *Chem. Phys. Lett.* **1991**, *185*, 23.
17) Hallmark, V. M.; Chiang, S. Surface Science **1994**, *329*, 255.
18) Biscarini, F.; Bustamante, C.; Kenkre, V. M. *Phys. Rev. B* **1995**, *51*, 11089.
19) Chi, L. F.; Anders, M; Fuchs, H.; Johnston, R. R.; Ringsdorf, H. *Science* **1993**, *259*, 213.
20) Hara, M.; Sasabe, H. in *Introduction to Molecular Electronics*; Editors, Petty, M. C.; Bryce, M. R.; Bloor, D. Ed.; Edward Arnold: London, UK, 1995; pp 243-260.
21) Biscarini, F.; Zamboni, R.; Samorí, P.; Ostoja, P.; Taliani, C. *Phys. Rev. B* **1995**, *52*, 14868.
22) Muccini, M.; Mahrt, R. F.; Lemmer, U.; Hennig, R.; Bässler, H.; Biscarini, F.; Zamboni,R.; Taliani, C. *Chem. Phys. Lett.* **1995**, *242*, 207.
23) Zhao, M. T.; Singh, B. P; Prasad, P. N. *J. Chem. Phys.* **1988**, *89*, 5535.
24) Periasamy, N.; Danieli, R.; Ruani, G.; Zamboni, R.; Taliani, C. *Phys. Rev. Lett.* **1992**, *68*, 919.
25) Egelhaaf, H.-J.; Bäuerle, P.; Rauer, K.; Hoffmann, V.; Oelkrug, D. *Synth. Met.* **1993**, *61*, 143.
26) Hotta, S.; Waragai, K. *Adv. Mater.* **1993**, *5*, 896.
27) Porzio, W.; Destri, S.; Mascherpa, M.; Brückner, S. *Acta Polymer.* **1993**, *44*, 266.
28) Horowitz, G.; Bachet, B.; Yassar, A.; Lang, P.; Demanze, F.; Fave, J-L.; Garnier, F. *Chem. Mat.* **1995**,.*7*, 1337.
29) Buongiorno Nardelli, M.; Cvetko, D.; De Renzi, V.; Floreano, L.; Gotter, R.; Morgante, A.; Peloi, M.; Tommasini, F.; Danieli, R.; Rossini, S.; Taliani, C.; Zamboni, R. *Phys. Rev. B* **1996**, *53*, 1095.
30) Servet, B.; Horowitz, G.; Ries, S.; Lagorsse, O.; Alnot, P.; Yassar, A.; Deloffre, F.; Srivastava, P.; Hajlaoui, R.; Lang, P.; Garnier, F. *Chem. Mat.* **1994**, *6*, 1809.
31) Torsi, L.; Dodalapur, A; Rothberg, L. J.; Fung, A. W. P.; Katz, H. E. *Science* **1996**, *272*, 1462.
32) Marks, R. N. ; Biscarini, F.; Zamboni, R.; Taliani C. *Europhys. Lett.* **1995**, *32*, 523.
33) Floreano, L.; Prato, S.; Cvetko, D.; De Renzi, V.; Morgante, A.; Modesti, S.; Biscarini, F.; Zamboni, R.; Taliani, C. *Phys. Rev. B* **1997**, submitted.
34) Lovinger, A.J.; Davis, D.D.; Ruel, R.; Torsi, L.; Dodabalapur, A.; Katz, H.E. *J. Mater. Res.* **1995**, *10*, 2958.
35) Ostoja, J.; Guerri, S.; Impronta, M.; Zabberoni, P.; Danieli, R.; Rossini, S.; Taliani, C.; Zamboni, R. *Adv. Mat. Opt. Electr.* **1992**, *1*, 127.
36) Dediu, V.; Kursumovic, A.; Greco, O.; Biscarini, F.; Matacotta, F. C. *Phys. Rev. B* **1996**, *54*, 1564.
37) Tersoff, J.; Denier van der Gon, A. W.; Tromp, R. M. *Phys. Rev. Lett.* **1994**, *72*, 266.

38) Biscarini, F.; Greco, O.; Lauria, A.; Samorì, P.; Taliani, C.; Zamboni, R. *Polymer Preprints* **1996**, *37*, 618.
39) Marks, R. N.; Biscarini, F., Muccini, M.; Virgili, T.; Zamboni, R.; Taliani, C. *Phil. Trans. Roy. Soc. A* **1997**, *1725*, 763.
40) Biscarini, F.; Greco, O.; Lauria, A.; Zamboni R.; Taliani, C. *Mol. Cryst. Liq. Cryst.* **1996**, *290*, 203.
41) Voss, R. F. In *Scaling Phenomena in Disordered Systems* ; Editors, Pynn R.; Skjeltorp A., Ed.; NATO ASI series; Plenum Publ. Corp: New York, 1985; pp 1-16.
42) Vicsek, T. *Fractal Growth Phenomena* ; World Scientific, Singapore 1989.
43) Family, F. *Physica A* **1990**, *168*, 561.
44) Krim J.; Palasantzas, G. *Int. J. Mod. Phys.* B **1995**, *9* , 599.
45) Krim, J.; Heyvaert, I.; Haesendonck, C. V.; Bruynseraede, Y. *Phys. Rev. Lett.* **1993**, *70*, 57.
46) Vázquez, L.; Salvarezza, R. C.; Ocón, P.; Herrasti, P.; Vara, J. M.; A. J. Arvia, *Phys. Rev. E* **1994**, *49*, 1507.
47) Weber, W.; Lengeler, B. *Phys. Rev. B* **1992**, *46*, 7953.
48) Mitchell, M. W.; Bonnell, D. A. *J. Mater. Res.* **1990**, *5*, 2244.
49) Dumas, Ph.; Bouffakhredine, B.; Amra, C.; Vatel, O.; Andre, E.; Galindo, R.; Salvan, F.; *Europhys. Lett.* **1993**, *22*, 717.
50) Collins, G. W.; Letts, S. A.; Fearon, E. M.; McEachern, R. L.; Bernat, T. P. *Phys. Rev. Lett.* **1994**, *73*, 708.
51) Biscarini, F.; Samorì, P.; Greco, O.; Zamboni, R. *Phys. Rev. Lett.* **1997**, *78*, 2389.
52) Press, W. H.; Flannery, B. P.; Teukolsky, S. A.; Vetterling, W. T. *Numerical Recipes: The Art of Scientific Computing*; Cambridge University Press: Cambridge, UK, 1990.
53) Wolf, D. E.; Villain, J. *Europhys. Lett.* 1990, *13*, 389.
54) Lai; Z.-W.; Das Sarma, S. *Phys. Rev. Lett.* **1991**, *66*, 2348.
55) Israelachvili, J. *Intermolecular and Surface Forces* ; Academic Press: San Diego, CA, 1992.
56) Deer, W. A.; Howie, R. A.; Zussmann J. *An introduction to the Rock-Forming Minerals*; Longman House: Burnt Mill Harlow, UK, 1992.
57) Degli Esposti, A.; Moze, O.; Taliani, C.; Tomkinson, J. T.; Zamboni, R.; Zerbetto, F. *J. Chem. Phys.* **1996**, *104*, 9704.
58) Gavezzotti, A.; Filippini, G.*Synth. Met.* **1991**, *40*, 257.

Chapter 10

Atomic Force Microscopy of Soft Samples: Gelatin and Living Cells

Jan Domke[1], Christian Rotsch[1], Paul K. Hansma[2,] Ken Jacobson[3], and Manfred Radmacher[1,4]

[1]Lehrstuhl für Angewandte Physik, Ludwig-Maximilians Universität München, 80799 München, Germany
[2]Department of Physics, University of California, Santa Barbara, CA 93106
[3]Department of Cell Biology and Anatomy, University of North Carolina, Chapel Hill, NC 27599

We have investigated soft samples with the atomic force microscope. Gelatin films can be patterned with the AFM using high loading force. Scanning the gelatin for a long time results in wrinkles, roughly equally spaced and perpendicular to the fast scan directions. The characteristic distance between adjacent wrinkles does not depend on the experimental conditions, however it varies with the molecular weight of the sample. Gelatin immersed in mixtures of water and propanol is a polymeric gel with a Young's modulus comparable to values typically found in living cells. Therefore films of gelatin are ideal test samples to characterize the measurement of the elastic properties of thin and soft samples. For this purpose we investigated shallow wedges of gelatin with a thickness below 1 μm and a slope of only a few degrees. These investigations show that there is a dependence of the Young's modulus on the thickness of the film such that these ultrathin films can be 5 times stiffer than the bulk value of the same sample. We also investigated the Young's modulus of living fibroblast cells. We measured local variations of the Young's modulus, for instance at stress fibres, by AFM with a lateral resolution of better than 100 nm.

The AFM (Atomic force microscope) (1; 2) has been used widely to investigate soft polymeric and biological samples (for reviews see: (3) and (4)). The AFM can not only be used to obtain surface topographs of the sample, it is also a powerful tool to measure and apply very small forces. Examples are mapping of adhesion forces with a lateral resolution of only a few nanometers (5; 6), mapping the electrostatic interaction (7) and probing the local elasticty of cells (8-10). Recently it has become possible to manipulate single molecules such as to detect the activity of single enzyme molecules (11) or to measure specific binding forces between individual ligand/receptor pairs (12; 13) or to stretch single molecules (14). Our main interest is the investigation of living cells focusing on their elastic properties. However, to understand the results with this rather new technique better it is

[4]Corresponding author.

necessary to apply it first to a simpler system. Gelatin seemed a promising candidate because of the ease of preparation, availability and the large amount of knowledge about this polymer gained because of its widespread use in the food and photographic industries (15; 16). Gelatin has also been investigated with the AFM (4; 17; 18) (a review about AFM investigations of biopolymers is: (19)). The softness of gelatin can be adjusted by immersing gels in propanol and mixtures of propanol and water (20). Thus we were able to gain a polymeric sample with a tunable Young's modulus comparable to that of the cytoskeleton of living cells (10 to 100 kPa) (9). This enabled us to address one point, which proved to be problematic already in the first investigations of the softness of living cells: the sample's thickness (8). Since the peripheral parts of cells are generally very thin (only a few hundreds of nanometers), an AFM tip probing and compressing the cell can easily "feel" the underlying stiff substrate and thus measure too large a value of the Young's modulus. To check this potential artifact we prepared gelatin films of defined thicknesses below 1 μm and determined the thickness dependence of the Young's modulus.

The investigation of living cells with the AFM became possible several years ago (21; 22) (for review see (23)). It has been recognized that the AFM is a promising tool to follow dynamic processes in living cells, as has been shown by several groups (24-27). An impressive, recent application of AFM is the investigation and imaging of the local events during exocytosis (28). The AFM was used first by Weisenhorn et al. to characterize the elastic properties of cells (29). Hoh and Schoenenberger investigated the changes of these elastic properties during fixation with glutaraldehyde (30). Radmacher et al. were the first to analyze the elastic properties of living cells qualitatively as a function of position on the cell (8; 10).

Materials and Methods

Gelatin. For the surface modification experiments gelatin from pork skin (purchased from Fluka, Buchs, CH) was dissolved in pure water (Milli-Q quality, Millipore Systems, Molsheim, Fr) at 40-50°C at a concentration of 50 mg/ml. Freshly cleaved mica was mounted in a home built spincoater. A small droplet of the gelatin solution was applied to the mica while the spincoater was already spinning and kept spinning for 30 more seconds. Samples were freshly prepared and used the same day. Two different types of gelatin were used, 60 Bloom and 300 Bloom corresponding to molecular weights of 20-25 kD and 50-100 kD, respectively. The unit bloom measures the so called gel-strength and is defined using the compressibility of the gel. A cylindrical block of gel (6 2/3 % by weight in water) with a diameter of 0.5 inch is loaded by a cylindrical punch. The mass in gramms needed to compress the gel by 4 μm is equal to the gel strength in Bloom. For the elasticity investigations the gelatin was dissolved in 20 ml of pure water at 50°C at a concentration of 5 mg/ml. We added 10 droplets of 2 mg/ml Coomasie blue to be able to locate the edge of the droplet easily using an optical microscope. Two droplets of this solution were put on a cleaned glass cover slip. This droplet was first air dried for several hours and then immersed in propanol for investigation with the AFM. We added water to the propanol to change the elastic properties of the gelatin during the time-course of the experiment.

Fibroblasts. 3T3 fibroblasts were cultured following standard procedures. The cells were cultured in DMEM (Dulbeccos Modified Eagle's Medium) supplemented with 10% fetal calf serum and 1 % penicillin and streptomycin (all chemicals were purchased from Sigma). 25 mM Hepes were added to increase pH stability while the cells were investigated in the AFM. Usually the cells were cultured in T-flasks and seeded in Petri dishes 1-2 days before the experiment was performed.

Atomic Force Microscopy. AFM was performed using a commercial instrument (Bioscope, Digital Instruments, Santa Barbara CA) combined with a modified optical microscope (Zeiss Axiomat, Zeiss, Oberkochen, FRG). The combination of an optical and an atomic force microscope helped to position the tip on the area of interest, in this case either the edge of the gelatin droplet or a particular cell. We used oxide sharpened silicon nitride cantilevers for this investigation with a force constant of 20 mN/m (DNPS tips from Digital Instruments, Santa Barbara, CA) or 8 mN/m (MicroLevers, Park Scientific, Santa Clara, CA). The force constant was determined by measuring the amplitude of the thermally driven motion of the free cantilever (31).

Analyis of force curves. Force curves were analyzed as described previously (20). By comparing force curves taken on the soft sample with the curves taken on the substrate, the indentation versus loading force relation can be calculated. This relation was then fitted to the Hertz-model describing the elastic indentation of a homogeneous sample with a cone-like tip (32). For fitting purposes we have used a value of 0.5 for the Poisson ratio and 18 degrees for the opening angle of the cone (i.e. the AFM tip).

Results

Surface modifications of gelatin films. Thin gelatin films were prepared by spin coating as described above and then imaged in air. By applying a high loading force we could create characteristic wrinkles on the surface (33). These wrinkles formed always perpendicluar to the fast scan direction and had a typical periodicty. Figure 1 shows a time series of this process using a gelatin of 300 Bloom, corresponding to 50-100 kD molecular weight. The time delay between each image is 3 minutes. We show only every 12th image from the entire series, so the process is very slow and can be followed easily by AFM. In comparison figure 2 shows the same experiment done with a gelatin film of different molecular weight: 60 Bloom, corresponding to 20-25 kD molecular weight. The pattern always forms perpendicular to the fast scan direction, as is illustrated in figure 3. First we have patterned four areas using different scan directions (figure 3a, 90°, 0°, 45° and -45°, starting top left and listing clockwise, areas 1 through 4, see legend in fig. 3d). Then we have patterened areas #5 and #6 at zero scan angle. It was surprising that when zooming into the area where the patterns overlap we see that the created patterns do line up (fig. 3c). Here we have patterned the lower half (#5) of this image first and then patterned the top half (#6) while scanning from top to bottom. Therefore it is very surprising that the wrinkles do line up so nicely.

The observed pattern creation with the AFM has the following properties: the distance between adjacent wrinkles seems to be independent of all instrumental parameters like scan rate, loading force, scan orientation and tip shape (data not presented here). The rate of formation of these patterns in terms of height of wrinkle created per time is dependent on the scan parameters that the higher the force the faster the wrinkles grow in height as could be expected. Since all images are performed here in the constant deflection mode higher scan rates also result in higher loading forces. This is true because the feedback tries to keep the deflection constant, but it can only react with a finite response time, given by the sensitivity of the electronics, the feedback and the response times of the cantilever and the piezo translator. Therefore the feedback will react with a time delay at sharp edges and will also overshoot in this somewhat late reaction. This will become more prominent when scanning faster.

Figure 1
A time series of the process of creating wrinkles with a high molecular weight gelatin sample (300 Bloom). The images were taken at a loading force of about 5 nN, the scan rate was about 20 lines per second. The AFM was continuously scanning the frame from top to bottom and then from bottom to top again. Here we show only every 12th scan (every 6th down scan) of this time series. (adapted from ref. (33))

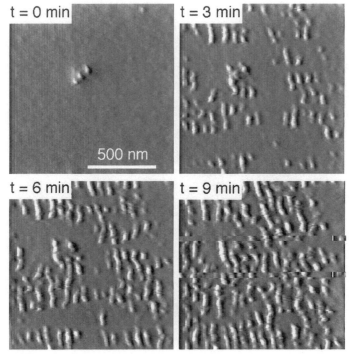

Figure 2
A time series of the wrinkling process of a low molecular weight gelatin sample (60 Bloom). The images were taken at a loading force of about 5 nN, the scanrate was about 20 lines per second. Here we present every 12th scan of this time series (only down scans shown). Note that the distance between adjacent wrinkles has become larger in comparison to those shown in figure 1 although the size of the molecule (the molecular weight) has become smaller. (adapted from ref. (33))

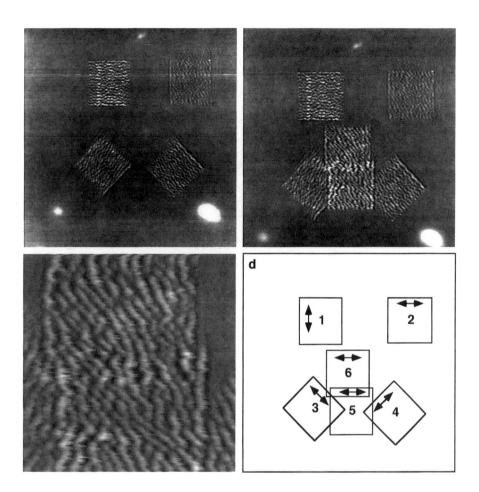

Figure 3
A gelatin film (60 Bloom) imaged in air with the AFM. The overview (a) shows that the sample has been scanned at higher force in four smaller areas (1μm size) with different scan angles (90°, 0°, 45° and -45°, starting top left and listing clockwise, areas 1 through 5, see legend in fig. 3d). While scanning the sample at high force wrinkles are created which are oriented perpendicular to the fast scan direction. In fig. 3b two more areas of the same sample spot have been wrinkled (area 5 and 6, see fig. 3d). A close up (fig. 3c) of the overlap region of these two new areas shows that the wrinkles of the two areas line up with each other. Amazingly the wrinkles in the lower area (#5) were created first while scanning from top to bottom and then the wrinkles in the top area (#6) were created while also scanning from top to bottom. (adapted from ref. (33))

The width of the wrinkles is dependent on the molecular weight of the sample: the higher the molecular weight of the sample the smaller the features. This counter intuitive dependence has been found previously in a similar study with a different polymeric system: polystyrene (34). Although the molecular weight influences the patterns, it is probably not the size of the molecule determining the size of the wrinkles directly but some material property which is dependent on the molecular weight. A likely material property which may determine the wavelength of the wrinkles is the elasticity (35), especially since the patterns are reminiscent of Schallamach-waves (36). Such patterns have also been created and observed with the AFM on Langmuir-Blodgett films (37). Generally this phenomen might be very useful for nanometer-scale lithography (38-40).

Measuring the elastic properties of thin gelatin films. We have investigated wedge shaped films of gelatin to characterize the elastic properties of ultrathin soft samples. A typical image of these gelatin films immersed in 50% propanol with water is shown in figure 4. The image shows that the film exhibits a very homogeneous linear slope over a large range of 100's of μm and more (data not shown here), which is 11 degrees in this case. The border of the gelatin film is very well pronounced. Therefore it was easy to take a reference force curve on the substrate and then by moving the AFM tip along a line several force curves ontop of the gelatin. From the heights of the individual contact points in each force curve we could determine the actual thickness of the film at the location where we have taken the force curve. Figure 5 shows three force curves, which have been taken at the location of the three arrows in figure 4. These force curves show a slightly different slope, corresponding to a different Young's modulus at these spots. Because we have taken many force curves over the entire sample we know that the variations in the slope of these curves is not just a variablity in the gelatin film, but it is an effect of the sample's thickness. In figure 6 we have analyzed 32 force curves taken along one line perpendicular to the edge of the gelatin film and calculated the Young's modulus for each curve. The lower trace shows the height profile and the upper trace the corresponding Young's modulus. The Young's modulus starts around 1 MPa at the edge of the film and drops to roughly 20 kPa at thicknesses of more than 1 μm.

In these experiments the maximum indentation of the AFM tip into the soft gelatin film was about 150 nm. However, we have only analyzed the part of the force curve corresponding to deflections between 110-150 nm. Thus the maximum indentation of the part of the force curve analyzed is between 100 and 150 nm. Therefore it is not surprising that the tip will penetrate or compress the film entirely when its thickness is comparable or even less than 100 nm. This explains the very high values of apparent Young's modulus of 1 MPa right at the border of the gelatin film. However, we would expect that if the film becomes much thicker than the indentation of 100 nm the Young's modulus should become independent of the thickness of the film. Figure 6 shows that the Young's modulus is approaching a constant value at higher thicknesses. However, there is still a slight dependence of the Young's modulus even at thicknesses of more than 1 μm. Therefore we believe that the change in apparent Young's modulus in the thickness range of 500 to 1500 nm shows a real variation of the elastic properties rather than the artifactual effect of the underlying stiff substrate.

The elasticity of a crosslinked polymeric network is determined by the available conformational space, as suggested by simple models such as the Gaussian chain mode (41). A surface like the substrate in our case could significantly change the number of available states of conformational modes for a polymer which could result in a very different elastic modulus. This interesting effect of ultrathin polymeric films is now accessible for studies with the AFM. The results we have obtained up to now, however, need to be extended by investigating different

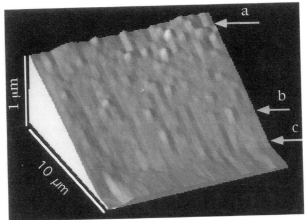

Figure 4
AFM image of a thin wedge shaped film of Gelatin immersed in 50%
propanol with water. This film was used to characterize the influence of the
thickness of the film on the measured elastic properties. We have taken three
force curves at the positions indicated by the arrows.

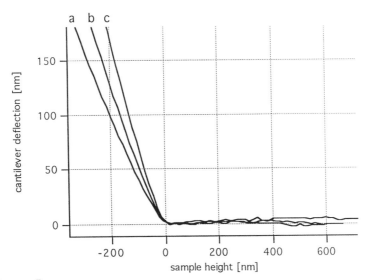

Figure 5
Three force curves taken at the positions denoted by arrows in the previous
figure. The different slopes of the three curves show that there is a
dependence of the elastic response of the sample on the thickness of the film.
The three force curves have been shifted horizontally such that the points of
contact coincide. This helps in comparing the slopes of the different force
curves visually.

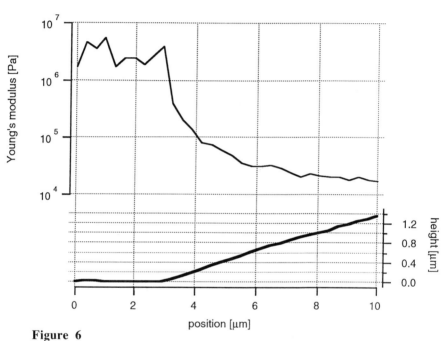

Figure 6

We have taken 32 force curves, while scanning the AFM tip along one line such the AFM tip has been moved from the substrate to the top of the gelatin film. The data were then analyzed. Thus we obtained values for the sample height (lower trace) and the sample elastic modulus (upper trace). To the left is the substrate and at x = 3 μm the gelatin film starts. As can be seen the Young's modulus, as determined with the AFM, starts with a very high value of around 1 MPa at the edge of the film and then drops to around 20 kPa at thicker parts.

polymers (different molecular weight) at different elasticities. Both variations can be easily done using gelatin and will be accessed by us in future work.

Measuring the elastic properties of living cells. Figure 7 shows constant height topographs of living fibroblasts. We have taken a two-dimensional array of force curves while raster-scanning the AFM's tip over the surface of the cell (force mapping) (5). Since each force curve gives a relation between the cantilever's deflection, which is proportional to the loading force and the sample height we can reconstruct topographic images from the two-dimensional array of force curves. On a stiff sample, as is the substrate here, the cantilever will follow directly the vertical motion of the sample. Thus, the deflection will be proportional to the sample height. However, with soft samples there will be a deviation from this linear behaviour due to elastic indentation of the sample, as is the case here with the cells. Thus at high loading forces the soft parts of the cell will be strongly compressed while the stiffer parts of the cell will be compressed less. Therefore in the topograph calculated at high loading force the stiffer areas of the cell will appear elevated, whereas the topograph at zero loading force will show the "real" topography of the cell. Stiff components of the cell are, for instance, bundles of actin filaments, so called stress fibres. The idea of the force generated contrast in AFM imaging as described here is depicted in figure 8. To illustrate this point figure 9 shows a typical contact mode AFM image of a living fibroblast. Fig. 9a is the height image showing the overall topography and fig. 9b is the deflection image showing the small corrugations. Many cytoskeletal elements are clearly depicted in these images taken at a force around 1 nN.

To quantify the local elastic properties of cells, we can calculate the Young's modulus from each force as has been described in previous work (8; 9; 20). Figure 10 shows the elasticity map of the same fibroblast as has been presented in figure 7.

We have measured the elastic modulus of a variety of different cells (figure 11). In general we have observed the following pattern: the softest area of cells, like at the nucleus, shows the lowest elastic modulus of around 1 kPa. Peripheral parts, as lamellipodia show a higher modulus around 10 kPa. The stiffest areas of the cell correspond to dense cytoskeletal structures like stress fibres with Young's moduli of 100 kPa or even higher (this work and (10)). We could also show that in the lamellipodial parts the elastic modulus is to a large degree determined by the actin cytoskeleton, which can be demonstrated by applying drugs degrading actin filaments (10). The measured values match very nicely theoretical estimates of the Young's modulus of actin networks (42) and indicate that in many situations the actin cytoskeleton provides the dominant component of the cell's stiffness as its elastic modulus greatly exceed that of the cell membrane (43).

Conclusions

We have demonstrated here that the AFM can be used to investigate soft samples, as demonstrated with gelatin films and living cells. The AFM can be used to determine the sample's topography but in addition it can be used to apply well defined forces with a precision of approximately 20 pN with very high spatial resolution which depends on the sample stiffness. On stiff samples like calcite crystals the lateral resolution of the AFM is better than 1 nm (44); On soft samples, like cells, the resolution is reduced to 10's or even 100's of nm (9). Therefore the elasticity of cells can be determined with resolution of better than 100 nm with the AFM. Since typically 20 force measurement can be performed per second, dynamic processes and changes in elasticity can be monitored with a temporal resolution of a tenth of a second. Thus, the AFM is a unique technique to map the mechanical response of living cells with a force and spatial resolution

188

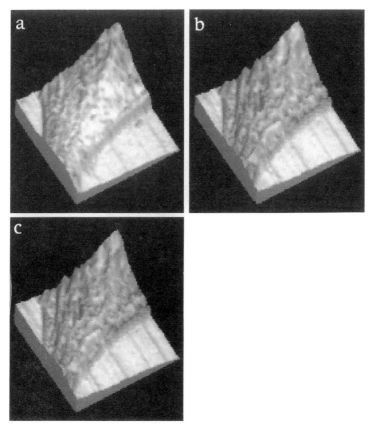

Figure 7

Constant height images of a living fibroblast. We have taken an array of 64 *
64 force curves on top of this cell. From the force curves the height values
were extracted corresponding to a given loading force (0, 0.75 and 1.5 nN,
fig. 7a, b and c respectively). The resulting height map was then modeled and
rendered to give a three-dimensional impression. The cell appears rather
smooth at zero loading force and very corrugated at the higher force values.
This is because at these high force values the membrane and the cytosol can
be compressed, but the underlying cytoskeleton resists the pressure of the
cantilever. Typically AFM contact mode images are taken at around 1 nN,
thus they will reveal an image similar to fig. 7b.

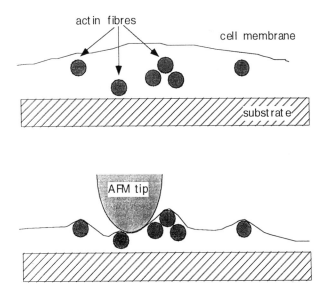

Figure 8
A simplified model of image generation on cells with the AFM. In contact
mode imaging of cells the tip is typically done by applying a force around 0.5
- 1nN. At this loading force the softer parts, as the glykocalix and the
membrane will be deformed and the force can only be counter balanced by
stiffer structures like the underlying cytoskeleton. Therefore the AFM does
not pick up the real topography, rather the apparent topography. This
apparent topography will depend on the loading force resulting due to locally
varying deformation of the sample. This idea has already been brought
forward in the earliest publications on investigations of living cells with the
AFM (22).

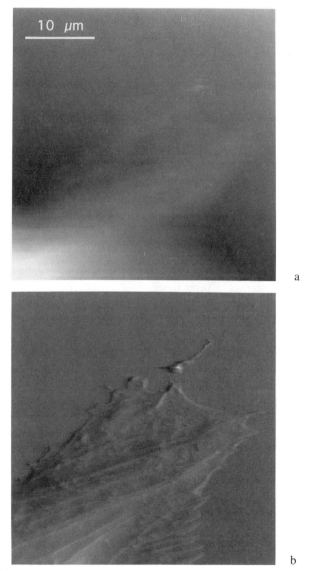

Figure 9
Contact mode AFM image of a living fibroblast. Figure 9a is the height mode image corresponding to the overall topography, figure 9b is the deflection image showing the small corrugations. This image shows very clearly cytoskeletal structures like stress fibres.

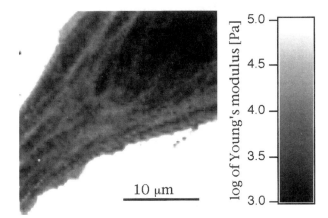

Figure 10
The same force data of figure 7 can be analyzed to yield the local Young's modulus. The Young's modulus is then displayed here on a logarithmic scale to show the large variations from around 1 kP at softer parts up to 50 kPa ontop of the stress fibres.

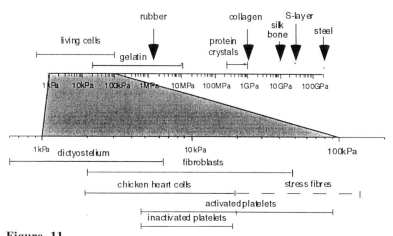

Figure 11
Overview of the elastic properties of cells and other materials. Cells show a Young's modulus between 0.5 kPa and 200 kPa depending on the area and the cell type. (Lower scale shows expanded range of elastic moduli for cells) This is much softer than other biological materials like collagen or silk (1 GPa and 10 GPa respectively) and like engineering materials such as steel (200 GPa). The stiffest biological material, to our knowledge, is a part of the cell wall of certain bacteria, the so called S-layer, which possesses a Young's modulus of 30 GPa. (the data are compiled from several sources: platelets (8), chicken heart cells (10), fibroblasts (this work), dictyostelium (unpublished results, Radmacher et al.), rubber, collagen, silk, bone & steel (45), gelatin (20), protein crystals (46) and bacterial cell walls (47).

previouisly not accessible. Undoubtedly the AFM will help us to gain new insights in the field of nanostructured polymeric samples and living cells.

Acknowledgment

This work was supported by the Deutsche Forschungsgemeinschaft as part of the Schwerpunkt für "Neue mikroskopische Techniken für die Biologie und Medizin" (MR, JD & CR) and as part of the SFB 266 (KJ), by the NIH under grant GM35325 (KJ) and by the Materials Research Division of the National Science Foundation under grant # NSF-DMR 9622169 (PKH). Equipment was supplied by Digital Instruments.

Literature Cited

(1) Binnig, G.; C. F. Quate and C. Gerber *Phys. Rev. Lett.* **1986**, 56,930
(2) Rugar, D. and P. K. Hansma *Physics Today* **1990**, 43,23-30
(3) Shao, Z. and J. Yang *Quarterly Reviews of Biophysics* **1995**, 28,195-251
(4) Haugstad, G.; W. L. Gladfelter; E. B. Weberg; R. T. Weberg and T. D. Weatherill *Langmuir* **1994**, 10,4295-4306
(5) Radmacher, M.; J. P. Cleveland; M. Fritz; H. G. Hansma and P. K. Hansma *Biophys. J.* **1994**, 66,2159-2165
(6) Radmacher, M.; M. Fritz; J. P. Cleveland; D. R. Walters and P. K. Hansma *Langmuir* **1994**, 10,**3809-3814**
(7) Rotsch, C. and M. Radmacher *Langmuir* **1997**, in press
(8) Radmacher, M.; M. Fritz; C. M. Kacher; J. P. Cleveland and P. K. Hansma *Biophys. J.* **1996**, 70,556-567
(9) Radmacher, M. *IEEE Eng. in Med. and Biol.* **1997**, 16,47-57
(10) Hofmann, U. G.; C. Rotsch; W. J. Parak and M. Radmacher *J. Struct. Biol.* **1997**, in press
(11) Radmacher, M.; M. Fritz; H. G. Hansma and P. K. Hansma *Science* **1994**, 265,1577-1579
(12) Florin, E.-L.; V. T. Moy and H. E. Gaub *Science* **1994**, 264,415-417
(13) Hinterdorfer, P.; W. Baumgartner; H. J. Gruber; K. Schilcher and H. Schindler *Proc. Natl. Acad. Sci. USA* **1996**, 93,3477-3481
(14) Rief, M.; F. Oesterheld; M. Berthold and H. E. Gaub *Science* **1997**,
(15) Clark, A. H. and S. B. Ross-Murphy *Adv. Polymer Sci.* **1987**, 83,57-192
(16) Djabourov, M. *Contemp. Phys.* **1988**, 29,273-297
(17) Haugstad, G.; W. L. Gladfelter and R. R. Jones *J. Vac. Sci. Technol. A* **1996**, 14,1864-1869
(18) Haugstad, G.; W. L. Gladfelter; E. B. Weberg; R. T. Weberg and R. R. Jones *Langmuir* **1995**, 11,3473-3482
(19) Haugstad, G. *Trends in Polymer Sci.* **1995**, 3,353-359
(20) Radmacher, M.; M. Fritz and P. K. Hansma *Biophys. J.* **1995**, 69,264-270
(21) Radmacher, M.; R. W. Tillmann; M. Fritz and H. E. Gaub *Science* **1992**, 257,1900-1905
(22) Henderson, E.; P. G. Haydon and D. S. Sakaguchi *Science* **1992**, 257,1944-1946
(23) Henderson, E. *Prog. Surf. Sci.* **1994**, 46,39-60
(24) Fritz, M.; M. Radmacher and H. E. Gaub *Exp. Cell Res.* **1993**, 205(1),187-190
(25) Oberleithner, H.; G. Giebisch and J. Geibel *Pflügers Archiv - european journal of physiology* **1993**, 425,506-510
(26) Fritz, M.; M. Radmacher and H. E. Gaub *Biophys. J.* **1994**, 66,1328-1334
(27) Schoenenberger, C.-A. and J. H. Hoh *Biophys. J.* **1994**, 67,929-936

(28) Schneider, S. W.; S. W. Sritharan; J. P. Oberleithner and B. Jena *PNAS* **1997**, 94,316-321
(29) Weisenhorn, A. L.; M. Khorsandi; S. Kasas; V. Gotozos; M. R. Celio and H. J. Butt *Nanotechnology* **1993**, 4,106-113
(30) Hoh, J. H. and C.-A. Schoenenberger *J. Cell Sci.* **1994**, 107,1105-1114
(31) Butt, H.-J. and M. Jaschke *Nanotechnology* **1995**, 6,1-7
(32) Johnson, K. L. *Contact Mechanics*; Cambridge University Press: Cambridge, 1994.
(33) Radmacher, M. and P. K. Hansma *Polymer Preprints* **1996**, 37,587-588
(34) Meyers, G. F.; B. M. DeKoven and J. T. Seitz *Langmuir* **1992**, 8,2330-2335
(35) Elkaakour, Z.; J. P. Aimé; T. Bouhacina; C. Odin and T. Masuda *Phys. Rev. Lett.* **1994**, 73,3231-3234
(36) Schallamach, A. *Wear* **1971**, 17,301-312
(37) Hui, S. W.; R. Viswanathan; J. A. Zasadzinski and J. N. Israelachvili *Biophys. J.* **1995**, 68,171-178
(38) Jin, X. and W. N. Unertl *Appl. Phys. Lett.* **1992**, 61,657-659
(39) Mamin, H. J. and D. Rugar *Appl. Phys. Lett.* **1992**, 61,1003-1005
(40) Majumdar, A.; P. I. Oden; J. P. Carrejo; L. A. Nagahara; J. J. Graham and J. Alexander *Appl. Phys. Lett.* **1992**, 61,2293-2295
(41) Ward, I. M. and D. W. Hadley *An introduction to the mechanical properties of solid polymers*; John Wiley & Sons: Chichester, 1993.
(42) Satcher, R. L. J. and F. C. J. Dewy *Biophys. J.* **1996**, 71,109-118
(43) Sackmann, E. *Science* **1996**, 271,43-48
(44) Ohnesorge, F. and G. Binnig *Science* **1993**, 260,1451-1456
(45) Fung, Y. C. *Biomechanics - mechanical properties of living tissues*; Springer: New York, 1993.
(46) Morozov, V. N. and T. Y. Morozova *Journal of Biomolecular Structure & Dynamics* **1993**, 11,459-481
(47) Xu, W.; P. J. Mulhern; B. L. Blackford; M. H. Jericho; M. Firtel and T. J. Beveridge *J. Bacter,* **1996**, 178,3106-3112

Chapter 11

Atomic Force-, Confocal Laser Scanning-, and Scanning Electron Microscopy Characterization of the Surface Degradation of a Polymer

Ch. Bourban[1], J. Mergaert[2], K. Ruffieux[1], J. Swings[2], and E. Wintermantel[1]

[1]Department of Materials, Chair of Biocompatible Materials and Engineering, ETH Zurich, Wagistrasse 23, CH-8952 Schlieren, Switzerland
[2]Laboratorium vow Microbiology, Universiteit Gent, K.L. Ledeganckstraat 35, B-9000 Gent, Belgium

Changes of surface morphology of degradable polymers are currently characterized qualitatively by SEM. In order to study the degradation behavior of surface eroding polymers, the effective surface area should be quantified. Therefore, AFM and CLSM techniques were used to calculate the surface area of extruded films of (P(3HB-co-3HV)), used as a model and degraded in a bacteria culture for up to 95 hours. AFM and CLSM methods showed similar results at the beginning of degradation: the surface area decreased slightly during the first 6 hours of incubation time (2% mass loss) as biodegradation occurs preferably at places with a higher surface to volume ratio. Later on, the surface became rougher and both microscopy methods showed an increase of the surface area which reached a maximum after 30 to 47 hours of degradation (6 to 15% mass loss). The surface area measured with high resolution AFM showed an increase of about 54% while CLSM, having a lower resolution, showed an increase of about 26%. After 78 hours of degradation, (21% mass loss) no reliable measurements could be obtained anymore with AFM due to a too rough topography, while CLSM was not limited by surface roughness but by pore shape. The quantitative results regarding surface area correlated with qualitative SEM results and with the growth of the bacteria culture.

Microscopy methods for the characterization of surface phenomena during polymer degradation were compared with regard to their ability and sensitivity to assess, describe and quantify the degradation process. Biopolymer surfaces during degradation were quantified by three-dimensional microscopic methods: Atomic Force Microscopy (AFM) and Confocal Laser Scanning Microscopy (CLSM) and were

194

documented by SEM. PHB/V films degraded in pure bacteria culture up to 95 hours were used as model substances.

Materials and Methods

The PHB/V used (poly(3-hydroxybutyrate-co-3-hydroxyvalerate, Biopol, D611G containing 12 phr plasticizer and 1 phr boron nitride as nucleant, Zeneca Bioproducts, Billingham, UK) was a copolymer containing 12% hydroxyvalerate (HV). Its density was about 1.25 g/cm^3 and Tm was 158-162 °C. Films of this polymer were extruded to a film strip of 16 cm width and 0.16 mm ($\pm$0.04 mm) thickness. This thermoplastic polyester is mainly aliphatic and hydrophobic in character. It degrades biologically in the presence of specific bacteria or fungi (1, 2). Degradation occurs by surface erosion mainly through enzymatic hydrolysis, as simple hydrolysis in water is very slow (3). The high HV content of 12% gives the copolymer a suitable toughness to be used as a film, in contrast to the brittle, pure PHB.

Degradation Protocol. Small film samples were obtained using a hollow punch of 15 mm diameter. They were disinfected by washing in 70% ethanol for 10 minutes, and rinsed twice in sterile distilled water for 10 minutes. Film samples were given to 100 ml of liquid medium (4), which was inoculated with a culture of *variovorax paradoxus* PHA 243. This strain has been shown to grow on and degrade (P(3HB)) and P(3HB-co-3HV) in pure culture on polymer overlay plates (5). The culture was incubated at 28°C in a rotary shaker at 125 rpm, samples were removed after about 6, 23, 30, 47 and 78 hours, washed, and dried to constant weight in vacuum. In parallel, disinfected control samples were incubated in uninoculated sterile 0,01 M Na_2HPO_4-KH_2PO_4 buffer, pH 6,8. Growth of the culture was followed by enumeration of colony forming units (CFU) on nutrient agar (Oxoid), incubated at 28°C for 3 days. Samples were characterized, on the upper side of the film only, by AFM, CLSM and SEM, and compared to the nonbiodegraded control samples.

Atomic Force Microscopy (AFM). AFM investigations were performed with a Nanoscope Dimension 3000 with a Scanning Probe Microscope IIIa controller (Digital Instruments). The cantilevers used were 125 μm long with a 10-15 μm pyramidal tip. The scans (50 x 50 μm, 512 x 512 points) were performed in tapping mode with a G type head. At least 3 scans were performed for each sample. The following parameters were chosen: scan rate: 0.8 Hz for flat samples to 0.35 Hz for rougher samples, setpoint: 1.1-2.5 V, drive frequency: 232-239 kHz, drive amplitude: 30-200 mV. Surface area measurements were performed using the area measurement tool integrated in the software (DI Nanoscope v. 3.24).

Confocal Laser Scanning Microscopy (CLSM). Samples were sputter coated with platinum for 600-800 seconds at 10-12 mA and 0.05 mbar in a Balzers SCD 004 Sputter Coater, as previous attempts with coated surfaces seemed to give more reproducible and reliable results than uncoated PHB/V surfaces. CLSM scans were performed with a Zeiss Axiovert 100 LSM with green HeNe-laser light (543 nm, 0.5 mW).

Each three-dimensional image was calculated from about 24 to 30 two-dimensional cross-sections regularly distributed along a depth of 16-18 μm. This depth matched the maximum distance observed between the highest peaks and deepest accessible pores. Each two-dimensional cross-section represented the average of 4 scans. Lateral dimensions of the scans were 80 x 80 μm. Scan parameters were kept constant for each sample as far as possible: zoom 8x, pinhole 10 or adapted to the sample, attenuation: 100 or 30. The final three-dimensional images were median filtered.

Surface area calculations were performed with a work station equipped with 2 C++ programs. The first one decoded CLSM image parameters and data files and the second calculated the surface area using a mesh of triangles over the surface (400 x 400 points). The data of at least 3 area measurements were calculated.

Scanning Electron Microscopy (SEM). Documentation of the PHB/V films was performed with a Hitachi S41C at an acceleration voltage of 5 kV, since higher voltages would thermally modify the polymeric sample. Thus, the highest possible magnification was limited to about 5'000 x. Working distances were typically 4-9 mm and the same platinum sputter coated samples as used for CLSM were analyzed.

Results

No relevant topography changes due to the incubation medium could be observed in the control samples. Thus, changes observed with inoculated samples were due to biodegradation by the PHA 243 strain.

AFM Scans. AFM pictures of the investigated PHB/V films are shown in figure 1. On the reference film surface, spherulites with diameter of 4 to 8 μm were visible. After 6 hours of degradation, the surface topography was flatter. Later on, the appearance of a roughness that increased regularly up to 78 hours was revealed by AFM monitoring. At certain locations on samples degraded for 78 hours, the maximum vertical oscillation range of the tip in tapping mode (about 6 μm) was reached, so that deeper pores could no longer be monitored using AFM. It was possible to observe the star shape of PHB/V spherulites probably due to preferential degradation of the amorphous polymer phase (fig. 2). The spherulite diameter was about 5-7 μm.

CLSM Scans. The degradation behavior as observed by AFM was confirmed by CLSM scans: a continuously increasing roughness appeared in function of degradation time. In contrast to AFM, CLSM allowed the monitoring of the surface area independently of the roughness and pore depth of the topography. The limitation of CLSM method was not the pore depth but the growing internal porosity, which cannot be reached by laser light. Neither could pores of very small diameters be monitored due to interference with laser green light frequency (543 nm). Arbitrarily

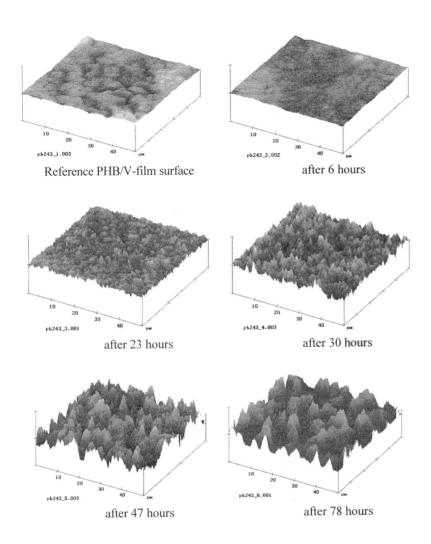

Reference PHB/V-film surface   after 6 hours

after 23 hours   after 30 hours

after 47 hours   after 78 hours

Figure 1. AFM images (50 x 50 µm) of biopolymer films during biodegradation by PHA 243 culture (x to z ratio is 2). After 6 hours, unevenness of the surface was slightly flattened by biodegradation. Later on, biodegradation increased the surface roughness significantly until the depth of the pores was too large to be monitored by AFM.

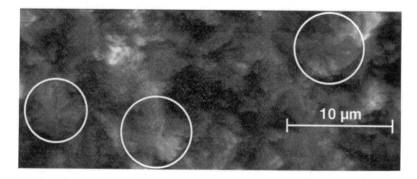

Figure 2. AFM image of the PHB/V film after 47 hours biodegradation by PHA 243 (not shaken). The star shape of spherulite lamellae with about 5-7 μm diameter (see circles) can be seen.

chosen sections showed a maximum distance between highest peaks and deepest pores of about 7.7 μm for the reference biopolymer surface to 18.4 μm after 47 hours biodegradation.

Surface Area Measurements. Results of the surface area measurements performed on the degraded films by AFM and CLSM are shown in figure 3. The y-axis shows, in percent ratio, the surface of the effective surface area divided by the projected rectangular area of scans as well as CFU, weight loss (±1%) and tensile strength (DIN 53 455, test shape nr. 3, 10 mm/min, n=10). 100% corresponded to the value of the PHB/V-film before degradation. As measured by AFM, the surface area slightly decreased (98.8 ±0.7%) during the first 6 hours. The CLSM average seemed to confirm this tendency despite a large standard deviation. First, the degradation process smoothed out unevenness caused by to the manufacturing process (texture, scratches). As biodegradation went on, both microscopy techniques indicated an increase in surface area which reached a maximum between 30 and 47 hours. This increase correlated with the increase in colony forming units (CFU), indicating an increase in bacterial activity. Later on, as described above, results of both microscopy techniques were no longer reliable due to growing internal porosity. A change in the curvature of the AFM and CLSM curves is likely to occur after about 26 hours, which corresponded with an increase in bacterial activity after a lag period.

SEM Documentation. SEM pictures of the films described the effect of biodegradation on the morphology of the biopolymer surface as a function of time in figure 4. On the reference topography, particles, probably due to the film processing, are visible which may contribute to an artificial increase of the surface area. After 6 hours degradation time, it seemed that spherulites got more visible locally, indicating a preferred degradation location at the spherulite boundaries (arrows in fig. 4, 6 hr.). The spherulites had a diameter of about 3-6 μm. After 23 hours, degradation extended

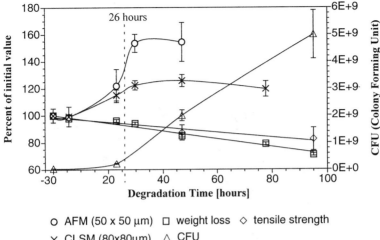

Figure 3. Surface area measurements by AFM and CLSM correlated with bacterial activity (CFU), weight loss and tensile strength of PHB/V films degraded in PHA 243 culture. The start of the significant increase in bacterial activity correlated well with the increase in surface area. Splines connecting points were arbitrarily chosen to better show tendencies.

to the spherulites themselves and an homogeneous rough surface topography appeared. Internal porosity seemed to take place locally already between 23 and 30 hours. After 30 hours, star shaped spherulite lamellae, became visible (circles in fig. 4, 30 hr.). Between 47 and 78 hours, a less homogeneous structure with an open porosity appeared. Pore diameters increased from 2-4 to 5-7 μm.

Discussion

Due to their physical principles, none of the microscopic methods used was able to reach and describe the internal porosity. As depicted by SEM, the limit between increasing roughness and internal porosity seemed to be reached within 23 and 30 hours of degradation. The start of internal porosity could not be observed by AFM or CLSM. It seemed that the point of inflection (at 26 hours) of AFM and CLSM curves corresponded with an increase of bacterial activity as shown by CFU-measurements. Since a decrease of surface area at that degradation stage was difficult to explain, it was assumed that the point of inflection corresponds to the start of internal porosity.

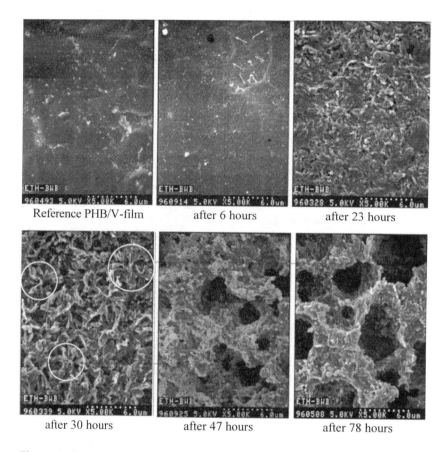

Reference PHB/V-film   after 6 hours   after 23 hours

after 30 hours   after 47 hours   after 78 hours

Figure 4. SEM pictures of PHB/V films from 0 to 78 hours degradation time in pure PHA 243 culture (magnification: 5'000x). The reference topography was rather flat except some nodes and dust particles. During degradation, spherulite boundaries got more visible and surface roughness increased.

Thus, AFM and CLSM area measurements were reliable only between 0 to 26 hours, in this particular study. AFM surface measurements were not only limited by the internal porosity but by the surface roughness, as the cantilever oscillation range was 6 μm.

The performed CLSM pore depth measurements, compared to results obtained by AFM, led to significant differences: as described by AFM, pore depth reached about 6 μm after 78 hours, while undegraded films scanned by CLSM showed a distance between highest peaks and deepest pores as high as 8 μm. As a control, surfaces providing an almost flat surface (a glass surface, an uncoated and a platinum coated tape) were analyzed by CLSM. Surprisingly, their maximum roughness was quite high: instead to be around 0 μm, they reached 3 to 5 μm. The method used seemed therefore to systematically deliver too large pore depths. Thus, CLSM depth values are to be treated cautiously as: 1) they may vary locally, 2) platinum coating of the sample may cause additional laser reflections which could be wrongly registered as surface points, 3) small pore diameters may interfere with laser light frequency, resulting in incorrect measurements. 4) It was also observed that, depending on the scan range used, the observed values of pore depth could vary. For these reasons, the CLSM method, as used in this work, allowed no reliable pore depth values but only comparisons between different stages of degradation.

The calculated surface area depended directly on the resolution of the microscopy technique: AFM had a higher resolution than CLSM, so that the calculated surface area was larger. Differences in surface measurements between AFM and CLSM may also be due to the platinum coating used for CLSM investigations, as AFM samples were not coated. Alternative surface measurement method using gas adsorption (BET-method) has been employed to assess the internal porosity. Intrinsic problems of BET-method are 1) a large quantity of material, and a large surface area in particular, must be available and 2) the temperature which is applied to deadsorb gas molecules from the surface prior to measure their N_2 adsorption is limited for polymers.

Conclusion

It could be determined by the three applied microscopy technique, that polymer degradation was due to the bacterial strain PHA 243, as control samples did not show relevant surface topography modifications. Each microscopy method used could assess an increasing roughness with increasing degradation time. The application range of AFM and CLSM techniques for the particular PHB/V samples used in this work was limited from 0 to 26 hours degradation time mainly due to internal porosity appearance.

The AFM technique facilitated surface area measurements as a function of degradation time. It could be assessed that first the surface area decreased slightly during the first 6 hours and then increased continuously up to 23 to 30 hours of degradation. After about 26 hours of biodegradation, the maximum rate of surface area increase due to degradation was reached. Afterwards, AFM no longer delivered reliable results due to 1) the limited cantilever oscillation range, which was not able to monitor the increasing roughness and pore depth and 2) the growing internal porosity. CLSM measurements permitted the quantification of the surface area but showed less sensitivity to the roughness and extent of pore depth than AFM. While area measurements correlated with AFM tendencies, calculated surface areas were lower due to the lower resolution of the CLSM technique. Limitations of this technique were resolution, pore shape and internal porosity of the degraded samples.

SEM images allowed documentation of surface topography phenomena such as increasing roughness, spherulites or pores appearance and diameter measurements, as well as homogeneity and changes in porosity.

Microscopy methods, such as AFM and CLSM, proved to be more sensitive at the very beginning of biodegradation (until 26 hours of degradation) when compared to weight loss or tensile strength measurements. The surface area increase correlated well with culture growth.

Acknowledgments. European Community project: AIR2/CT93-1099: "Biodegradability of biopolymers: prenormative research, biorecycling and ecological impact", Swiss Federal Office for Education and Research, Project Nr. 93.0279, K.J. Kjoller (Digital Instruments, Santa Barbara, USA), D. Perels and A. Halter.

Literature Cited

(1) Brandl H.; Mayer J.; Wintermantel E. *Can. J. Microbiol.* **1995,** *41*, pp. 1-11.
(2) Mergaert J.; Anderson C.; Wouters A., Swings J.; Kersters K. *FEMS Microbiol. Lett.*, **1992,** *52*, pp. 317-322.
(3) Doi Y., *Microbial Polyesters*, VCH Verlag: New York, US, 1990, p. 156.
(4) Delafield F.P.; et al. *J. Bacteriol.*, **1965,** *90*, pp. 1455-1466.
(5) Mergaert J.; et al. *Appl. Environ. Microbiol.*, **1993,** *59*, pp. 3233-3238.

POLYMER MORPHOLOGY AND STRUCTURE: MOLECULAR FILMS AND INTERFACES

Chapter 12

Scanning Force Microscopy of Surface Structure and Surface Mechanical Properties of Organotrichlorosilane Monolayers Prepared by Langmuir Method

Atsushi Takahara, Shouren Ge, Ken Kojio, and Tisato Kajiyama[1]

Department of Materials Physics and Chemistry, Graduate School of Engineering, Kyushu University, Hakozaki 6-10-1, Higashi-ku, Fukuoka 812-81, Japan

Organochlorosilane monolayers were polymerized on the water surface and immobilized onto the silicon wafer surface by Langmuir method. The electron diffraction (ED) pattern of the n-octadecyltrichlorosilane(OTS) monolayer revealed that OTS molecules were regularly arranged in a hexagonal array with the (10) spacing of ca. 0.42 nm. On the other hand, ED study revealed that the [2-(perfluorooctyl)ethyl]trichlorosilane (FOETS) monolayer was amorphous state at room temperature. The high-resolution atomic force microscopic (AFM) image of the OTS monolayer in a scan area of 10x10 nm^2 exhibited the individual methyl groups of which packing was a hexagonal array in a similar molecular arrangement concluded by the ED study. AFM observation of the (OTS/FOETS) mixed monolayer revealed that the crystalline OTS circular domains of ca. 1-2 µm in diameter were surrounded by a sea-like amorphous FOETS matrix, even though the molar fraction of OTS was above 75%. Phase separation was observed for monolayer prepared from FOETS and non-polymerizable and crystallizable amphiphile such as lignoceric acid (LA). The phase separation of the (alkylsilane/fluoroalkylsilane) mixed monolayer might be attributed to both the crystallizable characteristics of alkylsilane molecules and faster spreading of FOETS molecules on the water surface. The mixed monolayer of crystalline alkylsilane (OTS) and amorphous alkylsilane (n-dodecyltrichlorosilane, DDTS) formed phase-separated structure on the water surface because of the crystallizable characteristics of OTS. Lateral force microscopic (LFM) observation revealed that the order of the magnitude of lateral force generated against the silicon nitride tip was crystalline Si substrate >crystalline n-triacontyltrichlorosilane (TATS) domain> crystalline n-dococyltrichlorosilane (DOTS) domain> amorphous FOETS matrix > crystalline OTS domain>amorphous DDTS matrix. On the other hand, scanning viscoelasticity microscopic observation revealed that the order of the magnitude of modulus was Si substrate > crystalline OTS domain > amorphous FOETS matrix.

[1]Corresponding author.

Langmuir-Blodgett (LB) films and self-assembled monolayers (SAMs) have attracted growing attention lately, since these systems are believed to have technological potential in molecular electronics, optical device and biomedical applications (1,2). In order to realize their potential, it is necessary to achieve the precise structural and morphological controls and enhance the environmental stability of the surface structure of the monolayers. Organochlorosilane monolayer is a novel monolayer system which can be polymerized and immobilized onto a substrate surface with hydroxyl groups. The organochlorosilane monolayers have been prepared by a solution adsorption method (chemisorption) (3,4). However, since structure control is difficult via the chemisorption, a new method of preparing the organochlorosilane monolayer, namely, the Langmuir method was developed(5-9). By transferring the monolayer onto the substrate, by Langmuir method, a structurally controlled monolayer can be prepared. The chemisorbed monolayers such as organosilane on silicon and alkanethiol on gold, are physically robust. Generally speaking, monolayers of conventional amphiphiles are thermally and environmentally unstable, due to the lack of strong interaction between monolayer and substrate.

The surface properties, such as wettability, frictional property, chemical reactivity, biocompatibility, permeability, charge storage capacity and electrical response can be controlled by the design of surface aggregation structure, such as mixing two or more monolayer components. Various mixed monolayers, prepared by chemisorption from solution (10,11) and LB transfer from the air-water interface (12,13), have been reported. However, precise structure characterization has not been realized until the development of a scanning force microscope(SFM). SFM is one of the new scanning probe microscopy techniques which has become an important method for investigating materials' surface morphology with high resolution (14,15). Traditional microscopes use waves, such as light or electrons, and suitable imaging optics to create a two-dimensional projection of optical properties of the object. On the other hand, the SFM image is created on the basis of the various forces acting between a sharp cantilever tip and the sample surface, such as, van der Waals, electrostatic, frictional(lateral), and magnetic forces. Thus, the SFM offers the opportunity for gathering various kinds of information on those forces previously impossible to obtain from small areas of a sample. In this study, two-dimensional molecular aggregation structure of the n-octadecyltrichlorosilane (OTS) monolayer prepared by the Langmuir method from the air-water interface was investigated on the basis of transmission electron microscope (TEM) and AFM. The (alkylsilane/fluoroalkylsilane) mixed monolayers were prepared as a two-dimensional phase separation model surface. The phase-separated structure of the (alkylsilane/fluoroalkylsilane) mixed monolayer was investigated by AFM. The mechanism of the phase separation of the (alkylsilane/fluoroalkylsilane) mixed monolayer was discussed. Also, the surface mechanical properties of the (alkylsilane/fluoroalkylsilane) mixed monolayers were investigated by lateral force microscopic (LFM) and scanning viscoelasticity microscopic (SVM) measurements.

Experimental

Preparation of (alkylsilane/fluoroalkylsilane) mixed monolayers.
n-Triacontyltrichlorosilane (TATS, $CH_3(CH_2)_{29}SiCl_3$), n-dococyltrichlorosilane (DOTS, $CH_3(CH_2)_{21}SiCl_3$), OTS($CH_3(CH_2)_{17}SiCl_3$), n-hexadecyltrichlorosilane (HDTS, $CH_3(CH_2)_{15}SiCl_3$), n-dodecyltrichlorosilane (DDTS, $CH_3(CH_2)_{11}SiCl_3$), stearic acid (SA, $CH_3(CH_2)_{16}COOH$), lignoseric acid (LA, $CH_3(CH_2)_{22}COOH$) and [2-(perfluorooctyl)ethyl]trichlorosilane (FOETS, $CF_3(CF_2)_7CH_2CH_2SiCl_3$) were used to prepare the (alkylsilane/fluoroalkylsilane) mixed monolayers. SA and LA are

not polymerizable and also, non-reactive to the silicon substrate. FOETS has a lower surface free energy than that of alkylsilanes and LA. Organosilanes were purified by vacuum distillation. A (alkylsilane/fluoroalkylsilane) mixture solution in mixture solvent of benzene and 1,1,2-trichloro-1,2,2-trifluoroethane (TCTFE) ($v_{benzene}/v_{TCTFE}$ = alikylsilane/FOETS molar ratio) was prepared with a concentration of 2×10^{-3} M. The mixture solutions of organosilane compounds were spread on the pure water surface at a subphase temperature, T_{sp} of 293 K. The surface pressure-area (π-A) isotherms were measured with a microprocessor-controlled film balance system. In order to polymerize the monolayers, they were kept on the water surface under a certain surface pressure for 30 minutes. The monolayers were transferred by an upward drawing method onto the substrate surface with Si-OH groups and immobilized through condensation reaction.

Scanning force microscopic observation. Topographic images of the monolayer surfaces were taken with an atomic force microscope (AFM, SPA300, Seiko Instruments Industry, Co., JAPAN). The high-resolution AFM imaging was operated in water under constant height mode. The AFM observation of mixed monolayer surfaces was operated under the constant force mode, in air at room temperature, using a 20 μm x 20 μm scanner and a silicon nitride tip on a cantilever with a small bending spring constant of 0.022 N m^{-1}. The imaging force was in a repulsive range from 0.1 to 1 nN. The lateral force microscope (LFM) is a modified version of AFM which can be operated as a regular imaging device or as a force measuring device. Both the vertical and torsional motions of the cantilever were detected by the reflected laser beam. The vertical and torsional motions were proportional to the normal force (AFM image) and the lateral force (LFM image), respectively. In order to obtain the maximum LFM output voltage as torsional motion, the cantilever was scanned along the vertical direction to its axis. The two-dimensional image of dynamic viscoelastic functions was also obtained by utilizing a forced oscillation AFM, that is scanning viscoelasticity microscope (SVM)(*16-18*). The magnitude of strain was modulated sinusoidally by applying sinusoidal voltage generated by a frequency generator. The modulation frequency was 5 kHz. The z-sensitivity of the piezo-scanner was 4.46 nm V^{-1} for a 20 μm x 20 μm scanner. The measurement of the surface dynamic viscoelasticity was carried out in air at room temperature under the repulsive force of 0.021 nN.

Results and discussion

π-A isotherm and TEM observation of OTS and FOETS monolayers. Figure 1 provides a schematic representation of the polycondensation and immobilization process of the alkylsilane monolayer on a water surface. The chlorine groups of organotrichlorosilane on the water surface were substituted by hydroxy groups. At a certain surface pressure, the hydroxy groups in organochlorosilane molecules react with those in the adjacent molecules, in the case of a highly condensed monolayer, resulting in the formation of a polymerized monolayer. The polymerized monolayer was easily transferred onto a glass plate or silicon wafer by the Langmuir method and the residual hydroxyl groups could be strongly interacted with silanol groups on the glass or silicon wafer surface. Thus, stable immobilized monolayers were obtained. The transfer ratios of the OTS, FOETS and (OTS/FOETS) mixed monolayers were ca. 1.0. This indicated that the substrate surface was almost completely covered with monolayer.

Figure 2 shows the surface pressure-area (π-A) isotherm for the OTS monolayer spread on the pure water surface at T_{sp} of 293 K, as well as the bright-field images and electron-diffraction (ED) patterns of the monolayers which were

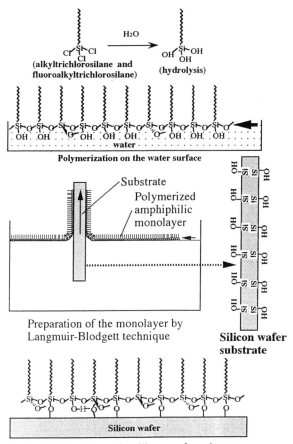

Polymerization on the water surface

Substrate
Polymerized
amphiphilic
monolayer

Preparation of the monolayer by
Langmuir-Blodgett technique

**Silicon wafer
substrate**

Immobilization on the silicon wafer substrate

Figure 1.   Schematic representation of the polymerization and immobilization process of the organosilane monolayer.

transferred onto the hydrophilic SiO substrate at surface pressures of 10, 35, and 50 mN m[-1]. The π-A isotherm reveals a rapid rise of surface pressure with a decrease in surface area. The molecular occupied area was 0.24 nm^2 molecule[-1] which was obtained on the basis of π-A isotherm measurements. The molecular aggregation structure in the OTS monolayer was investigated on the basis of the TEM observation and the ED pattern. Since the monolayer was transferred onto a hydrophilic substrate, the monolayer crystal system on the water subphase might maintain on the substrate surface(19). The ED pattern exhibited a hexagonal crystalline arc. This indicated that the crystalline OTS domains were gathered without any complete sintering at the domain interfaces. Also, the ED pattern showed that the (10) spacing of the OTS crystal is ca. 0.42 nm. The crystalline arc of ED pattern indicates the formation of larger two-dimensional crystalline domains over a range corresponding to an electron beam diameter of 10 μm. The (10) spacing of OTS agrees with that of the stearic acid (SA) monolayer. However, a change in the ED pattern of the OTS monolayer during a compressing process was completely different from that for the SA monolayer. The ED pattern of the SA monolayer at 0 mN m[-1] was a crystalline Debye ring and those at 24 and 40 mN m[-1] were crystalline hexagonal spots. This change of the ED pattern from the crystalline Debye ring to the crystalline hexagonal spots for the SA monolayer apparently indicates that the SA crystalline domains on the water surface were fused or recrystallized at the monolayer domain interface owing to sintering characteristics caused by compressive strain on stress, resulting in the formation of larger two-dimensional crystalline domains. In contrast, the ED patterns of the OTS monolayer at 10, 35 and 50 mN m[-1] were crystalline hexagonal arcs which were broadened along an azimuthal direction. This result also indicates that the crystallographical axes of the OTS monolayer were not aligned along a certain crystalline direction. Since the OTS molecules were polymerized on the water subphase, molecular aggregation rearrangement of the OTS molecules was fairly difficult and then, it seems reasonable to conclude that the sintering behavior at the monolayer domain interface did not easily proceed over a wide range of the OTS monolayer during a surface compression process. Barton et al. observed the formation of hexagonally packed polymerized OTS on a water surface via an in-plane surface X-ray diffraction measurement (20). The formation of the linear polymer is impossible for OTS having three equivalent hydroxy groups. It is possible to form a two dimensionally cross-linked monolayer on a water surface. The bright-field image for the OTS monolayer transferred at every surface pressure, exhibited the fairly uniform, smooth and continuous morphology. On the other hand, the bright-field image for the barium stearate and the fatty acid monolayers at a higher surface pressure (palmitic acid monolayer at 30 mN·m[-1]) showed a heterogeneous aggregation, which was composed of partially patched domains, due to the collapse of the monolayer. The homogeneous surface of the OTS monolayer even at higher surface pressure (50 mN·m[-1]) indicates that the polymerization enhances the mechanical stability of monolayer.

Figure 3 shows the high pass filtered high-resolution AFM images for the crystalline OTS monolayer transferred onto a silicon wafer substrate by Langmuir method at a surface pressure of 15 mN m[-1] and at T_{sp} of 293 K. The (10) spacing estimated on the basis of the two-dimensional fast Fourier transform (2D-FFT) of the image shown in Figure 3 was ca. 0.42 nm. This magnitude agreed with the (10) spacing which was calculated from the ED pattern of the crystalline OTS monolayer shown in Figure 2 . Therefore, the coincidence of the (10) spacings calculated on the basis of ED pattern and the high-resolution AFM image apparently indicates that the higher portions (the bright potions) in the AFM images of Figure 3 represent the individual methyl group of OTS molecules in the monolayer. The occupied area estimated from the ED pattern and the high-resolution AFM was 0.20 nm^2 molecule[-1].

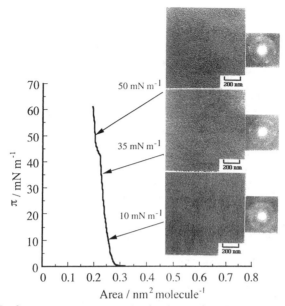

Figure 2. Surface pressure-area(π-A) isotherm for the OTS monolayer on the pure water surface at Tsp of 293 K, and the bright field images and electron diffraction (ED) patterns of the monolayers which were transferred onto the hydrophilic SiO substrate at the surface pressures of 10, 35, and 50 mN m^{-1}.

Figure 3. The high resolution high pass filtered AFM images for crystalline OTS monolayer on a scan area of 10x10 nm^2. The monolayer was prepared onto a silicon wafer substrate by Langmuir method at a surface pressure of 15 mN m^{-1} at the subphase temperature of 293 K. Note the periodic arrangement of the molecules with a hexagonal array.

The magnitude of the occupied area is slightly different from that of 0.24 nm^2 molecule^{-1} from π-A isotherm shown in Figure 2. Generally, the cross sectional area of linear alkyl chain is 0.19 nm^2 molecule^{-1} and the one for the OTS monolayer on the water subphase based on X-ray diffraction measurements is 0.20 nm^2 molecule^{-1}.(20) The molecular cross sectional area estimated by the ED pattern and the high-resolution AFM corresponded to these values. This indicates that the OTS molecules orient almost perpendicular to the silicon wafer substrate. On the other hand, the occupied area estimated by π-A isotherm is larger than the cross sectional area of methylene chain and the one for the OTS monolayer on the water subphase. It is considered that some defects are present, which may increase an apparent limiting area of OTS molecules.

It was reported that the high-resolution AFM image of the OTS monolayer prepared by a chemisorption method gave the occupied area of 0.227±15 nm^2 molecule^{-1} for the OTS monolayer(21). Also, the n-octadecyltriethoxysilane (OTE) monolayer prepared by the chemisorption was observed in molecular level and it was reported that the magnitude of short range ordering of the OTE monolayer was 0.8 nm.(22) In contrast, the magnitude of the molecular occupied area and the (10) spacing for the OTS monolayer prepared by Langmuir method were 0.20 nm^2 molecule^{-1} and 0.42 nm, respectively as mentioned above. The AFM image in this study clearly revealed the ordering of OTS molecules up to about 10 nm. Therefore, it seems reasonable to conclude that the OTS monolayer prepared by this method is more closely packed and has a homogeneous surface structure than those prepared by a chemisorption.

Figure 4 shows the π-A isotherm for the FOETS monolayer spread on the pure water surface at T_{sp} of 293 K, as well as the bright-field images and ED patterns of the monolayers, which were transferred onto the hydrophilic SiO substrate at surface pressures of 20 and 40 mN m^{-1}. The molecular occupied area of the FOETS monolayer was 0.31 nm^2 molecule^{-1}, which was obtained on the basis of π-A isotherm measurements. This shows that molecular cross-sectional area of the FOETS monolayer is larger than that of the OTS one, which reflects the large cross-sectional area of fluoroalkyl group compared with that of alkyl group. The bright field image of the FOETS monolayer showed a homogeneous surface and the ED pattern of the monolayer exhibited an amorphous halo at each surface pressure at 293 K. Since the fluoroalkyl chain length is not long enough to crystallize, the FOETS monolayer is amorphous state at 293 K.

Phase-separated structure of mixed monolayers. The mixed monolayer was prepared as a two-dimensional analogue of phase separation in a polymer blend. The π-A isotherm of Figure 5(a) is for the (OTS/FOETS) mixed monolayer (50/50 molar) on the water surface. The molecular-occupied area of 0.28 nm^2 molecule^{-1} for the (OTS/FOETS) mixed monolayer (50:50 molar) is almost equal to the average of the molecular occupied area for the OTS and FOETS monolayers considering the fraction of OTS. Figure 5 (a) shows the AFM image of the (OTS/FOETS) mixed monolayer (50:50 molar), which was transferred onto the silicon wafer substrate by Langmuir method at the surface pressure of 25 mN m^{-1}. The brighter and the darker portions correspond to the higher and the lower regions of the monolayer surface, respectively. The (OTS/FOETS) mixed monolayers were in a phase-separated state (Figure 5), and circular flat-topped domains of ca. 1-2 μm diameter , which were surrounded by a sea-like and flat region were observed. Since the area occupied by the circular flat-topped domains increased with an increase in the OTS content, it is expected that the circular flat-topped domain corresponds to the OTS domain. The domains might be formed due to the difference in the surface free energy between OTS and FOETS, and also, the crystallization of OTS would induce the phase separation.

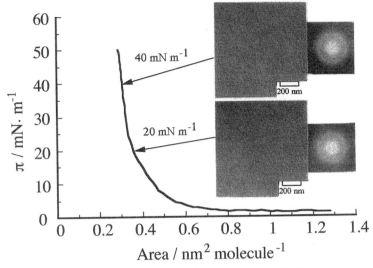

Figure 4.   Surface pressure-area(π-A) isotherm for the FOETS monolayer on the pure water surface at Tsp of 293 K, and the bright-field images and electron diffraction (ED) patterns of the monolayers which were transferred onto the hydrophilic SiO substrate at the surface pressures of 20 and 40 mN m^{-1}.

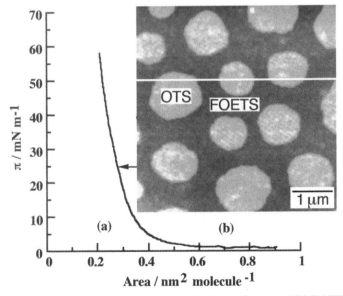

Figure 5.   (a) Surface pressure-area(π-A) isotherm for the (OTS/FOETS) mixed monolayer (50:50 molar) on the pure water surface at Tsp of 293 K, and (b) top view AFM image of the (OTS/FOETS) mixed monolayer (50/50 molar) prepared on the silicon wafer by Langmuir method at the surface pressures of 25 mN m^{-1}.

Figure 6 (a) shows the height profile along the line in AFM image in Figure 5(b). The height of circular flat-topped domains was 1.1-1.3 nm higher than the surrounding flat monolayer. Figure 6 (b) provides a schematic representation of the surface structure profile for the immobilized (OTS/FOETS) mixed monolayer. Since the difference in molecular lengths between OTS and FOETS is ca. 1.3 nm, it is apparent that the higher, circular domains and surrounding flat regions are composed of OTS and FOETS molecules, respectively. Since destruction of the surface structure has never been detected, even when dozens of AFM scans were done, it is reasonable to consider that the monolayer is remarkably well anchored to the substrate surface. Again, this demonstrates the exceptional high stability and cohesion of these monolayers to the substrate, due to both polymerization of organotrichlorosilane monolayer and the strong interaction between the monolayer and the substrate.

Figure 7 shows the AFM images (scanned area 5 x 5 μm^2) of the (OTS/FOETS)(25/75), (OTS/FOETS)(50/50) and (OTS/FOETS)(75/25) mixed monolayers, respectively, which were transferred onto the silicon wafer substrate by Langmuir method at the surface pressure of 25 mN m^{-1}. The AFM image is given in a top-view presentation. The area fraction of the circular flat-topped domains was in good agreement with the area fraction of OTS component, which was calculated on the basis of the limiting area (molecular-occupied area) of π-A isotherm. In other words, the (OTS/FOETS) mixed monolayer formed a phase-separated structure on the water surface and were transferred onto the silicon wafer surface quantitatively. Generally, polymer blends or copolymers are expected to show a surface enrichment of one component, in order to minimize interfacial free energy between the surface and the exterior environment. Even for compatible blends, the mean-field theory requires a surface enrichment of one component in order to minimize interfacial free energy. Therefore, it is very difficult to control the surface structure of polymer-blend films systematically on the basis of only the bulk composition. The definite surface composition, corresponding to the feed composition and the stable surface structure of the organochlorosilane monolayers, indicate that each monolayer is well anchored to the substrate surface. Though a scanning operation was done repeatedly for AFM observation with a repulsive force larger than 10^{-9} N, the OTS, FOETS and their mixed monolayers were not damaged by the tip.

Therefore, it seems reasonable to conclude that the remarkably stable surface and strong cohesion of these monolayers to the substrate are due to both the wide-area two-dimensional polymerization of these organotrichlorosilane monolayers and the formation of chemical bonding between OTS or FOETS molecules and the substrate surface. As shown in Figure 7, OTS formed circular domains even if the molar percent of OTS was 75%. It is apparent from the ED pattern in Figure 7 that the OTS domain was in a crystalline state, since the magnitude of the spacing corresponds to the (10) one of the OTS monolayer. Therefore, the crystallization of OTS molecules may be an important factor for the phase separation.

In order to confirm this conclusion, the alkylsilanes with various alkyl chain lengths were used to prepare the (alkylsilane/fluoroalkylsilane) mixed monolayers. Figure 8 shows the AFM images of the (HDTS/FOETS)(50/50) and the (DDTS/FOETS) (50/50) mixed monolayers which were transferred onto the silicon wafer surface by Langmuir method at the surface pressure of 25 mN m^{-1} at 293K. Irregular and non-circular domains were formed for the (HDTS/FOETS) mixed monolayer, because the crystallizability of HDTS molecules was lower and also, its polymerization reaction was faster than OTS, due to the shorter alkyl chain length of HDTS. Moreover, the phase-separated domain was not observed in the case of the (DDTS/FOETS) mixture, because the DDTS monolayer was in an amorphous state at 293 K.

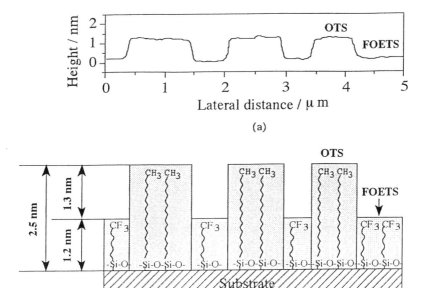

(a)

(b)

Figure 6. (a) Height profile along line in AFM image in Figure 5 (b), and (b) the surface structure model for the immobilized (OTS/FOETS) mixed monolayer.

214

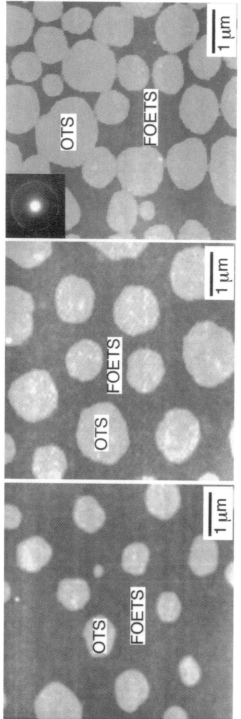

Figure 7. AFM images (scanned area 5x5 µm^2) of the (a) (OTS/FOETS)(25/75), (b) (OTS/FOETS)(50/50) and (c)(OTS/FOETS)(75/25) mixed monolayers, respectively, which were transferred onto the silicon wafer substrate by Langmuir method at the surface pressure of 25 mN m^{-1}. ED is inserted in Figure 7(c).

The AFM observations for the mixed monolayers (Figures 7 and 8) apparently indicate that the crystallization of alkylsilane plays an important role in the phase separation process of the (alkylchlorosilane/fluoroalkylsilane) mixed monolayer. Therefore, the surface structure of the (crystalline alkylchlorosilane/amorphous alkylchlorosilane) mixed monolayer, such as, (OTS/DDTS) was investigated. The AFM observation revealed that the (OTS/DDTS) mixed monolayer formed a phase-separated structure as shown in Figure 9. Since the OTS molecules can independently form the crystal phase on the water surface at 293 K in the amorphous DDTS matrix, the (crystalline alkylsilane/amorphous alkylsilane) mixed monolayer forms the phase-separated structure, which indicates that the crystallization plays an important role in phase separation.

In order to confirm the role of crystallization and polymerization at an air-water interface, the (OTS/FOETS) mixed monolayer was prepared by chemisorption. The clear phase-separated structure was not formed in the case of the (OTS/FOETS) mixed monolayer prepared by the chemisorption from the mixed solution of OTS and FOETS, because the formation of chemisorbed monolayer proceeds through a molecularly random adsorption process in solution. The formation of binary monolayers of OTS and 11-(2-naphthyl) undecyltrichlorosilane (2-Np) by backfilling of partial monolayers of one component, as a result of exposure to solution of other component has been reported (23). However, precise structural control has not been achieved by the chemisorption. The advantage of the method proposed in this study is one of the promising structural control method on the basis of phase separation and crystallization kinetics.

A phase-separated monolayer can also be prepared from both FOETS and the non-polymerizable and crystallizable amphiphile such as SA and LA. Figure 10 shows the 3-D representation AFM topographic image of (a) the (LA/FOETS) mixed monolayer, (b) the extracted (LA/FOETS) mixed monolayer and (c) the (OTS/FOETS) mixed monolayer prepared by the chemisorption of OTS onto the Si part of the FOETS monolayer. The (LA/FOETS) mixed monolayer was in a phase-separated state in a similar fashion to the (OTS/FOETS) monolayer in Figure 5(b). We may conclude from Figure 10 that the circular domains are composed of LA molecules, since the circular flat-topped domains were preferentially extracted with hexane as shown in Figure 10(b). Also, the FOETS matrix was not extracted with hexane because FOETS molecules were immobilized onto the silicon wafer surface by Si-O-Si covalent bond and hydrogen bonding among OH groups. The ED pattern of the (LA/FOETS) monolayer revealed a crystalline diffraction from LA domains. AFM observation revealed that the circular domain of LA was higher than FOETS by ca. 2 nm. The height difference in Figure 10(a) corresponds to the difference in the chain length between LA and FOETS molecules. The LA phase disappeared after repeated scanning of cantilever tip in the repulsive-force region, during AFM imaging, because of the mechanical instability, due to the absence of covalent bonding between LA and Si-substrate. The patterned monolayer surface with high surface free energy (bare Si wafer) and low surface free energy (FOETS) of Figures 10(b) was obtained after the removal of LA. The depth of hole is ca.1.4 nm, which corresponded to the thickness of the FOETS monolayer. Since the bare-Si surface with Si-OH groups was exposed to the surface, the Si phase can easily be backfilled by another organosilane through chemisorption from its solution. Thus, various surface modifications are possible by chemisorption to the FOETS monolayer with bare Si holes. The (OTS/FOETS) mixed monolayer was prepared by the chemisorption of OTS onto the Si part of the FOETS monolayer. The surface structure of Figure 10(c) was very similar to the phase-separated structure observed for the (OTS/FOETS) mixed monolayer directly prepared by Langmuir method. The OTS phase is mechanically and chemically very stable, as a consequence of the polymerization and anchoring to the Si-wafer.

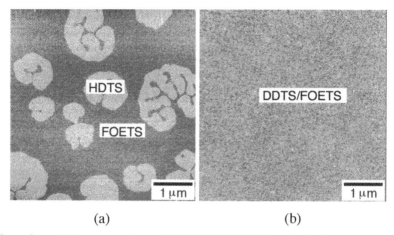

(a)   (b)

Figure 8.   AFM images of (a) (HDTS/FOETS)(50/50) and (b) (DDTS/FOETS) (50/50) mixed monolayers which were transferred onto the silicon wafer substrate by Langmuir method at the surface pressure of 25 mN m^{-1} at Tsp of 293K.

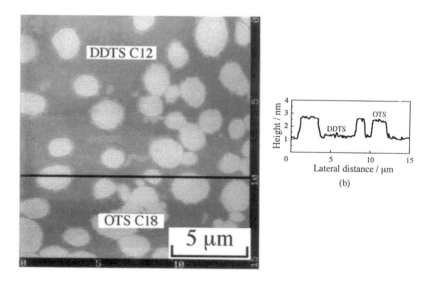

Figure 9.   (a) AFM image and (b) height profile along the line in AFM image for the (OTS/DDTS) (50/50) mixed monolayer which was transferred onto the silicon wafer substrate by Langmuir method at the surface pressure of 13 mN m^{-1} at 293K.

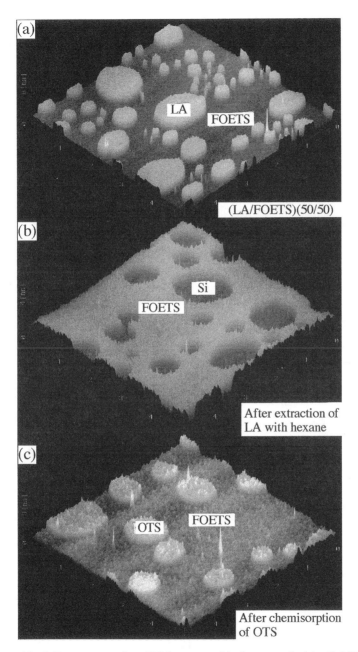

Figure 10. 3-D representation AFM topographic images of (a) (LA/FOETS) mixed monolayer, (b) extracted (LA/FOETS) mixed monolayer and (c) (OTS/FOETS) mixed monolayer prepared by chemisorption of OTS onto the Si part of FOETS monolayer.

However, the domain height of OTS phase in Figure 10(c) was less than that observed for the (OTS/FOETS) mixed monolayer prepared by the Langmuir method in Figures 5 and 6. Moreover, the surface roughness of the OTS domains prepared by chemisorption was more distinct than the OTS ones prepared by Langmuir method. These results indicate that the OTS monolayer prepared by chemisorption, is less densely packed compared with the one prepared by the Langmuir method. However, the above-mentioned procedure to backfill the Si part of FOETS monolayer by chemisorption of organosilane is applicable to the preparation of two-phase monolayers of which the constituents have different surface free energies or chemistries.

Surface mechanical properties of (alkylsilane/fluoroalkylsilane) mixed monolayer. Polymerization of organosilane molecules and immobilization of the monolayer to the substrate are very effective for improving the stability of the monolayers toward mechanical, thermal, and environmental attacks. The lateral force microscope (LFM) and the scanning viscoelasticity microscope (SVM) offer novel methods to estimate the mechanical properties of monolayer surface with high resolution. Figure 11 shows the LFM images for the (OTS/FOETS) mixed monolayer (a) and the (Si/FOETS) monolayer (b), respectively. The contrast in LFM image corresponds to the magnitude of output voltage from the photodiode, which is proportional to the magnitude of lateral force. The bright part in Figure 11 corresponds to the region with the higher lateral force, while the dark part does to that with the lower lateral force. Then, Figure 11 clearly exhibits that the lateral force between the silicon nitride tip and the Si substrate, the FOETS matrix and the OTS domain decreases in this order. Figure 12 shows the variations of the magnitudes of lateral force against alkyl chain length at room temperature. The magnitudes of lateral force for Si and FOETS were also plotted in Figure 12. It is clear that the order of the magnitude of lateral force generated against the silicon nitride tip increased with alkyl chain length. Also, it was clarified that the magnitude of lateral force were Si substrate > crystalline TATS domain > crystalline DOTS domain> amorphous FOETS matrix > crystalline OTS domain > amorphous DDTS matrix. It has been reported that polytetrafluoroethylene (PTFE) shows lower frictional coefficient compared with polyethylene (PE) (24) due to the lower surface free energy of PTFE and smooth profile of PTFE molecule. However, it should take into consideration that the aggregation structure and molecular orientation against interface is quite different between monolayer and polymer. Since the alkylsilane and the fluoroalkylsilane monolayers were immobilized onto the substrate with the stable orientation of methyl or fluoromethyl groups to the air-monolayer interface and also, the scanning direction was perpendicular to the molecular chains in the monolayers, the frictional characteristics of the alkylsilane and the fluoroalkylsilane monolayers should be fairly different from PE and PTFE, respectively. Also, Overney et al. (25) and Briscoe et al. (26) reported that the lateral force and the shear strength on the alkyl surface were lower than that on the fluoroalkyl surface. These agree with the result for the (OTS/FOETS) mixed monolayer. However, the TATS domain with longer alkyl chain in the (TATS/FOETS) mixed monolayer showed higher lateral force compared with the FOETS matrix. Since the TATS molecules have longer alkyl chain in comparison to OTS molecules, it should have higher crystallizability, elasticity or shear strength. This indicates that the lateral force depends not only on the chemical composition but also on the crystallizability, elasticity and shear strength. Since the magnitude of the lateral force corresponds to the net force including adhesive force, externally applied forces, elasticity and the shear strength, the further systematic studies on lateral force should be done for organosilane monolayer in an air atmosphere and a liquid environment.

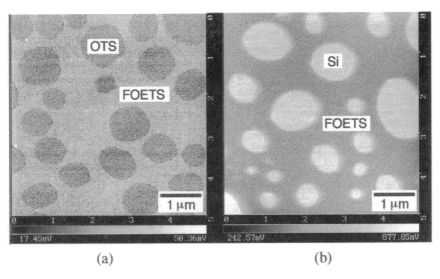

(a)   (b)

Figure 11. LFM images for the (OTS/FOETS) mixed monolayer (a) and the (Si/FOETS) monolayer surface (b) which was prepared from the (LA/FOETS) mixed monolayer.

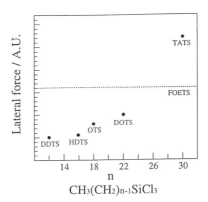

Figure 12. The variations of the magnitudes of lateral force against alkyl chain length at room temperature.

The immobilized organosilane monolayers were more stable in comparison with the monolayer composed of conventional amphiphilic molecules, because of the formation of covalent bond and hydrogen bond between substrate and monolayer. Since surface mechanical stability is required for the surface viscoelastic measurement, the organosilane monolayer is suitable for the surface viscoelastic measurement. The cyclic compressive strain must be applied to the monolayer surface under a repulsive force region for the surface dynamic viscoelastic measurement. In the case of the mechanically unstable SA monolayer, the monolayer was completely swept out during first scanning, then, the image of viscoelasticity could not be obtained. The modulated response force was detected by the deflection of the cantilever. The deflection signal was filtered by a band pass filter to obtain the dynamic mechanical functions and also, was simultaneously filtered by a low pass filter to obtain an AFM image. Figure 13 shows the images on the real part of dynamic modulus for (a) (OTS/FOETS) and (b) (Si/FOETS) mixed monolayer surfaces. The linear response of force against the magnitude of dynamic strain amplitude was observed up to the modulation amplitude of ca. 2 nm. Therefore, the image of the real part of modulus was obtained under the dynamic modulation of 0.89 nm along the z-axis. The brighter and the darker portions correspond to the higher and the lower values of apparent elastic modulus on the monolayer surface, respectively. Since the OTS region was in a crystalline state at 293 K, the circular flat-topped OTS domains showed higher modulus than that of the amorphous FOETS matrix as shown in Figure 13. Also, the Si substrate showed higher modulus compared with the FOETS monolayers as shown in Figure 13. Moreover, the SVM observation of the OTS monolayer with some defects (holes) exposed bare Si substrate shows that the modulus of Si substrate was higher than that of the OTS monolayer. It is clear that the order of the magnitude of modulus was Si substrate > crystalline OTS domain > amorphous FOETS matrix.

Conclusion

The TEM and high-resolution AFM analyses clarified that the crystalline OTS monolayer prepared by Langmuir method has a highly oriented alkyl chain perpendicular to the substrate. High-resolution AFM observation revealed the closely packed OTS molecules in a hexagonal unit cell. On the other hand, ED pattern revealed that FOETS was in an amorphous state. The (alkylsilane/fluoroalkylsilane) mixed monolayers prepared by Langmuir method showed a phase-separated structure. The surface structure and properties were characterized by SFM. Formation of phase-separated structure was governed by the factors such as the difference in the surface free energy between alkylsilane and fluoroalkylsilane and crystallinity of the alkylsilane. Especially in the case of the (crystalline alkylsilane/ amorphous alkylsilane) mixed monolayer such as (OTS/DDTS), the crystallization of the OTS chain was a decisive factor of the OTS and DDTS phase separation. Also, the surface mechanical properties of the (alkylsilane/fluoroalkylsilane) mixed monolayers were evaluated successfully. The LFM observation revealed that the order of the magnitude of lateral force generated against the silicon nitride tip was crystalline TATS domain with longest alkyl chain > crystalline DOTS domain > amorphous FOETS matrix > crystalline OTS domain >amorphous DDTS matrix. The result of SVM observation revealed that the crystalline OTS domain had higher modulus than that of the amorphous FOETS matrix. This information would be useful for the surface structural designs on the monolayer surface.

Acknowledgment. This work was supported in part by a Grant-in-Aid for Scientific Research No.08751052 and 08878175 from the Ministry of Education, Science, and Culture of Japan and by the Cosmetology Research Foundation.

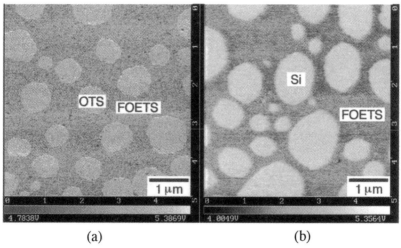

(a)   (b)

Figure 13. Images of real part of dynamic modulus for the (OTS/FOETS) mixed monolayer (a) and the (Si/FOETS) monolayer (b) which was prepared from the (LA/FOETS) mixed monolayer at 5 kHz.

222

Literatures cited
1. Tredgold, R. H. *Order in Thin Organic Films*; Cambridge University Press: New York, NY, **1995**.
2. Ulman, A. *An Introduction to Ultrathin Organic Films, from Langmuir-Blodgett to Self-Assembly*; Academic Press: San Diego, CA, **1991**.
3. Maoz, R.; Sagiv, J. *V. Colloid. Interface. Sci.*, **1984**, 100, 465.
4. Tillman, N.; Ulman A.; Penner T. L. *Langmuir*, **1989**, 5, 101.
5. Ge, S.-R.; Takahara, A.; Kajiyama, T. *J. Vac. Sci. Technol.* A, **1994**, 12, 2530.
6. Ge, S.-R.; Takahara, A.; Kajiyama, T. *Langmuir*, **1995**, 11, 1341.
7. Takahara, A.; Kojio, K.; Ge, S.-R.; Kajiyama, T. *J. Vac. Sci. Tech.* A,**1996**, 14 1747.
8. Kajiyama, T.; Ge, S.-R.; Kojio, K.; . Takahara, A. *Supramolecular Science*, **1996** 3, 123.
9. Ariga, K.; Okahata, Y. *J. Am. Chem. Soc.*, **1989**, 111, 5618.
10. Sagiv, J. *J. Am. Chem. Soc.*, **1980**, 102, 92.
11. Prime, K. L.; Whitesides, G. M.; *Science*, **1991**, 252, 1164.
12. Frey, W. ; Schneider, J.; Ringsdorf, H., Sackmann, E. *Macromolecules*, **1987**, 20, 1312.
13. Thibodeaux, A. F.; Radler, U.; Shashidhar, R.; Duran, R. S. *Macromolecules*, **1994**, 27, 784.
14. Marti, O. ; Amrein, M. , Eds., *STM and SFM in Biology*; Academic Press: New York, NY **1993**.
15. Wiesendanger, R. *Scanning Probe Microscopy and Spectroscopy, Methods and Applications*, Cambridge University Press, Cambridge, **1995**.
16. Kajiyama, T.; Tanaka, K.; Ohki, I.; Ge, S.-R.; Yoon, J.-S.; Takahara, A. *Macromolecules*, **1994**, 27, 7932.
17. Kajiyama, T.; Ohki, I.; Tanaka, K.; Ge, S.-R.; Yoon, J.-S.; Takahara, A. *Proc.Japan Acad.*, **1995**, 71B, 75.
18. Takahara, A.; Tanaka, K.; Ge, S.-R.; Kajiyama, T. In *Surface Science of Crystalline Polymers*, Terano, M; Yui, N., Eds.; Kodansha Scientific:Tokyo Japan, **1996**, pp.1-16.
19. Kajiyama, T.; Oishi, Y.; Uchida, M.; Morotomi, N.; Ishikawa, J.; Tanimoto, Y. *Bull. Chem. Soc. Jpn.*, **1992**, 65, 864.
20. Barton, S. W.; Goudot, A.; Rondelez, F. *Langmuir*, **1991**, 7, 1029.
21. Fujii, S.; Sugisawa, S.; Fukuda, K.; Kato, T.; Seimiya, T. *Langmuir*, **1995**, 11, 405.
22. Xiao, X. D.; Liu, G.; Charych, D. H.; Salmeron, M. *Langmuir*, **1995**, 11, 1600.
23. Mathauer , K.; Frank , C. W. *Langmuir*, **1993**, 9, 3446.
24. Pooley, C. M.; Tabor, D. *Proc. Roy. Soc. London*, **1979**, A329, 251.
25. Overney, R.; Meyer, E; Frommer, J; Brodbeck, D; Luthi, R; Howard, L.; Guntherodt, H.-J.; Fujihira, M; Takano, H; Gotoh, Y. *Nature*, **1992**, 359, 349.
26. Briscoe, B. J.; Evans, D. C. B. *Proc. R. Soc. Lond.*, **1982**, A380, 389.

Chapter 13

Optical and Structural Properties of Langmuir–Blodgett Films at the Air–Water and the Air–Solid Interface

L. M. Eng

Institute of Quantun Electronics, ETH Hönggerberg, HPF E7,
CH-8093 Zürich, Switzerland

We report on novel investigations of non-linear optical 2-docosylamino-5-nitropyridine (DCANP) Langmuir-Blodgett (LB) films by means of polarisation, second harmonic and scanning force microscopy (SFM). Direct optical measurements at the air/water interface showed that the molecules reoriented when depositing a LB film onto a solid substrate by the Langmuir-Blodgett method. Using the Langmuir-Schäfer method for horizontal dipping, however, resulted in the same surface structure being transferred onto substrates as was found for the DCANP film floating on a water subphase. Furthermore, the DCANP LB film was improved in film quality by adding 10-20% of arachidic acid (AA) to the pure DCANP phase. As shown by SFM, this results in a smooth surface structure free of defects while no significant changes in the optical properties of the mixed LB film were found. Such LB films therefore show improved structural and optical properties for optical wave guide applications.

Additionally, a novel force microscope was set up which is able to image the conformation of the chromophores of a LB monolayer floating at the air/water interface. Combined operation of both this new force microscope and the optical methods offers the possibility for the simultaneous inspection of both the structural and optical properties of the floating Langmuir films down to the molecular scale.

Introduction

For the application of non-linear optical materials in wave guide structures, filters and switches, an essential prerequisite is the overall homogeneous and constant properties of the material in use. It is not only necessary that the non-linear optical material has a high hyperpolarisation, but also that its physical structure is smooth on a scale much smaller than the typical wavelength. Recently, we showed that scattering losses at defects of 10-100 nm in diameter significantly attenuate the intensity of the propagating electromagnetic wave (~20 dB/cm) [1]. Thus we are interested in improving the material quality for the assembling of a usable wave guide structure. In a *bottom-up* approach the construction of a perfect monomolecular layer is desired using LB molecules. Not only do our organic films offer a big variety for chromophore syntheses, but also a wave guide structure can be controlled on a molecular level.

Here, we investigated the effect of LB film transfer onto a solid substrate, a crucial problem with these type of molecular structures. The influence of asymmetric forces on the molecules can be seen in direct space using our new set-up of second harmonic (SH) and polarisation microscopy [2]. Furthermore, film quality of freely floating Langmuir films was controlled using a novel SFM specially designed for the inspection of liquid/air interfaces.

Results and Discussion

Fig. 1 presents the DCANP monolayer film structure (a) when floating on the water subphase on a LB trough, (b) deposited onto a Pyrex substrate by perpendicular dipping (LB method), and (c) horizontally dipped by the Langmuir-Schäfer (LS) method. Note that in the LS method, the horizontally mounted sample substrate slightly touches the floating LB molecules until they interact with the atoms at the sample surface. Therefore, the structure of the Langmuir films transferred by the LS method is quite similar to the floating monolayer (see Fig. 1a and 1c). In both cases the DCANP film appears in so-called spherulites of a mixed crystalline and amorphous structure [3].

Perpendicular dipping (Fig. 1b) results in preferential arrangement of the chromophore molecules along the dipping direction. This is due to the force gradient induced by capillary and meniscus forces at the water/substrate interface [3]. The induced non-linear optical anisotropy measured parallel and perpendicular to the dipping direction is 10:1.

Fig. 2 presents the optical set-up as used to measure the reorientation of the DCANP LB molecules when deposited onto solid substrates. As seen the molecules are randomly oriented when floating on the water subphase (left). However, when interaction between the molecular dipoles of the chromophores and the surface atoms occurs, the molecules change their conformation and arrange along the dipping direction (right).

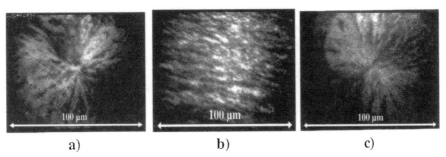

a)　　　　　　　　b)　　　　　　　　c)

Figure 1. Domain structure of a DCANP LB film (a) floating on the water
subphase, (b) deposited perpendicular onto a Pyrex substrate by
the LB method, and (c) horizontally dipped by the LS method.

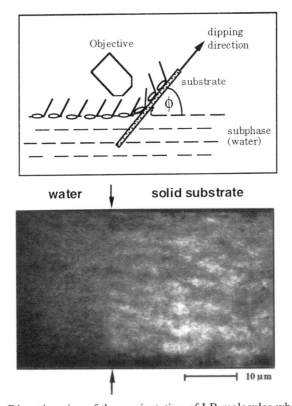

Figure 2. Direct imaging of the reorientation of LB molecules when being
deposited onto a solid substrate. While the molecules are
randomly oriented when floating on the water subphase (left
part), elongated domains appear on the solid substrate (right part)
arranged parallel to the direction of dipping.

The mechanical film properties of the DCANP LB film were further improved by adding a known percentage of arachidic acid (AA) to the DCANP phase [4]. Fig. 3 shows two SFM images representing (a) the pure DCANP LB film and (b) the mixed DCANP/AA structure containing 20% of AA in the DCANP phase. While the pure DCANP film (Fig. 3a) undergoes a structural phase transition with the LB film shrinking laterally by 10-15%, the mixed LB film presents a smooth sample surface topography containing only a minor number of defects. The latter sample is observed to be stable for several weeks and the rms surface roughness being constant at 0.1 nm_{rms} measured over 1 μm^2. We therefore succeeded in optimising a LB film with respect to its structure and transferring it onto a solid substrate.

Fig. 4 illustrates the behaviour of the non-linear optical susceptibility d_{33} as a function of AA concentration. As seen from this figure, the non-linear optical intensity is not affected too much for AA concentrations smaller than 20%. For higher concentrations, SFM inspection shows that the mixed LB film segregates into the two components forming islands of AA molecules in the DCANP phase. The phase separated film has a lower optical signal since the AA islands induce scattering losses of a propagating electromagnetic wave, in the same way as did structural defects in the pure DCANP film (see Fig. 2a). The optical intensity drops to zero for AA concentrations bigger than 30 %.

To complement our optical techniques (polarisation and second-harmonic microscopy) which were already set up on top of the Langmuir-Blodgett trough we constructed an underwater scanning force microscope (SAFM) [5] which is able to image the structural conformation of LB molecules floating at the liquid/air interface, i.e. a Langmuir layer. The SAFM was constructed as a standalone, remote-controlled and water-tight scanning force microscope [5].

Fig. 5 shows a constant height SFM picture of the freely floating AA Langmuir film. To perform this experiment AA was chosen as a LB molecule since it is known to exhibit a very well ordered and solid phase. AA molecules were spread onto the water subphase (containing 10^{-3} mol of CdCl for stabilisation) and then compressed to 30 mN/m. As seen from Fig. 5 the AA headgroups are well aligned along distinct directions with the most pronounced of them crossing the picture diagonally from the bottom-left to the top-right corner. The unit cell formed by the headgroups of the floating AA film was measured to 0.5 x 0.5 nm^2 and agrees well with the values found for AA films transferred to solid substrates.

The results reported above clearly show that SFM can equally be used for the inspection of liquid/solid interfaces as well as liquid/air interfaces. With the invention of the SAFM it is possible to answer some important scientific questions which are relevant for molecular binding and agglomeration processes of individual molecules at this interface. Because van der Waals, electrostatic, solvation and steric interaction forces between the floating Langmuir layer and the SFM tip are essentially the same as those acting between adjacent headgroups within a layer, we should expect to find many correlations between interlayer and intralayer forces.

Fig. 6 finally presents a series of force - distance (F-s) curves recorded with the SAFM at various interfaces. While Fig. 6a) presents the reference curve recorded

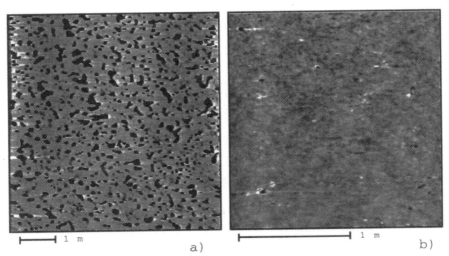

a)   b)

Figure 3. (a) 3.7 x 3.7 μm² SFM image of the DCANP LB film deposited
onto a mica substrate. Ageing results in film shrinking of 10-
15%. The molecular height between dark and grey levels
measures 2.1 nm.
(b) 2 x 2 μm² SFM image of the mixed DCANP/AA monolayer.
At 20% AA in the DCANP phase, the surface has a very smooth
appearance showing very few defects.

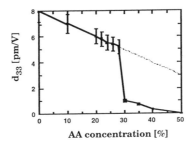

Figure 4. Non-linear optical susceptibility d_{33} of a mixed DCANP/AA LB
film as a function of AA concentration.

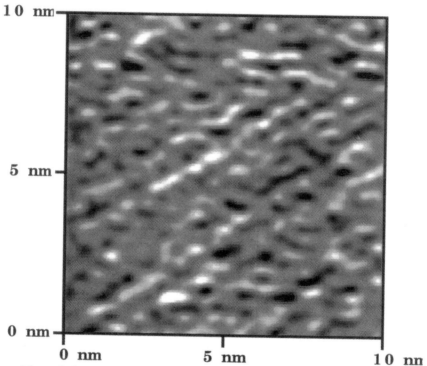

Figure 5. SAFM picture of the freely floating AA monolayer taken over an image size of $10 \times 10 \ nm^2$. The headgroups of the AA molecules are aligned along specific directions indicating good short range ordering.

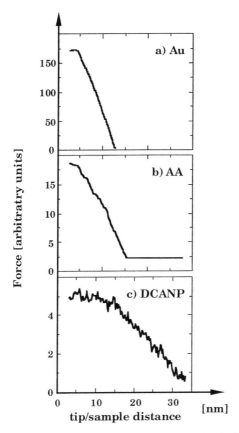

Figure 6. Force - distance (F-s) curves recorded with the SAFM at various interfaces (k = 0.64 N/m) : (a) polycrystalline gold sample, (b) floating AA LB film, and (c) floating DCANP LB film. Note the similarities in F-s curves between Au and AA.

on a polycrystalline gold sample, Fig. 6b) and Fig. 6c) illustrate the behaviour for two floating LB films : a densely packed AA LB film and a fluid-like DCANP LB film, respectively. All curves show the repulsive force interaction exerted between tip and sample. While the z-values are accurately read and represented in a nm-scale (horizontal axes), force numbers are not calibrated (vertical axes). Nevertheless, a relative comparison between the three F-s curves is possible :

First, both the AA and the Au curve are very similar in shape. In fact, the AA LB film was compressed into its solid phase in order to deduce the lateral packing of molecules with the SAFM (see Fig. 5). The only difference stems from the force values recorded for AA and Au, which are smaller by a factor of 10 when comparing Fig 6b) with Fig. 6a). Recording the same curves with a cantilever of a very soft spring constant (k = 0.03 N/m) instead of k = 0.64 N/m (not shown [6]), the two force plots become even more similar with the force values deviating only by a factor of 2. A floating AA LB film therefore may be regarded as forming a sold interface with a very high elastic modulus in the same way as does a solid substrate.

Secondly, comparison of Fig. 6b) and Fig. 6c) shows the pronounced difference between the solid- and the liquid-like LB film. While force values deviate by a factor of 2-3 only (even for a strong cantilever spring constant k = 0.64 N/m) interaction forces are found to be longer ranged extending over more than 30 nm. Clearly the DCANP LB film contains rather strong dipolar headgroups which significantly influence the force interaction between tip and sample. We therefore expect the SAFM to be sensitive enough to record both dipolar and van der Waals interactions in floating LB films.

Acknowledgement

We are pleased to express our gratitude to Ch. Seuret, P. Günter, and F. Grünfeld. Financial support by the Swiss National Science Foundation under grant # 30520 is greatly acknowledged.

References

1. Ch. Bosshard, M. Küpfer, M. Flörsheimer, and P. Günter, *Thin Solid Films* **1992**, *210/211*, 153.
2. M. Flörsheimer, D. Jundt, H. Looser, K. Sutter, M. Küpfer, and P. Günter, *Ber. Bunsenges. Phys. Chem.* **1994**, *98*, 521.
3. L.M. Eng, M. Küpfer, Ch. Seuret, and P. Günter, *Helv. Phys. Acta* **1994**, *67*, 757.
4. B. Schmidt, *Kraftmikroskopische Untersuchungen an DCANP und VECANP Langmuir-Blodgett Filmen,* Diploma thesis, ETH Hönggerberg, Zürich, Switzerland ; **1993**.
5. L.M. Eng, Ch. Seuret, H. Looser, and P. Günter, *J. Vac. Sci. Technol B* **1996**, March/April, 1386.
6. Ch. Seuret and L.M. Eng, to be published.

Chapter 14

Atomic Force Microscopy of Langmuir–Blodgett Films Polymerized as a Floating Monolayer

Jouko Peltonen and Tapani Viitala

Department of Physical Chemistry, Åbo Akademi University,
Porthaninkatu 3-5, FIN-20500 Turku, Finland

These topographical studies demonstrate that linoleic acid Langmuir-Blodgett films polymerized as floating monolayers are not ideally flat and defect-free. Both the monomeric and UV-polymerized films contain hole-defects. As a result of the polymerization part of the film folds up from the film plane. This effect is strongly surface pressure dependent and can be inhibited by polymerizing the floating monolayer at a pressure lower than 10 mN/m.

In order to improve the mechanical, thermal and chemical stability of organic Langmuir-Blodgett films, polymeric materials have been widely applied (1-51). Either preformed polymers may be used (1-27) or the monolayer can be formed of amphiphiles with reactive functional groups available for polymerization after the film formation (28-51).

Several different types of surface-active polymers have been employed, one of the most typical being poly(methacrylate) (or poly(acrylate)) and its numerous derivatives (3-10). The influence of the tacticity on the film forming properties has been studied (4-5) and the rheological behaviour has been compared with other types of polymers, e.g. poly(vinyl acetate) (10) which represents another class of widely studied vinyl-based polymers (10-13). Homogeneous multilayer structures of alternating anionic and cationic polyelectrolytes have been successfully formed by using e.g. poly(vinyl sulfate) and poly(allylamine) (13) or poly(styrenesulfonate) and poly(allylamine) (14), respectively. So called "hairy-rod" or "rigid-rod" polymers introduced by Wegner represent a novel class of polymeric materials that, as a result of deposition onto a solid substrate, mimic properties of liquid crystalline (e.g. nematic) order and as such are potential candidates for LB film applications as sensors and optical devices (15-20). The use of mechanically, thermally and chemically stable polyamides has attracted attention, not least due to the fact that after deposition, the polymer can be converted to the polyimide form by chemically or through heat treatment removing the hydrocarbon chains (21-27).

Since polymeric monolayers are occasionally difficult to process due to their high viscosity and inhomogeneous distribution at the interface, an alternative approach has been to study reactive monomers which are spread at the gas/liquid interface and the so formed monolayer is polymerized either prior to or after the deposition onto a solid substrate (28-47, 49-51). Diacetylenes and their various substituted derivatives with conjugated triple bonds responsible for the reactivity are the most commonly studied in this class (28-43). In ideal conditions, topochemical polymerization (41) takes place, i.e. the crystal structure remains unchanged and no area contraction occurs during the reaction. In reality, however, the resulting film is not a 2-D single crystal but consists of domains and thus also domain boundaries, representing discontinuities. The diacetylene monolayer may be polymerized as a floating monolayer or after deposition, even as a multilayer structure; the main prerequisite is that the monomers within a monolayer are packed in a solid state and in a suitable orientation for the neighbouring triple bonds to crosslink during UV-curing. Recently, however, it has been reported that the polymerization of diacetylenes may also be carried out in the liquid-crystalline state (42). The diacetylenes are interesting not only because of their ability to create stable polymeric monolayers but also due to their optical activity which can be further tuned by the choice of the side groups (28, 29, 32-35, 38, 39). Acyl chains incorporating diacetylenic groups have also been applied in lipid systems in order to create bilayer structures with enhanced stability (30, 31, 43, 44). The reaction in these systems is, however, no more a pure topochemical one and furthermore the polymerized film restricts using the system to mimic a natural biomembrane e.g. with respect to transport phenomena through such a rigid crosslinked structure.

The real time reaction of a Langmuir monolayer may also be initiated by a catalyst introduced into the liquid subphase. The group of Duran has published results on the polymerization of e.g. aniline monolayers where the reaction has been activated by an oxidizing agent (46, 47).

The present work concentrates on the topographical characteristics of linoleic acid films, both as a monomer and a crosslinked polymer. The reasons for the use of this particular acid are discussed together with a short review of the recent results on in situ polymerization and the related monolayer phase behaviour (49-54). Tsukruk and Reneker have recently reviewed the use of scanning probe microscopy (SPM) in the characterization of different types of thin films (48). Zasadzinski has reviewed the latest developments in SPM instrumentation and its practical applications (55). Here, by implementing tapping mode SPM, it was possible not only to reveal new information about the phase structure of the films, but also to image unstable samples, e.g. monolayers deposited at pH's corresponding to highly metastable structures or ones only weakly adsorbed to the solid substrate (or underlying monolayer). This set of data nicely supports both the surface balance data and the spectroscopic measurements recently carried out for the same film structures (51-53).

The Surfactant and Related Substances

Z,Z-octadecadi-9,12-enoic (linoleic) acid (LA, reagent grade 99% purity), n-hexane (PA grade, >99.5% purity) and $TbCl_3 \cdot 6H_2O$ (98% purity) were obtained from FLUKA and used without further purification. LA was dissolved in n-hexane to form a solution with a concentration of 1 mg/mL. The details of the monolayer formation and LB-film deposition are given elsewhere (*51*).

It is well known that the introduction of cis-type double bonds into the acyl chain increases the minimal molecular cross-sectional area and makes it difficult to obtain a condensed monolayer at the air/water interface, as compared with e.g. stearic acid. The problem is not only due to the steric hindrance when compressing molecules with nonlinear chain, but also one of energetics; the double bonds have a rather strong attractive interaction with the water subphase. The presence of only one double bond can decrease the interfacial energy of octadecane with water from 52 to 19 mJ/m, whilst the surface energy changes only slightly (*56*). It is well known that the degree of unsaturation together with the position and configuration (cis/trans) of the double bonds affects the melting point and the equilibrium spreading pressure (*57*). Thus, special attention has been paid to the condensation of the initially expanded linoleic acid monomer film through pH adjustment and suitable choice of composition of the liquid subphase (*51,52*). The use of $TbCl_3$ subphase at elevated pH enabled the generation of a phase transition from a liquid-expanded (LE) to a solid-expanded (SE) state monolayer (Figure 1). The acid molecules were fully ionized at pH 5, but the complete condensation took place over a very narrow pH-range of 6.8-6.9, where the conversion of Tb to its monohydroxy complex reached a maximum concentration, just prior to precipitation (*51*). FTIR-measurements have confirmed that the soap formation really concerns the (ionized acid)$_2$ - $Tb(OH)^{2+}$ complexation (*51,52*), as schematically illustrated in Figure 2. It is worth mentioning that on a pure water subphase or on a subphase containing any divalent metal ions, the LA monolayer remains in a liquid-expanded (LE) state, throughout the compression and at all pH values.

As may be seen in Figure 1, the extrapolated molecular area of the SE-phase still deviates from that of the tightly packed stearic acid monolayer, but the density was enough to reach an 'activation area' for the polymerization reaction to be successfully initiated. In fact, the polymerization of LA in the SE-phase has been found to be more effective than that of e.g. trans-unsaturated petroselaidic (trans-6-octadecenoic acid) or elaidic (trans-9-octadecenoic) acid, which can be compressed to a crystalline state (*49,50,58*). The utilization of the SE-phase instead of a fully crystalline state may provide an important advantage as compared with other polymerizable monolayers; the semicrystalline SE-phase with liquid or amorphous characteristics enables not only the monomers to reorganize during the crosslinking process but also e.g. functional molecules to distribute evenly within this monomer matrix by diffusion. Subsequently the system can be stabilized by crosslinking the reactive monomers, consequently freezing the whole matrix and inhibiting the diffusion and phase separation e.g. during the deposition of the monolayer.

The Polymerization Technique

The polymerization of the monolayers was initiated by exposure to UV-light from a 30 W low-pressure Hg lamp, as schematically shown in Figure 3. The reaction was carried out at a predetermined constant surface pressure (typically in the range 5-20 mN/m) and irradiation-induced changes in the mean molecular area and barrier speed were detected. This set of data enabled the modelling of the reaction kinetics in a similar way as

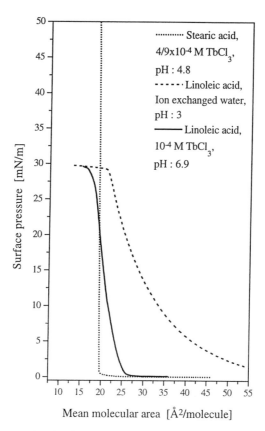

Figure 1. Compression isotherms of stearic acid and linoleic acid measured on different subphases as indicated in the inset.

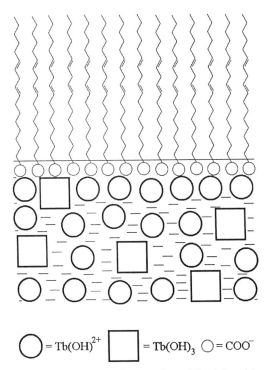

Figure 2. A schematic illustration of a condensed linoleic acid monolayer on a 10^{-4} M Tb subphase of pH 6.8.

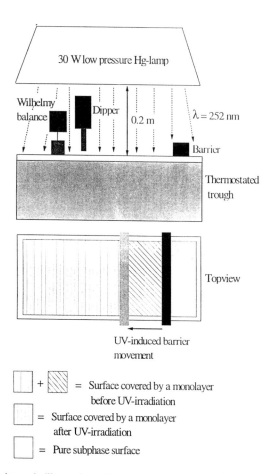

Figure 3. A schematic illustration of *in situ* polymerization of a floating monolayer.

reported by a few other groups (*46,47,59*). In our experiments, the main assumption was that the irradiation-induced change in mean molecular area was equal for each reaction step. The overall reaction has been found to follow first order kinetics (*51*), however, the best fit being obtained if the time-dependent change in mean molecular area is taken into account (Viitala et al., *J. Chem. Soc., Faraday Trans.*, submitted).

It may not be assumed that the UV-light will be absorbed by the monolayer and especially the double bonds since for LA they are non-conjugated. In dilute solutions or LB-multilayers representing the LE-phase (below the optimum pH of 6.8) UV-VIS spectra have revealed that the main absorption band of a "pure" or "isolated" monomer appears at (200 ± 7) nm, the exact energy being dependent on the concentration of the solution or the homogeneity of the film structure. A weak and broad band appeared centered at 233 nm. However, when the multilayer structure was prepared at pH 6.8, the band centered at 233 nm was stronger and more localized. It is suggested that this band corresponds to a charge transfer complex between individual monomers, strictly speaking the double bonds (Viitala et al., *J. Chem. Soc., Faraday Trans.*, submitted). The 233 nm band has been shown to be responsible for the reactivity of LA (e.g. it decreases in intensity and finally disappears during UV-curing), not least due to the fact that this band overlaped with the main emission band of the used UV-lamp centered at 252 nm. The rate of polymerization in the SE-phase was almost 10-fold as compared with the value observed for the LE-phase (*51,52*).

Since the UV-initiated reaction of LA has been found to be insensitive to oxygen and only weakly temperature-dependent, the polymerization mechanism has been suggested to be ionic (*51*)(cf. the earlier discussion on charge transfer complexation) rather than a radical polymerization which is more typical of vinyl-based polymers. This made control of the reaction easy since i) UV-irradiation is enough to initiate the reaction, no catalyst is needed, and ii) the reaction could be executed in normal air conditions, no safety gas was needed as is the case with e.g. conjugated dienoyl acid analogues for which the oxidative degradation may become a problem (*30*).

Topographical Studies

A NanoScope III multimode AFM (Digital Instruments, Santa Barbara, CA) was used for imaging, in the tapping mode. The 125 µm J-scanner with Si_3N_4 cantilevers supplied by the manufacturer was used. Freshly cleaved mica (S&J Trading, Glen Oaks, NY) was used as the substrate. All the measurements were carried out in ambient air conditions.

The Films Representing the LE-phase. The unstable nature of LA monolayers at any pH below that of 6.8 is demonstrated in Figure 4, where the structure of films deposited at pH 6.6 is shown, in the tapping mode AFM images. In Figure 4a the adsorbed monolayer appeared to be uneven with high (light) spot-like structures rising out of the film plane and being randomly distributed over the sample surface. The adsorbed monolayer was unstable and imaging with the AFM in contact mode frequently resulted in a modified surface and film desorption. The most dramatic effect was observed when a multilayer structure was prepared. It is obvious from the images representing a three-layer structure (Figures 4b & 4c), that the film does not exist as a pure monolayer but has mainly been converted to a lipid particle form. The phenomenon is especially well

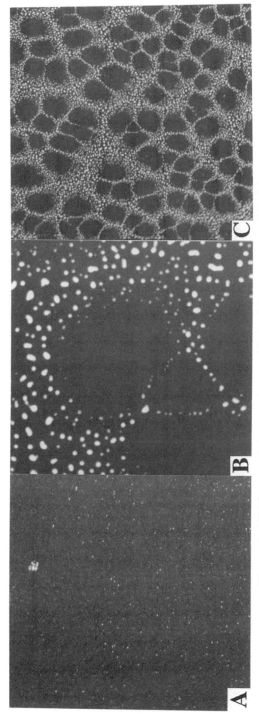

Figure 4. AFM images of a) a monolayer and b,c) a 3-layer structure of linoleic acid deposited on mica from a 10^{-4} M $TbCl_3$ subphase at pH 6.6. The image size is $10 \times 10 \ \mu m^2$ in a) and b) and $50 \times 50 \ \mu m^2$ in c) and dark-light (low-high) height scales are a) 10 nm and b,c) 100 nm.

demonstrated in Figure 4c in a 50 μm X 50 μm image where the interparticle attraction as evidenced by network formation is also visible. Grain size analysis of this particular but still very characteristic image yielded a mean grain size of 19 x 10^3 nm^2 with altogether ca. 6000 grains. The corresponding mean diameter of the particles is thus 160 nm which, however, includes the convolution effect between imaging tip and the imaged object. We have not measured the radius of curvature of our tips, but by using a reference value for the tip radius of curvature of 50 nm (60) and applying a simple deconvolution model (61,62) it was possible to roughly estimate the real diameter of the particles. The resulting mean particle diameter of 60 nm coincided with the observed average height value and indicated that the particles were close to spherical.

One possible reason for the particle formation is the elastic energy stored in the floating monolayer, which is released during the monolayer transfer. The exceptionally high elastic or bending energy is due to the asymmetric shape of the molecule arising from the relatively small polar head group in relation to the larger volyme of the nonlinear hydrocarbon chain. During the monolayer transfer, the release of this energy competes with the energy of adhesion to the underlying surface. The first monolayer still remained as such when adsorbed to the hydrophilic mica plate, however, being rougher than films made of saturated acid analogues (48,63,64,65). The subsequent layer-to-layer interaction seemed to be insufficiently strong to compete with the elastic energy responsible for the particle formation. Because the large mean molecular area in the LE-phase is mainly determined by the nonlinear acyl chain, the soap formation between the ionized acid groups and the Tb^{3+}-ions is partly inhibited, even if the ions still screen the charged monolayer. The resulting relatively loose structure may also allow the water molecules to penetrate to the polar head group region of the film and would explain the similarities between the particle formation and the wetting-dewetting phenomenon.

The trend with the subsequent layer adsorption was clear, as seen in Figure 5a representing an 11-layer structure. It was difficult to measure the polarity of the particles due to unstable contact angle measurements, but since the AFM-images captured for a fresh and e.g. one-week-old sample looked the same, thus the stable structure refered to aggregates of reversed micelle-like structures with the hydrocarbon chains in contact with the gas phase. Also here, the differences in resolution when measuring with either the contact mode or the tapping mode were obvious. A broader particle size distribution for a tapping mode image, such as Figure 5a, was observed, the smallest particles (diameter 8-12 nm) refering to single micelles. In the contact mode, the friction between the tip and the sample contributed to the imaging process, the lightest particles which were most weakly bound to the surface being moved by the tip and finally aggregating together with larger particles.

The possibility that the particles were formed as a result of modification of the layer structure by the imaging tip should be considered. Evidence for this not being the case was obtained by UV-VIS spectroscopy (Viitala et al., *J. Chem. Soc., Faraday Trans.*, submitted). All the spectra recorded for fresh 'low-pH' samples prior to any microscopic analysis showed considerable light scattering throughout a spectral range of 200-600 nm. As soon as the pH was raised to 6.8, this scattering disappeared and only the characteristic bands at 193 and 233 nm were present. This set of spectra coincided very well with the AFM images of Figures 4 and 5, especially when also considering the image in Figure 5b which represents a multilayer structure as deposited

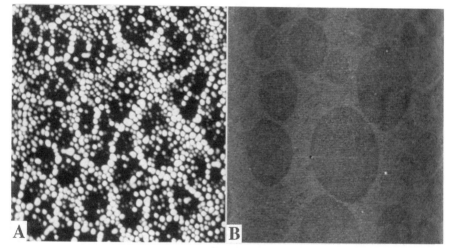

Figure 5. AFM images (20 x 20 μm^2) representing a multilayer structure (11 layers) of linoleic acid as deposited at a) pH 6.6 and b) pH 6.8. The height scales are a) 120 nm and b) 10 nm.

at pH 6.8; the surface was absolutely flat and smooth with no sign of particles. The tapping mode image also revealed that the film consisted of different phases. The average height difference between the domains and the surrounding phase was only a few angstroms. Furthermore, the film was smooth and the RMS (root mean square) roughness of a 11-layer structure in the 10 X 10 μm^2 area was typically 0.35 ± 0.05 nm.

There is an obvious analogy between the change in topography from strongly curved surfaces (pH 6.6 or lower) to a smooth LB film structure (pH 6.8-6.9) and the dramatically enhanced reaction rate of a Langmuir monolayer when screening the same pH interval. The small step upwards in the pH-scale changes the system radically because the concentration of $Tb(OH)^{2+}$ increases significantly and simultaneously the relative proportion of Tb^{3+} decreases due to gradually increasing precipitation. Thus a stable $Tb(OH)(Linoleic\ acid)_2$ complex network forms. In conclusion, the greater reaction rate, by an order of magnitude, at this particular pH strongly indicates that the monomers are now favourably oriented for a successful reaction.

Comparison of Monomeric and Polymeric Films. As already shown in Figure 2, the subphase at pH 6.8 contains not only $Tb(OH)^{2+}$ complexes but also solid particles from the precipitation process. $Tb(OH)_3$ solid particles adsorbed onto the mica substrate which was immersed and stored in the subphase (ca. 30 min) prior to deposition. Figure 6a shows that the mica surface was almost fully covered by the particles with an average deconvoluted diameter of 5 nm. The largest objects obviously represent aggregates.

The $Tb(OH)_3$ particles caused problems during the deposition of the first monolayer (upstroke of the mica plate) as confirmed by Figures 6b and 6c where the monomeric and polymerized linoleic acid monolayers on mica appear unexpectedly rough. The RMS roughness of all the three images of Figure 6 was in the same range, 1.2 ± 0.1 nm. We will not go into any detailed analysis of the film structure but it is worth noting that the polymer film seemed to have retained its film form better than the monomeric film. It should also be mentioned that when the deposition was continued for a sample such as that of Figure 6b, the multilayer structure finally smoothed out as in fact is demonstrated in Figure 5b. This is also evidence for the fact that the adsorption of solid Tb particles concerns only the mica substrate, no adsorption onto or transfer of the particles by the monolayer took place.

The reason for the adsorption of solid Tb particles was a result of a strongly hydrophilic and hydrated mica surface at pH 6.8 (*63,66*). The problem was overcome by first depositing a Tb-stearate monolayer on mica at pH 5. At this pH, no precipitation of $Tb(OH)_3$ takes place and the surface of a mica plate is less hydrated (*63*). As a result, a smooth layer of stearic acid rendered the substrate hydrophobic and electrically neutral. Hence, the subsequent monomer and polymer LA films appeared smooth and contained only few or no particles (Figure 7). Holes appear in both images. Since the diameter of the holes was 110 ± 40 nm, it is difficult to regard the holes as traces of Tb-particles, these had a mean diameter of 5 nm (Figure 6). Hence, they are considered to be defects in the monolayer in a similar way as reported for films of saturated fatty acids (*63-65,67*). Furthermore, the height profile of the surface revealed that the majority (about 3/4) of the holes corresponded to a single monolayer in depth (2.4 ± 0.3 nm) whereas the rest of the holes had a depth close to a bilayer thickness. This means that the uppermost layer bridged the holes of the underlying film with no

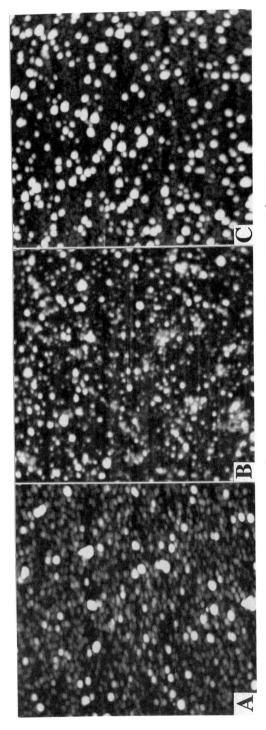

Figure 6. 1 x 1 μm^2 AFM images of a mica plate being stored in a 10^{-4} M $TbCl_3$-solution of pH 6.8 during 30 min. The same substrate was further covered by a b) monomeric or c) polymerized linoleic acid monolayer. The height scale in each image is 10 nm.

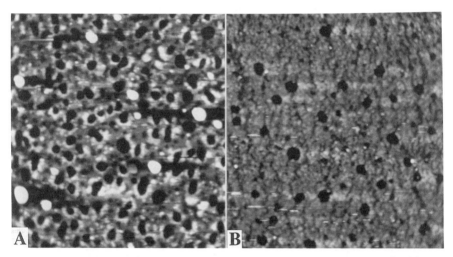

Figure 7. 2.5 x 2.5 μm² AFM images of a a) monomer and b) polymerized linoleic acid bilayer. The substrate was mica covered by a Tb-stearate monolayer. The dark-light Z-scale is 10 nm.

indication of rapture or bending of the cover film. In the monomer film, the holes appeared as craters with raised edges. This was not the case for the polymerized film, where the edges remained at the same level as the rest of the film surface. The RMS roughness of the polymer film is lower than that of the monomer film, however, in part due to the absence of particles in the polymer film.

Large-scale images of a monomeric and polymerized film are shown (Figure 8), in order to demonstrate that the topology of the made up film structures was not homogeneous over larger areas. Clear phase boundaries are visible in the otherwise smooth surface of a monomer film. The surface of the polymer film with the characteristics similar to that of Figure 7b is disrupted by a zone very much resembling a hyperbolic or saddle surface (68). It seems that the monolayer in this apparently higher phase has pleated, i.e. folded up, however, still appearing as a 2D continuous structure. The height difference between various phases was much more pronounced in the polymer film than in the monomer film, where the height difference was hardly measurable.

Interestingly, the folded structure was only found in the structures polymerized at high surface pressure. A close look at the isotherm (Figure 1) revealed that there is a kink at pressure 10 mN/m where the compressibility decreases. If the Langmuir monolayer was polymerized at a pressure lower than 10 mN/m, no folded phase was observed. Neither was there any phase contrast in the images captured for the low-pressure monomeric films. Above this critical pressure the monolayer is believed to be so tightly packed that the UV-irradiation induced area contraction in the hydrophobic chain region generates internal stress in the monolayer. Due to the crosslinked structure, however, the structure does not totally break during the monolayer transfer but instead appears as a partly folded film. The folded structure was visible in high-pressure samples irradiated for only 2 minutes, indicating that the folding was not connected to e.g. collapse or degradation induced by an overdose of irradiation. This set of data shows that the phenomenon of folding is strongly density-dependent.

Discussion

At present, it is not exactly known how the $Tb(OH)_3$ complexes affect the state of the dissociated monolayer. We believe that we have in the main been able to overcome the imbalance in the volumetric asymmetry of the studied surfactant, as schematically shown in Figure 9, by using Tb metal ions in the subphase at elevated pH where the $Tb(OH)^{2+}$ complexes contribute to the monolayer condensation through an electrostatic mechanism. The condensation may also be generated by e.g. using Al^{3+}-ions, but the coordination chemistry of Al is much more complicated than that of Tb and it is difficult to say which complex form primarily causes the condensation (52). No LE-SE phase transition could be generated when using pure metal ions such as Cd^{2+} or Mn^{2+} including Tb^{3+} of smaller volume, which are normally enough to condense a saturated fatty acid monolayer (53,54).

The data on the polymerization kinetics has shown that the final mean molecular area per monomer unit of the formed polymer is ca. 14 $Å^2$ which is clearly too small a value to be a realistic cross-sectional area of a tightly-packed hydrocarbon chain (51). Thus, we have speculated that part of the floating monolayer may have buckled out of

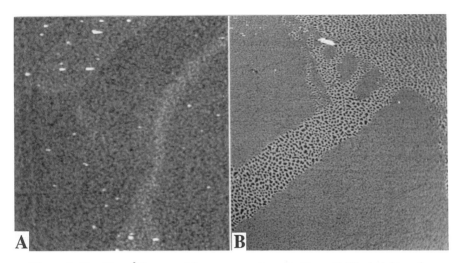

Figure 8. 10 x 10 µm² images of the same samples as in Figure 7. The height scale is 10 nm in both images.

Polymerization

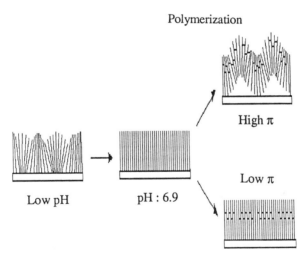

High π

Low π

Low pH   pH : 6.9

Figure 9. A schematic illustration on the effect of pH of the Tb-subphase on the condensation of a linoleic acid monolayer, and subsequently the effect of surface pressure on the topology of the polymerized film.

the film plane (*51*). The AFM images shown here regarding the folded phase nicely support this assumption and give further proof that the buckled (*69,70*) or curved polymer Langmuir monolayer appears folded after deposition. This strongly surface pressure dependent effect is schematically drawn in Figure 9.

It was only with the tapping mode that the phase contrast shown in Figures 5b and 8a could be obtained. The absence of the LE-SE phase coexistence region in the isotherm (*51,52*) was confirmed by fluorescence microscopy; no images could be obtained due to the lack of contrast. Both phases appearing in Figure 5b are furthermore so dense that neither of them can be considered as representing an LE-phase. The round or slightly distorted form of the domains (observe the similarities with Figure 8b) represents a special case of minimum line tension and thus line energy, with a minimal repulsive electrostatic interaction of the molecules within a domain (*71*). However, the round form was only characteristic of the multilayer samples; in bilayer structures (Figure 8a) the domain shape appeared to be much more distorted and irregular. This indicates that the structure becomes more homogeneous with an increasing number of layers in a multilayer structure. It should be further pointed out that no evidence of kinetic crystallization, reorganisation or overturning of molecules was visible in the multilayer structure, being further indication of a stable structure. These often problematic phenomena are frequently observed for saturated fatty acid soap films (*72, 73*).

Since the difference in topographical height between the two phases was very small (Figure 5b) and since no LE-SE phase coexistence was evident, the contrast between the domains and the surrounding phase is suggested to be due to the difference in local film density and thus stiffness (*74-76*). Within the limits of resolution, even the roughness was seen to be equal for the two phases. The specific response may be connected to the characteristic elasticity and/or tip-sample interaction forces within various surface regions (*74,75*). The most logical explanation for the observed contrast is that the force gradient is modified when going from one phase to another, affecting the resonance frequency and also the driving amplitude of the vibrating tip (*75*). Since a constant force mode was used, the described phenomenon finally affects the Z-piezo, i.e. height response, in order to maintain the initially selected setpoint force.

Concluding remarks

The presented results demonstrate the ability of the tapping mode of AFM to distinguish between the details of both topographical and mechanical properties of the studied soft monomer and polymer surfaces that nicely support the existing surface balance, spectroscopic and reaction kinetic data. However, by implementing the force modulation technique, more information about the rheological properties is to be expected, not only between but also within the monomer and polymer films.

Acknowledgments

Prof. Jarl Rosenholm, Dr. Hazel Watson, Dr. Mika Lindén, Dr. Serge Durand-Vidal and Dr. Ulrich Hofmann are acknowledged for useful discussions. We thank the Academy of Finland for financing this work.

Literature Cited

1. Embs, F.; Funhoff, D.; Laschewsky, A.; Licht, U.; Ohst, H.; Prass, W.; Ringsdorf, H.; Wegner, G.; Wehrmann, R. *Adv. Mater.* **1991**, *3*, 25-31.
2. Gaines, Jr., G.L. *Langmuir* **1991**, *7*, 834-839.
3. Hann, R.A. In *Langmuir-Blodgett Films;* Roberts, G.G., Ed.; Plenum Press: New York, NY, 1990; pp 68-78.
4. Brinkhuis, R.H.G.; Schouten, A.J. *Langmuir* **1992**, *8*, 2247-2254.
5. Ha, J.S.; Roh, H.-S.; Jung, S.-D.; Park, S.-J.; Kim, J.-J.; Lee, E.-H. *J. Vac. Sci. Technol. B* **1994**, *12*, 1977-1980.
6. Judge, M.D.; Gardin, G.P.;Thompson, E.; Lowen, S.V.; Holden, D.A. *J. Polymer Sci. A* **1991**, *29*, 1203-1206.
7. Kawaguchi, M.; Sauer, B.B.; Yu, H. *Macromolecules* **1989**, *22*, 1735-1743.
8. Peng, J.B.; Barnes, G.T. *Langmuir* **1991**, *7*, 3090-3093.
9. Lowack, K.; Helm, C.A. *Adv. Mater.* **1995**, *7*, 156-160.
10. Kawaguchi, M.; Nagata, K. *Langmuir* **1991**, *7*, 1478-1482.
11. Wang, L.-F.; Kuo, J.-F.; Chen, C.-Y. *Colloid Polym. Sci.* **1995**, *273*, 426-430.
12. Peng, J.B.; Barnes, G.T. *Colloids Surfaces A* **1995**, *102*, 75-79.
13. Lvov, Y.; Decher, G.; Möhwald, H. *Langmuir* **1993**, *9*, 481-486.
14. Decher, G.; Hong, J.D.; Schmitt, J. *Thin Solid Films* **1992**, *210/211*, 831-835.
15. Orthmann, E.; Wegner, G. *Angew. Chem.* **1986**, *98*, 1114-1115.
16. Duda, G.; Wegner, G. *Makromol. Chem., Rapid Commun.* **1988**, *9*, 495-501.
17. Schwiegk, S.; Vahlenkamp, T.; Wegner, G.; Xu, Y. *Thin Solid Films* **1992**, *210/211*, 6-8.
18. Schaub, M.; Mathauer, K.; Schwiegk, S.; Albouy, P.-A.; Wenz, G.; Wegner, G. *Thin Solid Films* **1992**, *210/211*, 397-400.
19. Sauer, T.; Caseri, W.; Wegner, G.; Vogel, A.; Hoffman, B. *J. Phys. D: Appl. Phys.* **1990**, *23*, 79-84.
20. Gupta, V.K.; Kornfield, J.A.; Ferencz, A.; Wegner, G. *Science* **1994**, *265*, 940-942.
21. Kakimoto, M.; Suzuki, M.; Konishi, T.; Imai, Y.; Iwamoto, M.; Hino, T. *Chem. Letters* **1986**, 823-826.
22. Suzuki, M.; Kakimoto, M.; Konishi, T.; Imai, Y.; Iwamoto, M.; Hino, T. *Chem. Letters* **1986**, 395-398.
23. Hirano, K.; Nishi, Y.; Fukuda, H.; Kakimoto, M.; Imai, Y.; Araki, T.; Iriyama, K. *Thin Solid Films* **1994**, *244*, 696-699.
24. C.E. Sroog In *Macromolecular Synthesis;* Moore, J.A., Ed.; John Wiley & Sons: New York, 1977; Coll. Vol. 1; pp 295-298.
25. Lupo, D.; Prass, W.; Scheunemann, U. *Thin solid Films* **1989**, *178*, 403-411.
26. Schoch, Jr., K.F.; Su, W.-F.A.; Burke, M.G. *Langmuir* **1993**, *9*, 278-283.
27. Yang, X.-M.; Min, G.-W.; Gu, N.; Lu, Z.-H.; Wei, Y. *J. Vac. Sci. Technol. B* **1994**, *12*, 1981-1983.
28. Tieke, B.; Lieser, G.; Wegner, G. *J. Polymer Sci.* **1979**, *17*, 1631-1644.
29. Tieke, B.; Lieser, G. *J. Colloid Interface Sci.* **1982**, *88*, 471-486.
30. Meller, P.; Peters, R.; Ringsdorf, H. *Colloid Pol. Sci.* **1989**, *267*, 97-107.

248

31. Göbel, H.D.; Gaub, H.E.; Möhwald, H. *Chem. Phys. Lett.* **1987**, *138*, 441-446.
32. Warta, R.; Sixl, H. *J. Chem. Phys.* **1988**, *88*, 95-99.
33. Ogawa, K.; Tamura, H.; Hatada, M.; Ishihara, T. *Langmuir* **1988**, *4*, 903-906.
34. Hasegawa, T.; Ishikawa, K.; Kanetaka, T.; Koda, T.; Takeda, K.; Kobayashi, H.; Kubodera, K. *Chem. Phys. Lett.* **1990**, *171*, 239-244.
35. Mino, N.; Tamura, H.; Ogawa, K. *Langmuir* **1991**, *7*, 2336-2341.
36. Goettgens, B.M.; Tillmann, R.W.; Radmacher, M.; Gaub, H.E. *Langmuir* **1992**, *8*, 1768-1774.
37. Tillmann, R.W.; Radmacher, M.; Gaub, H.E.; Kenney, P.; Ribi, H.O. *J. Phys. Chem.* **1993**, *97*, 2928-2932.
38. Deckert, A.A.; Horne, J.C.; Valentine, B.; Kiernan, L.; Fallon, L. *Langmuir* **1995**, *11*, 643-649.
39. Kim, T.; Crooks, R.M.; Tsen, M.; Sun, L. *J. Am. Chem. Soc.* **1995**, *117*, 3963-3967.
40. Sarkar, M.; Lando, J.B. *Thin Solid Films* **1983**, *99*, 119-126.
41. Wegner, G. *Z. Naturforschg.* **1969**, *24 b*, 824-832.
42. Tsibouklis, J. *Adv. Mater.* **1995**, *7*, 407-408.
43. Tillmann, R.W.; Hoffmann, U.G.; Gaub, H.E. *Chem. Phys. Lipids* **1994**, *73*, 81-89.
44. Nakaya, T.; Yanada, M.; Shibata, K.; Imoto, M.; Tsuchiya, H.; Okuno, M.; Nakaya, S.; Ohno, S.; Matsuyama, T.; Yamaoka, H. *Langmuir* **1990**, *6*, 291-293.
45. Cemel, A.; Fort, T.; Lando, J.B. *J. Polym. Sci. A-1* **1972**, *10*, 2061-2083.
46. Zhou, H.; Stern, R.; Batich, C.; Duran, R.S. *Makromol. Chem., Rapid Commun.* **1990**, *11*, 409-414.
47. Bodalia, R.; Duran, R. *J. Am. Chem. Soc.* **1993**, *115*, 11467-11474.
48. Tsukruk, V.V.; Reneker, D.H. *Polymer* **1995**, *36*, 1791-1808.
49. Peltonen, J.P.K.; He, P.; Rosenholm, J.B. *J. Am. Chem. Soc.* **1992**, *114*, 7637-7642.
50. Peltonen, J.P.K.; He, P.; Rosenholm, J.B. *Langmuir* **1993**, *9*, 2363-2369.
51. Peltonen, J.; He, P.; Lindén, M.; Rosenholm, J. *J. Phys. Chem.* **1994**, *98*, 12403-12409.
52. Lindén, M.; Györvary, E.; Peltonen, J.; Rosenholm, J.B. *Colloids Surfaces A* **1995**, *102*, 105-115.
53. Peltonen, J.; Lindén, M.; Fagerholm, H.; Györvary, E.; Eriksson, F. *Thin Solid Films* **1994**, *242*, 88-91.
54. Lindén, M.; Rosenholm, J. *Langmuir* **1995**, *11*, 4499-4504.
55. Zasadzinski, J.A. *Current Opinion Colloid & Interface Sci.* **1996**, *1*, 264-269.
56. Israelachvili, J.N. *Intermolecular and Surface Forces;* Academic Press: London, 1992; pp 312-319.
57. Jalal, I.M.; Zografi, G.; Rakshit, A.K.; Gunstone, F.D. *J. Colloid Interface Sci.* **1980**, *76*, 146-156.
58. Dérue, V.; Alexandre, S.; Valleton, J.-M. *Langmuir* **1996**, *12*, 3740-3742.
59. Rolandi, R.; Dante, S.; Gussoni, A.; Leporatti, S.; Maga, L.; Tundo, P. *Langmuir* **1995**, *11*, 3119-3129.
60. Hanley, S.J.; Giasson, J.; Revol, J.-F.; Gray, D.G. *Polymer* **1992**, *33*, 4639-4642.

61. Butt, H.-J.; Guckenberger, R.; Rabe, J.P. *Ultramicroscopy* **1992**, *46*, 375-393.
62. Zenhausern, F.; Adrian, M.; Heggeler-Bordier, B.; Eng, L.M.; Descouts, P. *Scanning* **1992**, *14*, 212-217.
63. Viswanathan, R.; Schwartz, D.; Garnaes, J.; Zasadzinski, J. *Langmuir* **1992**, *8*, 1603-1607.
64. Schwartz, D.K.; Viswanathan, R.; Garnaes, J.; Zasadzinski, J.A. *J. Am. Chem. Soc.* **1993**, *115*, 7374-7380.
65. Wolthaus, L.; Schaper, A.; Möbius, D. *J. Phys. Chem.* **1994**, *98*, 10809-10813.
66. Lyons, J.S.; Furlong, N.; Healy, T.W. *Aust. J. Chem.* **1981**, *34*, 1177-1187.
67. Chi, L.; Fuchs, H.; Johnston, R.; Ringsdorf, H. *J. Vac. Sci. Technol. B* **1994**, *12*, 1967-1972.
68. Hyde, S.T. *Colloids Surfaces A* **1995**, *103*, 227-247.
69. Milner, S.; Joanny, J.-F.; Pincus, P. *Europhys. Lett.* **1989**, *9*, 495-500.
70. Saint-Jalmes, A.; Graner, F.; Gallet, F.; Houchmandzadeh, B. *Europhys. Lett.* **1994**, *28*, 565-571.
71. Vanderlick, T.; Möhwald, H. *J. Phys. Chem.* **1990**, *94*, 886-890.
72. Vierheller, T.R.; Foster, M.D.; Wu, H.; Schmidt, A.; Knoll, W.; Satija, S.; Majkrzak, C.F. *Langmuir* **1996**, *12*, 5156-5164.
73. Györvary, E.; Peltonen, J.; Lindén, M.; Rosenholm, J.B. *Thin Solid Films* **1996**, *284-285*, 368-372.
74. Burnham, N. *J. Vac. Sci. Technol. B* **1994**, *12*, 2219-2221.
75. Chen, G.; Warmack, R.; Huang, A.; Thundat, T. *J. Appl. Phys.* **1995**, *78*, 1465-1469.
76. Overney, R.; Bonner, T.; Meyer, E.; Rüetschi, M.; Lüthi, R.; Howald, L.; Frommer, J.; Günterodt, H.-J.; Fujihira, M.; Takano, H. *J. Vac. Sci. Technol. B* **1994**, *12*, 1973-1976.

PROBING OF LOCAL SURFACE PROPERTIES OF POLYMERS

Chapter 15

Quantitative Probing in Atomic Force Microscopy of polymer Surfaces

Valery N. Bliznyuk, John L. Hazel, John Wu, and Vladimir V. Tsukruk[1]

College of Egineering and Applied Sciences, Western Michigan University, Kalmazoo, MI 49008

Quantitative measurements of friction, elastic, and shearing behavior on a sub-micron scale and *"multidimensional"* characterization of surface properties are crucial for studying multicomponent polymer systems. We discuss the latest developments in quantitative characterization of local surface properties by scanning probe microscopy. First, we analyze calibration of cantilever spring constants. Second, we focus on approaches for estimation of tip/surface contact area that is vital for calculation of specific parameters such as surface energy and elastic compliance. Third, we discuss a chemical modification of the tips for chemical force microscopy. Recent endeavors have resulted in substantial progress in all these fields. We are moving towards a new level of surface nanoprobing when a poorly characterized tool with unknown shape, mechanical parameters, and surface chemistry will be replaced by a well defined nanoprobe that allows quantitative nanoprobing of surface properties.

Quantitative measurements of friction, elastic, and shearing behavior on a sub-micron scale is crucial for studying surface properties of multicomponent polymers systems *(1-3)*. Promising results have been recently obtained using combined atomic force (AFM), chemical force (CFM), and friction force (FFM) microscopy techniques. A major advantage of these methods is the possibility of local testing of the surface physical properties in relation to the surface's topography and chemical composition. Apparently, the focus of SPM studies on polymer surfaces is gradually changing towards *quantification* of the surface measurements and a *"multidimensional"* characterization of surfaces. Just a few years ago, almost all publications in this field discussed polymer surface topography and microstructure. In contrast, now, more

[1]Corresponding authors. Fax: 616–387–6517; e-mail: vladimir@wmich.edu.

and more papers discuss surface topography in conjunction with friction properties, elastic behavior, adhesion, chemical composition, viscoelastic properties, conductive state, and thermal transformations. Such "multidimensional" collection of experimental data opens exciting opportunities for readdressing the traditional task of "structure-property relationships" of polymer surfaces on the molecular/submicron levels.

An example of "multidimensional mapping" of surface properties (topography, friction, adhesion, and compliance) is demonstrated for a composite polymer molecular layer in Figure 1 *(4)*. Langmuir film of polynaphthoylene benzemidazole/stearic acid complexes deposited on silicon wafer shows domain morphology with clearly visible polymer domains of 2 nm thick separated by interdomain boundaries (silicon oxide) on topographical mode (Figure 1a). Other scanning modes reveal lower friction properties, lower stiffness, and lower adhesion for the domains as compared to the silicon surface (Figure 1b -1d). Such concurrent information is invaluable for intelligent design of molecular films with controlled micromechanical properties *(4)*.

Currently, researchers use the third generation of scanning probe microscopes (SPM) with sophisticated modes and options which were not available on preceding versions. A general scheme of any SPM apparatus includes several major blocks which provide various functions: precise 3D movements of either a sample or a cantilever (piezoelements); SPM tip deflection (detection scheme); microfabricated probe (cantilever/tip); control of scanning parameters (electronic feedback); on-line and off-line analysis of 3D images (analyzing software) (Figure 2). Various aspects of reliable collection of experimental data are critical for quantitative measurements of surface properties: calibration, non-linearity, and creep of piezoelements; flexibility of tunneling, optical, and electric detection schemes; different designs of cantilevers (parallel beam, V-shaped) and tips (growth, aspect ratio, sharpness); different versions of electronic feedback/detection schemes (constant height, constant deflection, amplitude, phase) and modes of operation (contact, dynamic, lateral); on-line corrections (filters, gains, scanning parameters) and off-line analysis (sections, roughness, FFT, autocorrelation). Detailed recent reviews on these subjects can be found in the literature *(5-10)*.

In the present communication, we focus on one crucial part of this complex SPM scheme: the cantilever with integrated nanoprobe for correct measurements of surface responses (Figure 2). We review the current level of understanding of this critical component of the SPM instrument and some of the latest developments in quantitative characterization of local surface properties with emphasis on composite polymer molecular films. First, we analyze different approaches in calibration of cantilever spring constants that are used to convert tip deflections to a force scale. Second, we focus on estimation of contact area tip/surface that is vital for calculation of specific parameters such as surface energy or elastic compliance. This is especially important for compliant polymer surfaces where elastic deformation can be very substantial. Third, we discuss chemical modification of the tip surface and study of interactions between well defined surface chemical groups. Experimental results are illustrated by several examples of such studies undertaken in our research group.

Calibration of cantilever spring parameters

Knowledge of spring constants of the SPM cantilevers is critical for determination of absolute values of vertical and torsional forces acting on a tip. In linear approximation, conversion of cantilever deflections, Δx, to a force scale, $F = k_x \Delta x$, requires knowledge of k_x with a high precision. However, due to uncertain spring

Figure 1. Multidimensional approach in SPM representing concurrent imaging of composite monolayer of ladder polymer/stearic acid in different scanning regimes: by atomic force (a), friction force (b), force modulation (c), and adhesive force (d) modes *(4)*.

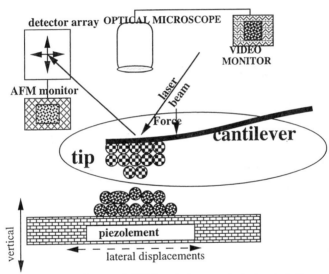

Figure 2. A general scheme of SPM technique and designated focus of this paper.

parameters, forces resulting from the tip deflections are not so easily determined and, therefore, *quantitative* force measurements are still challenging.

The absolute values of normal and lateral forces might be within ± 100% of nominal values quoted by a manufacturer *(11, 12)*. The V-shaped cantilevers are the most popular in SPM because of stable scanning (Figure 3). Simple "diving board" cantilevers (Figure 3) are also available which allows use of standard analytical solutions *(12)*. However, this type of cantilever does not provide stable scanning for variety of surfaces. Several experimental, analytical, and computational approaches were recently explored for estimation of normal (k_n) spring constants of the SPM cantilevers but only a few papers considered estimations of torsional (k_t) spring constants.

The first approach for calibration of the V-shaped cantilevers is a direct experimental measurement of a variation of resonant frequency for a cantilever loaded by microspheres *(12)*. Heavy microspheres (e.g., tungsten) are attached to the end of the cantilever and resonant frequency is measured as a function of applied mass. Extrapolation to a zero mass gives "effective mass" of the cantilever that can be used for evaluation of spring constants. Measurement of resonant frequency for unloaded cantilevers provides means for estimation of actual spring constants if calibration curves are available *(13)*.

However, accumulation of various uncertainties like the errors in microsphere weight or position might lead to a possible error of at least 50% in k_n. Alternative determination of the vertical spring constant by measurement of deflections between a cantilever and a wire or a pendulum results in large uncertainty as well *(14)*. Moreover, these approaches do not solve the problem completely. In fact, to calculate k_t from known k_n one has to use some analytical approximation anyway.

Second approach is an analytical one that uses simple theoretical equations (such as those derived from a parallel beam approximation (PBA)) for estimation of both k_n and k_t *(15)*. This scheme works nicely for simple parallel beam cantilevers where an exact solution for the spring constant exists *(10)*. However, analytical estimates for a "workhorse", V-shaped cantilevers, may vary widely due to different treatments of the models (e. g., replacement of the complicated V-shaped beam by a parallel beam) and are rarely confirmed by independent measurements. In addition, all analytical approaches use a "best guess" for material properties of microfabricated ceramic cantilevers. This raises additional concerns because the properties of ceramic materials with thickness below 1 μm (such as Young's modulus, E,) can be greatly influenced by surface effects and stoichiometry rendering bulk values grossly in error *(16)*. On the other hand, important geometrical dimensions of the microfabricated cantilevers (typical length of 100 - 200 μm, width of 10 - 30 μm, and thickness below 1 μm) are taken frequently from a manual despite possible variations of geometrical sizes within ± 5% from set to set.

Third, the most recent approach, relies on computer modeling of the SPM cantilever actual geometry by a finite element analysis (FEA) and numerical calculation of mechanical parameters for the V-shaped beams *(17)*. This approach can provide reliable values of spring constants for the SPM cantilevers. However, past FEA studies have not adequately modeled the composite nature of the SPM cantilevers (ceramics/gold) although they have a gold coating that significantly changes the mass and stiffness of the cantilever. In addition, experimental verification of material properties such as Young's modulus has not been implemented.

We briefly summarize our version of the later approach that helps to produce the most reliable solution. A complete description of the FEA calculation results can be found elsewhere *(11)*. Tip dimensions were measured by high magnification optical microscopy, and cantilever thickness, gold film thickness, and tip location and height were obtained from scanning electron microscopic and SPM images. Geometrical dimensions measured on several cantilevers randomly selected within the same wafer have been shown to be quite uniform. Cantilevers in our particular wafer have an overall thickness, t, of 0.60 ± 0.01 μm and a gold coating thickness of 47 ± 7 nm.

The change in stiffness due to the Au overlay was determined by analysis of the equivalent transformed section *(19)*. Then, we measured the lowest resonant frequency of real cantilevers and then adjusted the equivalent Young's modulus for the homogeneous plate elements so that the FEA model had a calculated resonant frequency matching the actual values. Our results for different types of cantilevers gave a Young's modulus for Si_3N_4 of 215 - 245 GPa for density of 3.2 g/cm³ (185 - 215 GPa for density of 3.0 g/cm³). These values are within the range of 135 to 305 GPa reported for different stoichiometry of Si_3N_4*(16)*.

After the value of Young's modulus had been found, the spring constant k_n was determined by FEA simulations with loads applied at the tip of the cantilever model. The torsional spring constant k_t was calculated from inclinations of the tip base plane caused by application of lateral forces to the end of the tip. We compared FEA-determined spring constants for parallel beam, ideal triangular, and real V-shaped (Digital Instruments, DI) cantilevers (Figure 3) to test the validity of various simplifications of the actual cantilever geometry. We also tested our FEA modeling against the exact analytical solution for the PBA (diving board) model. The FEA model gave vertical spring constant 0.279 N/m (for density of 3.2 g/cm³) that was fairly close to the exact solution *(12)* k_n = Et^3w/L^3 = 0.275 N/m (see Figure 3 for designations). Two analytical parallel beam approximations *(15, 20)* used for evaluation of triangular cantilevers gave k_n of 0.220 and 0.189 N/m that is on 6% - 25% below FEA results for either ideal triangular or real V-shaped cantilevers.

Estimation of a torsion spring constant k_t according to an analytical approach proposed in Ref. 21 gave ratio k_n/k_t = 4.0 x 10^{-9} that underestimates the torsion constant by about 45%. Modification of this approach that includes the supporting beam's connection at the tip resulted in different mode of bending deformation gave a new expression for k_n/k_t (for detailed discussion see Ref. 11):

$$\frac{k_t}{k_n} = \frac{4}{3} \frac{L^2}{\left[\cos^2\theta + 2(1+v)\sin^2\theta\right]}$$

that is close to the exact FEA solution (within 10%).

Example. Large differences in the friction force were observed between the Langmuir monolayers of fatty acids and underlying silicon oxide in several reports *(22-24)*. All results here and below were obtained on the Nanoscope III microscope (Digital Instruments). To quantify this difference we measured a friction loop, measured loading curves, and calculated friction coefficients using vertical and torsional spring constants discussed above for silicon wafers, stearic acid monolayer, and cadmium salt of stearic acid monolayers (Figure 3) *(23)*. As is clear from these data, the friction forces increase almost linearly with load and are much higher for silicon surface than for Langmuir monolayers. In addition, cadmium salt Langmuir monolayer possesses the lowest friction coefficient. Friction coefficient μ for the Langmuir monolayer is in the range 0.01 - 0.05 but reaches 0.2 for silicon oxide surface *(23)*.

Contact area and mechanical parameters

Knowledge of the tip end shape is critical for recognition of imaging artifacts that are produced by broken, asymmetric, blunt, or double SPM tips. On the other hand, estimation of tip radius is required for calculation of surface compliance, J, and surface energy, E, through work of adhesion, W, *(8)*. Importance of this information can be illustrated by variation of a formula for work of adhesion derived from pull-off forces, ΔF, according to: $W_{ad} = k\Delta F/3\pi R_c$ *(25, 26)*.

Numerical coefficient k in this equation varies from 0 to 2 depending upon the type of physical contact. For three major mechanical contacts, namely, Hertzian, JKR, and DMT, the values of k are presented in Figure 4 *(8, 25, 26)*. Different scenarios of mechanical contact can take place depending upon parameters of interacting surfaces (tip and substrate in SPM experiment) such as tip radius, adhesive forces, and elastic modulus. Hertzian behavior is valid for high loads and low surface forces. The JKR approach works well for high adhesive forces, large radii, and soft materials. Finally, DMT theory is a good approximation for hard surfaces, low adhesion, and small radii *(8, 25, 26)*.

Because it is not clear what situation corresponds to a particular mating pair, a proof of applicability of one of the approaches should be provided. It can be done by studying variation of friction response, F_f, or contact area, A, versus normal load, F_n : F_f (F_n) or A (F_n) *(8, 25)*. Variation of $F_f \propto F_n{}^n + F_o$ where n $\approx$ 2/3 (spherical approximation) and a non-zero intercept ($F_o \gg 0$) are an indication of the JKR behavior. To date, such analysis is done in just a few cases and all JKR, DMT, and Hertzian types of the contact behavior were observed for different mating surfaces *(18, 25-27)*.

As an example, we show the non-linear behavior of the F_f (F_n) curves at high loads which is consistently observed for composite molecular films with a rigid component such as chemically tethered fullerenes or ladder polymers *(4, 27)*. The analysis of F_f (F_n) data for Langmuir monolayers shows $F_f \propto F_n{}^{1/2} + F_o$ (~0) (Figure 5a) that corresponds to the Hertzian elastic behavior of a composite monolayer contacting a conical end of the SPM tip. Such behavior can be explained within the Hertzian theory of an elastic contact with finite (and small) adhesion contribution at zero separation.

The estimation of the actual contact area and an absolute value of thickness for soft molecular layers at various normal loads (and, therefore, the depth of the tip penetration) is crucial for the estimation of their elastic and shearing behavior. AFM technique has no means to measure absolute distance between the tip and the probing surface. This problem can be addressed by using intentionally incomplete molecular layers deposited on solid substrates *(23)*. The silicon substrate can be used as a reference level for the calculation of monolayer thickness under shearing stresses introduced by the sliding tip because its Young's modulus is orders of magnitude higher than that of the polymer layer. Having this reference surface, Young's modulus, contact area, and shear strength all can be estimated from the experimental data *(23)*.

To estimate tip radii of curvature, R_c, special reference samples with sharp (as compared to tip shape) features should be used. Several different standard samples have been proposed that include the edges of atomic planes, nanofibres, latex microparticles, whiskers single crystals, and gold nanoparticles *(6, 18, 28)*. Various procedures for extraction of tip shape information have been proposed ranging from "deconvolution" by using mathematical morphology formalism to simple analytical expressions assuming a spherical shape for tip end *(28,29)*. Depending upon application (loads, type of cantilevers) and a spatial scale of interest one can select any of these samples and apply one of the standard procedures for estimation of R_c.

We observed significant discrepancies between the diameter of the smallest gold nanoparticles reported and measured by TEM and their SPM measured height for standard sample of dispersion of gold nanoparticles of 5, 14, and 28 nm in diameter deposited on polylisine spin-coated film according to the procedure described in *(28)*. Usually, the height of nanoparticles was 1 - 2 nm smaller than expected due to the

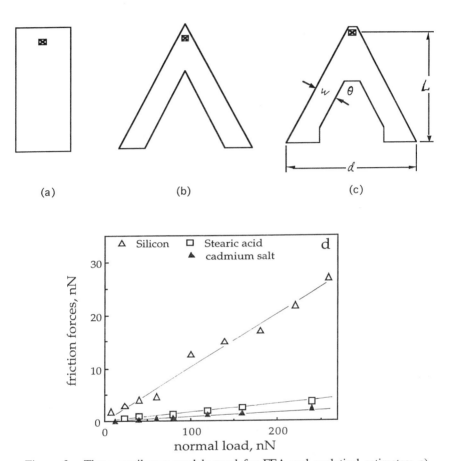

(a)   (b)   (c)

Figure 3. Three cantilevers models used for FEA and analytical estimates: a) parallel beam model; b) "ideal" triangular model; c) V-shaped DI model. Loading curves for silicon oxide surface in comparison with Langmuir monolayers of stearic acid and cadmium salt of stearic acid (d).

contact shapes at pull-off point

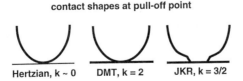

Hertzian, k ~ 0   DMT, k = 2   JKR, k = 3/2

Figure 4. Different version of mechanical contacts and numerical coefficients.

partial penetration of gold nanoparticles in the supporting soft film under local pressure exerted by the SPM tip. Tip contamination by loosely attached polymeric material during scanning is also a significant problem.

To avoid this problem, we used self-assembled monolayer with thiol or amine surface groups as a supporting functional sublayer (Figure 6a). In this way, we provided chemical tethering of gold nanoparticles to a very robust and thin (less than 1 nm) supporting functional sublayer that cannot be damaged by the scanning SPM tip. As a result, we achieved stable scanning in the tapping mode in air and the contact mode in water and unambiguous determination of tip radii. The radii were determined from the image cross-sections by applications of the analytical expression for a spherical tip scanning a spherical asperity: $R_c = W^2/8H$ (Figure 6b) (28). As we observed, tip radii varied widely in the range of 20 - 200 nm from tip to tip from the same batch. One example of AFM images obtained by scanning of a standard sample with tethered gold nanoparticles by a "blunt" tip in comparison with an ideal symmetrical (spherical) tip with a small radius is presented in Figures 6c, 6d.

Example. We observed significant compression of the fatty acid Langmuir monolayers under the SPM tip reaching 65% of initial thickness before irreversible damage occurred (23). The observed compression behavior was related to a collective tilt of the molecules under normal loads due to the formation of gauche conformers in initially extended alkyl chains. The compliance of the monolayers estimated using the Hertzian approximation for the spherical asperity - flat plane contact area was in the range from 0.3 to 0.7 GPa. Two very different types of the deformational behavior were observed for these monolayers (Figure 5b). At low normal loads (< 100 nN) we observed relatively stiff behavior and small deformations (<3%) with high elastic modulus. In contrast, significant elastic deformation (up to 35%) with very high compliance was observed at higher loads.

Adhesive forces and chemical sensing

Chemical sensing requires modification of the probe tip surface by the fabrication of a robust molecular layer firmly tethered to the tip's surface with appropriate terminal groups. The CFM technique allows discrimination of surface forces related to intermolecular interactions of different chemical groups and is promising for chemical sensing of surfaces on the submicron scale $(2, 21, 30\text{-}40)$. "Nanotitration" of functional surfaces and evaluation of local isoelectric points has been recently done for different self-assembled monolayers $(38 - 40)$.

Two different approaches use self-assembly to fabricate thiol or silane monolayers by chemisorption onto a gold coated tip or a bare silicon tip, respectively (Figure 7). To date, several examples of thiol modified tips with surface CH_3, COOH, CH_2OH, CH_2Br, CH_3, and NH_2 groups have been demonstrated $(30\text{-}35, 38)$. Selective chemical interactions between these surfaces were tested in air at various humidities, in water, and in solvents with different pH (see other papers in this volume). General trends were observed that correlate well with estimates made from known surface energies and individual molecular interactions. This behavior was tested for groups of molecules as small as several dozens. Friction measurements on mating chemical groups of different types were made and good correlation was observed between the level of adhesive forces and friction behavior $(21, 32)$. The role of concurrent intermolecular interactions and surface mechanical response on frictional behavior of soft surfaces was discussed recently (36). It was demonstrated that for some monolayers variations of the contact area due to different stiffness of microscale domains composed of different molecules can be responsible for apparent differences in the frictional behavior of elastic surfaces.

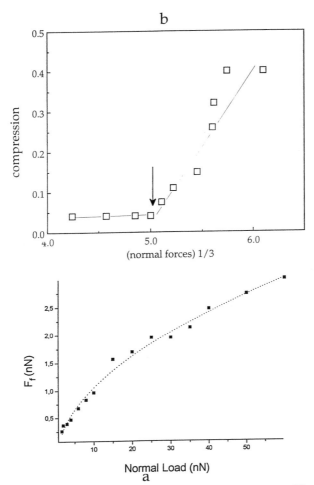

Figure 5. Nonlinear F_f (F_n) for composite monolayer *(27)* and $F^{1/2}$ fit (a) and compression behavior of Langmuir monolayer *(23)* (b).

For some applications involving significant local stresses (nanotribology and nanomechanics), thiol modification is not very suitable in its current design. The major concern is the weak adhesion between the ultrathin gold coating and silicon nitride surfaces that leads to rapid deterioration of the modified tips. In addition, deposition of relatively thick gold layer (50 - 100 nm) may result in substantial worsening of the tip end sharpness. Therefore, a direct modification of tip surface by silane chemistry is thought to be a valuable alternative. Applicability of this approach for tip hydrophobization was demonstrated by several groups very recently (see Refs. 31, 37, and other chapters in this volume).

We explore chemisorption of silane-based molecules with various functional terminal groups to modify surface properties of silicon and silicon nitride tips (Figure 7) *(37, 40)*. Silane compounds used for tip modification form robust uniform self-assembled monolayers (SAMs) chemically tethered to silicon oxide surfaces of Si_3N_4 as a result of adsorption and hydrolization of terminal $Si(Cl)_3$ or $Si(O-C_2H_5)_3$ groups. As chemical compounds for surface modification we use 3-aminopropyltriethoxysilanes (NH_2 terminated nucleophilic surface), 2- (4-chlorosulfonylphenyl) ethyltrimethoxysilane (SO_3 terminated hydrophilic surface), and octadecyltrichlorosilane (CH_3 terminated hydrophobic surface) (Figure 7).

We made quantitative SPM measurements of work of adhesion, adhesion hysteresis, effective "residual forces", and friction coefficients for four different types of modified tips and surfaces (see Figure 7 for chemical structures) *(40)*. Absolute values of work of adhesion between various surfaces, as probed by SPM in normal water, W_{ad}, are in the range of 0.5 to 12 mJ/m^2. Work of adhesion for different surfaces correlates with changes of solid-liquid surface energy estimated from macroscopic contact angle measurements. Friction properties vary with pH in a register with adhesive variation showing a broad maximum at the intermediate pH values for silicon nitride-silicon nitride pair. Similar broad maximum is observed in the acid range for NH_2 terminated surfaces and the basic range for SO_3 modified surfaces. The behavior observed can be understood on the basis of a double-layer theory considering changes of surface charge state determined by the zwitterionic nature of silicon nitride surfaces with multiple isolectric points *(40)*.

Example. Adhesive forces between modified tips and different surfaces can be deduced from pull-off force-distance data according to the well known approach (see above) *(21)*. Forces were measured for 4 different modified surfaces by 4 different modified tips. Elimination of capillary phenomenon by submerging the SPM tip in fluid reveals actual intermolecular forces responsible for tip-surface interactions. Measured adhesive forces are in the range 0.1 - 8 nN for different interacting modified tips and surfaces (Figure 8). A general trend in these variations is that virtually for all modified tips the highest and the lowest adhesive forces are observed for NH_2 (positively charged after protonation in acid conditions) and CH_3 terminated surfaces, respectively.

Work of adhesion for separating the surfaces and tip W_{ad} is estimated from experimentally measured ΔF using JKR model *(25, 26)*. Absolute values of W_{ad} are in the range 0.1 to 12 mJ/m^2 that covers the range estimated for similar surfaces from surface energies at solid-liquid interface and contact angle measurements *(21)*. Despite very different absolute values of work of adhesion, general trends are similar for all modified surfaces if considered only symmetrical surface pairs. We can conclude that the lowest work of adhesion is observed for both hydrophobic and hydrophilic surfaces and the highest W_{ad} is detected for NH_2 and SO_3 terminated surfaces under water.

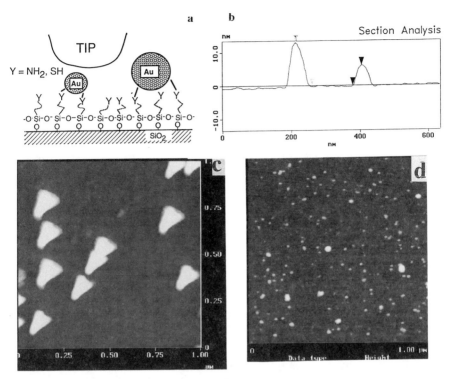

Figure 6. A general scheme of gold nanoparticles tethered to functional SAM (a); b) cross-section of a nanoparticle of 14 nm height; tethered nanoparticles imaged by blunt (c) and sharp (d) tips.

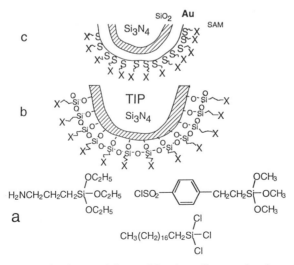

Figure 7. A general scheme of tip modification: silane molecules with different terminal groups (X, Y = Si-OH, NH_2, SO_3, and CH_3) (a); silanazation of a silicon nitride tip (b); and thiol-based modification of a gold coated tip (c).

264

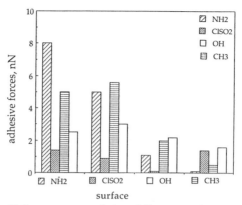

Figure 8. Pull-off forces measured for different surfaces (designated on a horizontal axis) for different tips (designated by various column patterns).

Conclusions

We discussed the current understanding of the SPM cantilever design for correct measurements of surface response and some of the latest developments in quantitative characterization of local surface properties. Different approaches in calibration of cantilever spring constants, estimation of contact tip/surface area, and study of interactions between well defined surface chemical groups were summarized and illustrated by the authors examples.

Recent efforts of a number of research groups resulted in substantial progress in all fields discussed above. SPM technique is moving very fast towards a new level of surface nanoprobing when instead of a poorly characterized and understood tool with unknown shape, mechanical parameters, and surface chemistry we will have a nicely designed nanoprobe with a wide range of controllable properties. This allows quantitative characterization of polymer surface properties on a submicron scale and opens a door for a unambiguous nanomechanical testing of surface properties.

Acknowledgments

Support for this research from The Surface Engineering and Tribology Program, The National Science Foundation, CMS-94-09431 Grant, Ford Motor Co., and Durametallic Corp. is gratefully acknowledged. The authors thank Dr. T. Bunning for SEM measurements and R. F. Lupardo (SAS) for donation of silicon wafers.

References

1. Bhushan, B.; Israelachvili, J.; Landman, U. *Nature* **1995,** *374,* 607.
2. Frommer, L. *Angew. Chem. Int. Ed. Eng.* **1992,** *31,* 1298.
3. Hues, S. M., Colton, R. J., Meyer, E. and Guntherodt, H.-J. *MRS Bull.* **1993,** *1,* 41.
4. Tsukruk, V. V., *Rubber Chem.&Techn.,* **1997,** accepted; Bliznyuk, V. N.; Everson., M., Tsukruk, V. V.; *J. Tribology,* **1996,** submitted.
5. Sarid, D. *Scanning Force Microscopy,* Oxford University Press, New York, 1991.
6. Magonov, S.; Whangbo, M.-H.; *Surface Analysis with STM and AFM,* VCH, Weinheim, 1996.
7. *Dimension^{TM} 3000 Scanning Probe Microscope Instruction Manual,* Digital Instruments Inc., 1995.
8. Burnham, N. A.; Colton, R. J.; Pollock, H. M. *Nanotechnology* **1993,** *4,* 64.

9. Aim, J. P.; Elkaakour, Z.; Gauthier, S.; Mitchel, D.; Bouhacina, T.; Curely, J. *Surface Science,* **1995**, *329,* 149.
10. Overney, R. M.; Meyer, E. *MRS Bull.* **1993**, *28(5),* 26.
11. Hazel, J.; Bliznyuk, V. N.; Tsukruk, V. V. *J. Tribology,* **1996**, in press.
12. Cleveland, J. P.; Manne, S.; Bocek, D.; Hansma, P. K. *Rev. Sci. Instrum.* **1994**, *64 (2),* 403.
13. Cleveland, J. P., personal communication.
14. Sader, J. E.; Larson, I.; Mulvaney, P.; White, L. R. *Rev. Sci. Instrum.* **1995**, *66 (7),* 3789.
15. Albrecht, T. R.; Akamine, S.; Carver, T. E.; Quate, C. F. *J. Vac. Sci. Technology.* **1990**, *A 8,* 3386.
16. Elliot, D. *Microlithography: Process Technology for IC fabrication,* Halliday Lith., NY, 1986.
17. Sader, J. E. *Rev. Sci. Inst.* **1995**, *66 (9),* 4583.
18. Carpick, R. W.; Agrait, N.; Ogletree, D. F.; Salmeron, M. *Langmuir,* **1996**, *12,* 3334.
19. Beer, F. P.; Johnson Jr., E. R. *Mechanics of Materials,* McGraw Hill, NY, 1981.
20. Butt, H. J.; Siedle, P.; Seifert, K.; Fendler, K.; Seeger, T.; Bamberg, E.; Weisenhorn, A. L.; Goldie, K.; Engle, A. *J. Microscopy* **1993**, *169 (1),* 75.
21. Noy, A.; Frisbie, C. D.; Rozsnyai, L. F.; Wrighton, M. S.; Lieber, C. M. *J. Am. Chem. Soc.* **1995**, *117,* 7943.
22. Lüthi, R.; Meyer, E.; Haefke, H.; Howald, L.; Gutmannsbauer, W.; Güntherodt, H.-L. *Science* **1994**, *266,* 1979.
23. Tsukruk, V. V.; Bliznyuk, V. N.; Hazel, J.; Visser, D.; Everson, M. P. *Langmuir* **1996**, *12,* 4840.
24. Tsukruk, V. V.; Bliznyuk, V. N.; Visser, D.; Hazel, J. *Tribology Letters* **1996**, *2,* 71.
25. Johnson, K. L.; Kendal, K.; Roberts, A. D. *Proc. R. Soc., London,* **1971**, *A324,* 301.
26. Derjaguin, B. V.; Muller, V. M.; Toporov, Y. P. *J. Coll. Interface Sci.* **1975**, *53,* 314.
27. Tsukruk, V. V.; Everson, M. P.; Lander, L. M.; Brittain, W. J. *Langmuir* **1996**, *12,* 3905.
28. Vesenka, J; Manne, S; Giberson, R; Marsh, T; Henderson, E, *Biophysical Journal* **1993**, *65 (9),* 992.
29. Goh, M. G.; Juhue, D.; Leung, O. M.; Wang, Y.; Winnik, M. A. *Langmuir* **1993**, *9,* 1319.
30. Frisbie, C. D.; Rozsnyai, L. W.; Noy, A.; Wrington, M. S.; Lieber, C. M. *Science* **1994**, *265,* 2071.
31. Nakagawa, T.; Ogawa, K; Kurumizawa, T.; Ozaki, S. *Jpn. J. Appl. Phys.* **1993**, *32,* L294.
32. Green, J.-B.; McDermott, M. T.; Porter, M. C.; Siperko, L. M. *J. Phys. Chem.* **1995**, *99,* 10960.
33. Berger, C. E.; van der Werf, K. O.; Kooyman, R. P.; de Grooth, B. G.; Greeve, J. *Langmuir* **1995**, *11,* 4188.
34. Sinniah, S. K.; Steel, A. B.; Miller, C. J.; Reutt-Robey, J. E. *J. Amer. Chem. Soc.* **1996**, *118,* 8925.
35. Frommer, J. E. *Thin Solid Films* **1996**, *273,* 112.
36. Barger, W.; Koleske, D.; Feldman, K.; Colton, R. *Polymer Prepr.* **1996**, *37(2),* 606.
37. Tsukruk, V. V.; Bliznyuk, V. N.; Wu, J.; Visser, D. *Polymer Prepr.* **1996**, *37 (2),* 575.
38. Vezenov, D. V.; Noy, A.; Frisbie, C. D.; Rozsnyai, L. F.; Lieber, C. M. *J. Am. Chem. Soc.* **1997**, *119,* 2006.
39. Noy, A.; Vezenov, D. V.; Lieber, C. M. *Annu. Rev. Mater. Sci.* **1997**, *27,* 381.
40. Tsukruk, V. N., Bliznyuk, V. N., *Langmuir,* **1997**, in press.

Chapter 16

Stretching a Network of Entangled Polymer Chains with a Nanotip

J. P. Aimé and S. Gauthier

CPMOH Université Bordeaux I, 351 Cours de la Libération, F–33405 Talence Cedex, France

Abstract.
Uniaxial extension of a network of entangled linear polymer chains (PDMS) is performed with an AFM. We show that whatever the shape, the size and the fragility of the network created between the tip and the silica grafted surface, the elastic responses always have the same structure just before the point at which the rupture happens. The molecular weight dependence is also investigated. The elastic responses are very well described with the tube model when the inextensibility is taken into account. Also we show that the capillary forces do have a sizeable effect at a small deformation when our estimation of the number of parallel chains contained in the network is around a hundred.

1-Introduction.

Beyond scanning a surface with a nanotip to assess the structure of a surface at the nanometre scale, AFM is also used to probe the sample properties. From that point of view the force - displacement curves and friction measurements are of particular interest to characterize surface modifications [1-8]. Several questions arise about the use of the theory which can be useful in describing the contact between a nanotip and a sample either soft or hard. The contact mechanic between two elastic solids[9,10], in which the two materials are considered as continuum elastic medium, is often used [3, 11, 12, 13]. But up to now there is no clear experimental evidence that these theories can be called up to interpret most of the situations encountered with an AFM. Dissipation and plastic deformation may also occur, which in many cases, make those theories less useful. Besides, a key difficulty with AFM is that the area of contact is an unknown parameter. Therefore, to interpret the experimental results, several assumptions have to be made which are not easy to verify. Nevertheless, there are ways that can be used to overcome this main difficulty, either because an accurate knowledge of the contact area between the tip and the sample is not required or because a comparison between several kinds of measurements on different samples allows the experimentalist to extract the relevant variations. One approach that has made progress in removing some uncertainty at the tip-sample interface is increased control over the chemical nature of the two interacting surface species [14, 15]. Another tact that minimizes the absence of knowledge on the contact area, is to investigate the influence of the tip velocity on the tribological behavior of grafted surfaces [13, 16].

Another key issue is the development of the force microscopes as biological sensors. For this purpose, the force-displacement curves are used as a way to quantify the interaction between host and guest sites [17, 18]. In that case histograms of rupture force are compiled giving the binding energy of the two interacting sites.

Following this aim one can expect to probe the elastic properties at the submicrometre scale and particularly for the case of polymer sample to assess the elastic behavior at the point at which a fracture happens in the network. The present paper aims to describe complex force-displacement curves in which a network of entangled linear polymer chain is involved. Because of the remarkable robustness and reproducibility of the elastic response before the rupture happens and the ability of the tube model to describe the elastic response, the whole force curve preceding the second rupture is interpreted within the framework of rubber elasticity. In the first part experimental results and treatment of the data are described, while the second part is devoted to a discussion in which the tube model introduced by S. F. Edwards takes a central part [19, 20].

2- Experimental results

The preparation of the silica surface and of the silica surfaces grafted with polydimethylsiloxanes (PDMS) is given in detail in preceding papers [16, 21]. We have chosen PDMS chains that have one function at both ends such that the chains can only be grafted at one or two locations, making for the latter case a bridge onto the silica surface. They are both monofunctional groups which avoids the possibility of creating a polymer network with chemical cross links. The chemical function is either a methoxy function or a silanol function. The molecular weights are M_W=27000 (terminated -OCH_3), M_W = 110000 and M_W = 310000 (terminated -$SiOH$). The pretreated silica wafers, after being hydroxylized in boiling deionized water for 20 min, are dipped in the polydimethylsiloxane solution with a concentration of 2.8 10^{-3} mol/l (M_W = 27000), 7.1 10^{-4} mol/l (M_W = 110000) and 2.2 10^{-4} mol/l (M_W = 310000). After the reaction, the treated wafers were rinsed with CH_2Cl_2.

With a Nanoscope III [22], the loading-unloading curve is obtained by monitoring the vertical displacement of the sample. The force-displacement curves obtained with this apparatus are widely described in the recent literature and will not be recalled here [1-3].

268

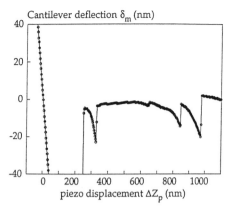

Figure 1 : Example of multiple rupture obtained with PDMS grafted onto silica.

The initial purpose of these curves was to calibrate the cantilever deflection using a known piezo vertical displacement [22]. As a consequence, only a limited number of data were available (no more than 512) with a restricted sampling time (100μs). To overcome these constraints, we built an electronic device which makes elementary algebraic operations. The signal is taken at the output voltage of the photodiode then is recorded on an oscilloscope (Fluke PM801). The modification allows us to have up to 8000 data points.

In figures 1 and 2 the force-displacement curves are displayed which show a complex behavior with several elastic responses and multiple ruptures. These multiple ruptures are never observed when hard surfaces, silica, mica, semiconductors or amorphous polymers are investigated with the nanotip. In this section we shall first give a qualitative description of the force curves, then we describe in detail the data treatment.

The first instability occurs when the vertical displacement of the piezoelectric actuator needed to unstick the tip from the surface is reached. By multiplying the vertical displacement with the announced cantilever stiffness [23], which may not be the correct one [21, 24], one gets the order of the magnitude of the pull-off force. The magnitude of the pull-off force is sensitive to chemical species [14] and to dissipation processes that might occur during the rupture of the contact [3, 10]. After few oscillations [21], the cantilever goes back to its equilibrium position at rest and should keep it when the sample moves further downwards. No further deflection of the cantilever is expected, because the tip-sample distance is greater than a hundred nanometers and the attractive interaction becomes negligible. Typically for a radius of the tip of 40 nm and an Hamaker constant of 10^{-20} Joule, the dispersive Van der Waals attraction is less than the thermal energy k_BT [25].

Two other possibilities must be considered when material makes a bridge between the tip and the grafted surface. One is when a perfect plastic flow takes place, in that case even if a material remains stuck at the tip we don't detect any mechanical response of the cantilever. The other is when the elastic force is not large enough to be detected by the microlever. When the tip is far away from the sample, the cantilever again bends downward which means that polymers remain grafted at the tip apex whatever the hypothesis we use to describe the part where no mechanical response is detected.

The question of an accurate determination of the domain size of the vertical displacement during which again an elastic response is observed is the key point to properly describe the force-displacement curve. The measure of the domain size depends on the sensitivity of the measurement. The criterion we have used is the following, the mechanical response becomes detectable and significant as soon as we are able to measure a slope of 3‰.. Over a piezo vertical displacement of 20 nm, this criterion leads, with an announced cantilever stiffness of 0.58 Nm^{-1} [23], to a sensitivity of 40 pN. The analysis of the data is focused on this part of the force-displacement curve, keeping in mind that the size of the domain during which an elastic response is observed is driven by the criterion described above. Figure 3 gives a sketch of a force-displacement curve defining the different lengths used in the paper.

In figure 2 are reported the typical shapes of the curve observed. The elastic response is non linear as shown with the increase of the slope as a function of the vertical piezo electric displacement. Before going further into the description of the data, it is worth recalling a few results coming from the framework of the continuum theory that describes the contact between two elastic solids in presence of adhesion. Following the usual analysis of a force - displacement curve, in the linear regime the relationship between the cantilever deflection δ_m and the vertical piezoelectric displacement ΔZ_p is given by [3]:

$$\delta_m = \frac{1}{1 + \dfrac{k_m}{k_s}} \Delta Z_p \qquad (1)$$

where k_m is the cantilever stiffness, and k_s the contact stiffness of the sample. For a stiffness k_s much larger than k_m the slope is close to one, while for a sample stiffness close to that of the cantilever the slope is equal to 0.5. In addition for a slope smaller or equal to 0.5, close to the instability, a non linear behavior is anticipated and the force-displacement curve ends with a rounded shape indicating that the contact stiffness decreases (figure 4). Here the behavior is non linear but with a convex curvature indicating an increase of the slope as ΔZ_p increases, that is just the opposite of what is predicted with the theory describing the contact of a tip with a semi infinite elastic medium.

As a first attempt to better understand the curves, a power law is used to fit the relationship between the cantilever deflection and the elastic displacement of the polymer :

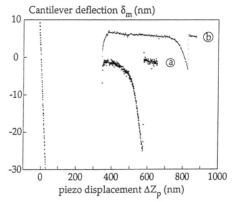

Figure 2 : Experimental force curves showing second ruptures which happen for different vertical displacements and different forces of rupture. For clarity the curve (b) is shifted upward.

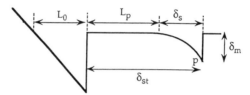

Figure 3: Sketch of a force - displacement curve showing the mean of the different parameters used : L_0 is the length necessary to pull the tip out of the surface; L_p is the length where no elastic response occurs or is detectable; δ_s is the length on which we measure an elastic response; $\delta_{st} = L_p + \delta_s$; p is the slope at the very end of the elastic response; δ_m is the cantilever deflection giving the force of rupture $f_{rupt} = 0.58 \ast \delta_m$.

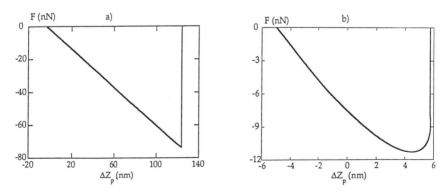

Figure 4 : Computed force curves. a) The stiffness of the sample k_s is larger than the one of the cantilever k_m, b) the stiffnesses k_m and k_s are approximately the same [3].

$$\delta_m = A\,\delta_s{}^\alpha \tag{2}$$

where the elastic displacement of the sample is normally given by $\delta_s = \Delta Z_p - \delta_m$ if the tip is attached at a spring. Because the spring is a microlever built in at one end, any vertical motion of the piezo actuator leads also to a lateral motion of the tip and the elastic displacement of the neck is given by $\delta_s = \Delta Z_p \backslash r\,(1 + \tan^2(\theta)) - \delta_m$, where θ is the angle made between the cantilever beam and the horizontal plane. For $\theta = 12°$, this makes a difference of 2.5% with the previous relation.

The values of the exponents α obtained for two molecular weights are reported in figure 5. They are centred around the exponent 3 with scattered values on the prefactor A. Therefore, we reduce the number of parameters and fix the exponent equal to 3. Typical results are reported in figures 6. The law with an exponent 3 describes quite well the elastic responses of the large molecular weights ($M_w = 110000$ and 310000) and reasonably well the ones obtained with the lower mass ($M_w=27000$). This result indicates that whatever the shape, the fragility and the length of the neck between the surface and the tip, the elastic responses remain identical. This can be shown in a more spectacular way by interpreting the non-linear behavior through consideration about change of the effective sample stiffness. The dimensional analysis tells us that A is the product of the reciprocal of a square length multiplied by the ratio of the stiffness of the neck and of the cantilever stiffness. Therefore it might be interesting to consider an effective stiffness expressed explicitly as a function of the elastic elongation of the polymer neck. We rewrite equation 2 as $k_m\,\delta_m = k_s(\delta_s)\,\delta_s$ with $k_s(\delta_s)$ given by $k_s(\delta_s)= k_s{}^0\,\delta_s{}^2$, where $k_s{}^0$ is the stiffness of the neck at the beginning of the measurable elastic response. Following this procedure, equation 1 can be used in introducing the measured non linear variation of the sample stiffness:

$$\delta_m = \cfrac{1}{1 + \cfrac{k_m}{k_s{}^0 \delta_s{}^2}}\,\Delta Z_p \tag{3}$$

or :

$$\delta_m{}' = \cfrac{1}{1 + \cfrac{1}{(\Delta Z_p{}' - \delta_m{}')^2}}\,\Delta Z_p{}' \tag{4}$$

where $\delta_m{}' = \sqrt{A}\,\delta_m$ and $\Delta Z_p{}' = \sqrt{A}\,\Delta Z_p$. The coordinate transformation and equation 4 predict that if the elastic response is correctly described with equation 2, the observed elastic responses as a function of the vertical piezoelectric displacement should appear identical. In other words whatever the particular properties of the polymer neck - fragility, stiffness, size ...-, which are not controlled in our experimental procedure, all the curves must be

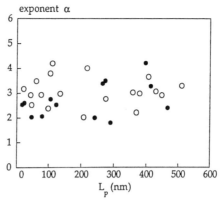

Figure 5 : Exponent α versus the length L_p. (●) M_w=27000 and (○) M_w=110000.

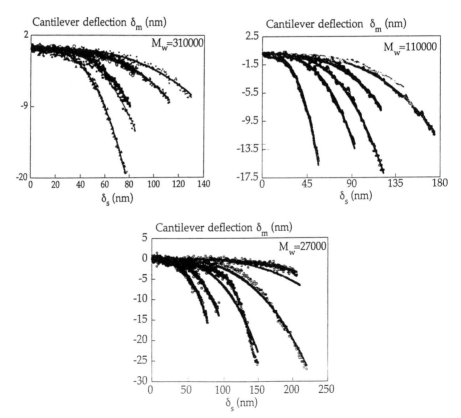

Figure 6 : Typicall ruptures with a measurable elastic response δ_s. For clarity, the curves are translated to zero. The red curves correspond to the fits with a power law with the exponent 3.

identical with the unique difference of the force at which the rupture happens. The results obtained in reduced coordinates $\delta_m' = F(\Delta Z_p')$ are reported in figures 7. The result is striking, it emphasizes that in spite of the very different conditions, the elastic responses are always the same, particularly when the experiments are performed on the large molecular weights. For example, after the tip has left the surface, the vertical displacement varies between 10 to 800 nanometers before the elastic response becomes detectable and, as shown below, this length has an influence on the sample stiffness but does not have any on the curve shape.

The evolution of the prefactor A as a function of the neck length is reported in figure 8. One can already note that, first a drastic decrease is observed as the length of the neck increases, second that the low molecular weight exhibits a more pronounced decrease. Since A contains the stiffness of the polymer network, such a variation suggests that the neck stiffness is decreasing as the length increases and that the way it decreases is molecular weight dependent. A more convincing attempt to extract the behavior of the polymer stiffness is to measure the slope in the quasi linear domain which happens just before the rupture. From a mechanical point of view, such a slope, as shown with the use of the equation 1, directly gives a measure of the stiffness of the neck.

$$k_s = k_m \frac{p}{1 - p} \qquad (5)$$

where p is the slope measured in the linear domain.

The results are given in figure 9a and reported as a function of the relative uniaxial extension λ_{max}. The uniaxial extension is defined as follows: L_0 is the height of the neck after the first instability and λ is given by $\lambda = (L_0 + \delta_{st})/L_0$, where δ_{st} is the sum of the length during which no mechanical response has been detected and the measured elastic deformation of the sample δ_s (figure 3). As for the A factor, a decrease of the stiffness is observed as the uniaxial extension increases. For $M_w = 310000$, k_s decreases slowly while the low molecular weight exhibits a pronounced decrease.

3- Discussion.

After the first instability, the force curve does not exhibit any mechanical response then it presents an elastic one. We shall consider that the first instability acts as a step like function which is applied to the microlever-polymer network system. As for the time analysis of the cantilever relaxation [21], the length L_0 corresponds to the magnitude of the step function. The relative deformation λ is calculated with L_0 as the length of the neck at rest as discussed above. So that we uniquely discuss the force curves as a function of the stretched length L_p measured from the origin up to the point at which the elastic response δ_s is detected and focus the analysis on the shape of the elastic response δ_s. Such an approach appears justified due to the remarkable

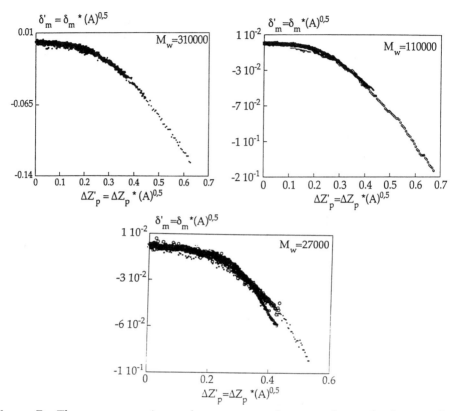

Figure 7 : The same experimental ruptures as the ones shown in figure 6 but displayed with the reduced coordinates (see text and equ. 4).

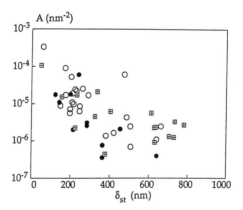

Figure 8 : Prefactor A (equ. 2) versus the neck length δ_{st} for three different weights of PDMS : (●) M_w=27000, (○) M_w=110000 and (⊞) M_w=310000.

robustness and reproducibility of the elastic responses (fig. 6 and 7).

The use of the power law does not bring much physical insight about the way the neck occurs and breaks. The main interest is to show that the elastic response remains identical in spite of the large variation of the polymer properties: wide range of stretched lengths, forces at break that range between 3 nm (1.8 nN) up to 30 nm (17 nN), wide range of neck stiffnesses. One way would then be to compare the elastic response to known models. The P. Pincus's conjecture leads to a non linear behavior [26], which, for polymer in good solvent or more generally for polymer chains that exhibit excluded volume effect, gives a force per chain that scales as $f_c \sim r^{1.5}$. As shown by our attempt to fit with a power law, the elastic behavior of our sample cannot be in agreement with this law. The random walk that describes the configuration of ideal polymers predicts a linear relationship between force and polymer elongation. Such a law cannot obviously match the data, as it corresponds to an asymptotic behavior which is not suitable as soon as the polymer elongation is within the range of the contour length of the polymer. For polymer chains with finite length, or in the domain of uniaxial extension in which the finite length of the macromolecule becomes a relevant parameter, an appropriate trial function is the Langevin function [27]. The Langevin function has been checked and figures 10 shows that the whole shape of the curve cannot be fit that way.

Estimation of the strength of the polymer network

Before going farther let us estimate the strength of the polymer network. To do so, we use the framework of rubber elasticity [27, 28]. A network of random coil chains gives $f_s \sim \Phi^2 N_c k_B T$, where Φ is the diameter of the sample, N_c the number of chains per unit volume, and $k_B T$ the thermal energy 4 10^{-21} J. A usual polymer density is around $\rho \sim 1.1$ g/cm^3. With $M_w = 110000$ we have $N_c \sim 6.02 \; 10^{24}$ per m^3, and with $\Phi=10$ nm a force $f_s \sim 2.4 \; 10^{-12}$ N. The value is rather low when it is compared with the values fitted with the tube model $f_s \le 10^{-10}$ N (see discussion below and figure 9b). To get the force of 10^{-10} N, we need a diameter of 70 nm which is a high value with a radius of the tip $R \sim 40$ nm. For a network of entangled chains, as soon as the entanglements between chains act as chemical cross links, we can replace the expression of the force by $f_s \sim \Phi^2 N_e k_B T$, where N_e is the number of entanglements per unit volume [27, 28, 29]. A critical mass $M_e \sim 10^4$ [30] above which the entanglements become efficient gives N_e ten times higher then N_c. Consequently, the force of 10^{-10} N is reached with a diameter 3 times smaller than the previous one which give a more reasonable value of the neck thickness.

One of our attempts to extract quantitative information was to use the linear domain just before the second instability to assess the stiffness k_s of the polymer network (equation 5). Actually, the quantities deduced are not strictly equivalent to a spring constant as the force-displacement curve exhibits a non-linear behavior, but their variations allow us to get a better idea about the way the polymer network behaves under a high deformation (figure 9a).

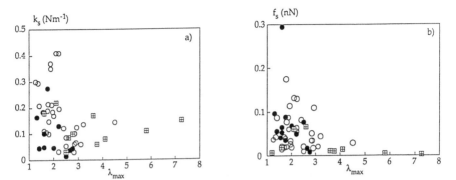

Figure 9 : a) Values of the stiffnesses of the neck k_s, deduced from the slope close to the second rupture (equ. 5), as a function of the maximal relative uniaxial extension λ_{max}. b) Calculated values of the strength of the neck f_s versus λ_{max} when the tube model is used. (●) $M_w=27000$, (○) $M_w=110000$ and (⊞) $M_w=310000$.

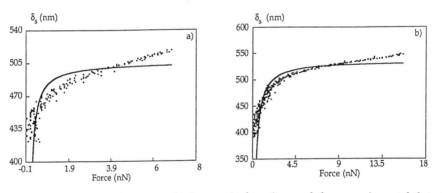

Figure 10 : Comparison between the Langevin function and the experimental data for $M_w=110000$ (fig. 10a) and $M_w=27000$ (fig. 10b). The Langevin function is given by δ_s = L {coth $(\frac{F}{B})$ - $\frac{B}{F}$} where the coefficient L has the dimension of a length (nm) and the coefficient B has the dimension of a force (nN). The fitting parameters used in fig. 10a and fig. 10b are L = 535 nm, B = 0.05 nN and L = 505 nm, B = 0.14 nN respectively.

There remains an important unknown parameter which is the shape of the polymer network between the tip and the grafted surface. Let us discuss the effect in terms of the neck stiffness, the stiffness is the product of an intrinsic property, the elastic modulus $G = N_e k_B T$ of the entangled polymer network, and a geometrical factor. We will discuss the two limiting cases sketched in figure 11:

i) the tip is stuck at a semi-infinite elastic medium. In that case the general theory describing the contact between two elastic solids applies [9, 10] from which the stiffness of the contact is given by $k_s \sim G\Phi$, where Φ is the diameter of the contact area between the tip and the sample. Typically Φ cannot exceed 10 nm for a radius of curvature of the tip apex of about 40 nm, which leads to values of G between 10^6 and 10^8 N m^{-2} for k_s ranging between 10^{-3} and 10^{-1} N m^{-1} (figure 9a). The hypothesis of a tip stuck at a semi infinite elastic medium gives values of G too large for an entangled polymer network at a small deformation, but as the slope is measured at a very high deformation such an assumption may become more acceptable [19, 20, 27]. Nevertheless because of two points, this approach is probably not correct. First there is an implicit contradiction, a linear slope just before the rupture means that the contact area remains constant which is not expected for a stiffness smaller than that of the cantilever (figure 4); second the polymer neck started with a height larger than a hundred nanometers which implies that this parameter must be included in any expression of the sample stiffness.

ii) The polymer network between the tip and the grafted surface is modelled as a cylinder of diameter Φ_1 and height L which gives $k_s \sim \dfrac{G\,\Phi_1^2}{L}$. The polymer neck may have a large thickness near the surface and ended at the tip with a diameter of 10 nm or less, thus an effective diameter Φ_1 around 30 nm can be a reasonable value as suggested with the numerical evaluation of the force. A simple numerical application with L = 200 nm, gives values of G half as large as the ones obtained previously.

Actually the real situation is somewhere between these two limiting cases, but we cannot give a reasonable guess of the shape of the neck and neither do we know to what extent the polymer chains remain stuck at the tip, keeping the contact area between the polymer and the tip constant throughout the uniaxial extension. Therefore any attempt to deduce the magnitude of G depends on the way the expression of k_s has been chosen.

Inextensibility and tube concept.

Our first attempt to describe the elastic response with a known model was to use the Langevin function. The Langevin function has often been used to describe the hardening of rubber at a high deformation [27]. The main physical idea is that hardening is due to the inextensibility of the chain at a high deformation because of the finite length of the polymer chain. The Langevin function fails to describe the elastic response (figures 10) while our attempt to fit with a power law shows the remarkable stability of the elastic behavior at a high deformation.

A somewhat different model to describe the effect of entanglements is the tube model [19, 20, 29] which also produces a singular behavior at a high deformation when inextensibility is considered [19, 20]. The tube model introduces topological constraint without placing it at a specific point. A given chain is forced to have its available configurations within a contorted contour and the structure of the tube gives a parameter equivalent to the average molecular weight M_e between entanglement points (figure 12). Within the framework of the tube model, the maximum of the uniaxial extension is given by the ratio of the tube diameter a and the Kuhn length b of the polymer. The length a is also a measure of the step length of the primitive path of the tube as defined by S. F. Edwards and Th. Vilgis [20] which provides the evaluation of λ_{max}.

$$\lambda_{max} = a/b \qquad\qquad 6$$

The entanglements are described as slip links and, if they are strong enough and act as chemical cross links, this gives a force in uniaxial extension [19, 20]:

$$f_c = N_c k_B T \, D \left[\frac{1 - \alpha^2}{(1 - \alpha^2 \phi)^2} - \frac{\alpha^2}{1 - \alpha^2 \phi} \right] \qquad\qquad 7$$

where N_c is the density of cross links, $\alpha = b/a$ and ϕ and D are given by :

$$\phi = \lambda^2 + \frac{2}{\lambda}$$

$$D = \frac{1}{2} \frac{d\phi}{d\lambda} = \lambda - \frac{1}{\lambda^2}$$

for slip links the leading term of the force is [19]:

$$f_{sl} \sim N_{sl} k_B T \, D \left[\frac{(1 - \alpha^2)(1 + \eta)}{(1 - \alpha^2 \phi)^2} \right] \cdots \qquad\qquad 8$$

where N_{sl} is the density of slip links, and η is a measure of the strength of the slip link. For $\eta = 0$, the entanglement acts as a cross link, while for large η the links slide almost freely and no elastic response is expected due to the entanglements.

From the structure of equation 7 or 8 one can see that the value of λ_{max} at which the rupture happens gives the order of magnitude of the value a/b that describes the topological constraint of the network. This is the reason why

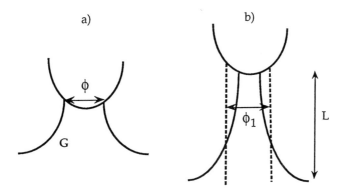

Figure 11 : Sketch of the shape of the polymer neck under the tip. a) The tip is stuck at a semi - infinite elastic medium, G is the elastic modulus of the sample. b) The neck is modelled as a cylinder with an average diameter Φ_1.

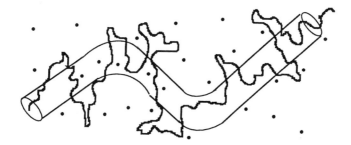

Figure 12 : Sketch of the tube model. The points illustrate the chains perpendicular to a reference chain represented by the solid line. The tube is due to entanglements, so that the polymer can only reptate along the tube [19, 29].

an accurate knowledge of the relative extension must be known. As we do not control the sample preparation, and cannot access the values of λ without making an hypothesis, we followed two kinds of fitting procedures, aiming to extract a general behavior rather than quantitative values. In a first attempt we set arbitrarily that over the total length of the neck, just before the point at which the elastic response becomes detectable, only 10% of the material contributes to the elastic extension. The second fit does not contain any constraint and consider that the polymer network is able to respond elastically throughout the material with an initial length at rest L_0 and $\lambda=(L_0+\delta_{st})/L_0$ (figures 13). In both cases the elastic domains of the network are considered as homogeneous and for the later case if crystallisation occurs, creating domains with a much higher modulus, this is not taken into account. For few cases, when the curves do exhibit a more complex behavior, for example with several ruptures (figure 1), the fitted were not included because λ is more difficult to evaluate.

Whatever the choice of the fit and consequently the value of the uniaxial extension λ_{max} at which the rupture happens, equation 7 is able to reproduce, with a very good agreement, the singular behavior of the curve. The relation between the measured λ_{max} and the deduced value a/b is reported in figure 14. The linear relationship emphasizes the pertinence of the fit procedure, only few data points disagreeing with the linear relation.

The prefactor in equation 7 is a measure of the strength of the neck and scales as $\phi^2 N_e k_B T/k_m \sim f_s/k_m$. The variations of f_s (figure 9b) and of the stiffness k_s (fig 9a) as a function of a/b or λ_{max} show an identical behavior. This is what can be expected as the two parameters are the product of a geometrical parameter and of an elastic modulus, but one is the result of the use of a function describing the whole part of the elastic response, while the second quantity was measured from a mechanical analysis of the curve in a small linear domain.

The key point is to discriminate the respective influence of the intrinsic properties, the elastic modulus of the network, and the geometrical factor. For example, for a cylinder shape, the force should also vary as $1/L$. It is not an easy task to try to separate these two contributions but we note that k_s and f_s vary much more rapidly than $1/\lambda$. In what follows, we shall focus the discussion on the network properties keeping in mind that a change in the geometry of the polymer network may also occur and have an influence on the force.

The decrease of the force as a function of λ_{max} is in good agreement with what we can conclude from the tube model. The topological constraint becomes less and less efficient as a/b increases and consequently the strength of the network decreases. Such analysis is in part supported by the influence of the molecular weight. For $M_w = 27000$, f_s and k_s show a more pronounced decrease and only a few results are obtained for $a/b \geq 3$, while value up to 5 have been obtained for $M_w = 110\,000$ and up to 7 for $M_w = 310000$.

Influence of the force of capillarity

For the highest molecular weight, not only higher deformations are measured but data are also more scattered and do not exhibit the strong decrease observed for the other two molecular weights. In addition, at small λ_{max} the forces exerted by the neck made by the chains of M_w= 310000 are smaller than that of the two other M_w. Typically the strengths either computed with equation 7 or deduced from the stiffness k_s are smaller when the rupture happens at $\lambda_{max} \leq 2$. If for ruptures happening at $\lambda_{max} < 2$ we make the hypothesis that the internal structure of the network are roughly the same whatever the value of M_w, then the difference in forces leads to the conclusion that the number of chains in parallel in the neck are 2 to 3 times less for the highest molecular weight and even a factor of ten if we consider the calculated values f_s (figure 9b). The assumption is reasonable since we expect that the probability of grafting the chains on the tip is molecular weight dependent. Each chain having uniquely one function at each end, the probability of grafting and therefore the diameter of the neck decreases as M_w increases. Such a remark is strongly supported by the shape of the force-displacement curve observed for M_w = 310000. In many cases after the first instability a noticeable mechanical response is observed (figure 13a). This force prevents the cantilever from reaching its position at rest and is interpreted as being due to the surface tension. The effect of the surface tension becomes significant when it balances the internal pressure inside the neck and we have:

$$P \, dV = \gamma \, dS \qquad\qquad 9$$

γ is about k_BT/a_u^2 [25, 31] where a_u is the average size of the molecular unit. The pressure P is given by $k_BT/N_e a_u^3$ [29], where N_e is of order 100. Taking the geometry of a cylinder, equation 9 becomes:

$$\frac{k_BT}{N_e a_u^3} \pi R^2 \, dL = \frac{k_BT}{a_u^2} 2 \pi R \, dL \qquad\qquad 10$$

the number of chains n_c in the section is given by $n_c a_u^2 \sim \phi_1^2 \sim R^2$ so that from equation 10, we obtain an order of magnitude of the number of chains per section below which the capillarity effect becomes measurable:

$$\sqrt{n_c} \sim N_e \qquad\qquad 11$$

with $N_e \sim 135$ (corresponding approximately to $M_e = 10^4$ and a molar mass of the unit m = 74 g) and $a_u = 0.5$ nm we have $n_c \sim 1.8 \ 10^4$. This is a high value compared to the one calculated with the diameter ϕ_1 estimated to get a force of about the 10^{-10} N. The number of chains per section is given by:

$$n_c \sim \left(\frac{\phi_1}{a_u}\right)^2 \qquad\qquad 12$$

282

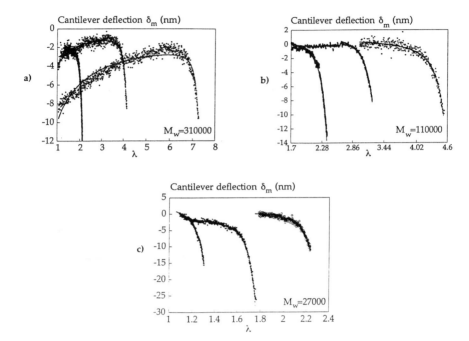

Figure 13 : Elastic responses expressed in function of λ. The theoretical curves, red lines, are calculated with the equation 7 for b) and c). A force of capillarity is added in the case of M_w=310000 (fig. 13a) with $f_{cap} = \Upsilon/\lambda^{0.75}$.

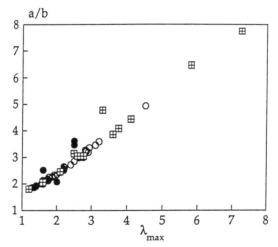

Figure 14 : Computed value of a/b versus λ_{max} (equ. 7). (●) M_w=27000, (○) M_w=110000 and (⊞) M_w=310000.

For most of the experiments performed with M_w = 27000 and 110000, after the first instability, the cantilever reaches a position close to the one at rest. The forces of capillarity can be neglected suggesting that n_c is larger than the value given by the criterion 11. The use of a unit length a_u = 0.5 nm gives a density ρ ~ 1 g/cm^3 which in turn leads to a diameter ϕ_1 ~ 25 nm to reach the strength f_s ~ 10^{-10} N, thus giving n_c ~ 2.5 10^3 which is an order of magnitude less than the value given by the criterion 11. For M_w = 310000, f_s is more likely to be around 10^{-11} N (figure 9b) which gives n_c ~ 2.5 10^2. Therefore as shown with these estimations and the shape of the curve (figure 13a), the situation being completely different for M_w = 310000, it becomes useful to evaluate the way the capillarity force varies as a function of λ.

To do so, again we consider the simple geometry of a cylinder and use the incompressibility rule of the rubber elasticity. Let us note λ_R = R/R0, the incompressibility of the neck gives λ_R = $1/\sqrt{\lambda}$. So that the force of capillarity varies as:

$$f_{cap} \sim \frac{\gamma}{\sqrt{\lambda}} \phi_0 \qquad\qquad 13$$

the total force of the neck is now given by the sum of the contribution of equation 7 and 13 :

$$f_s \sim f_{cap} + f_c \qquad\qquad 14$$

the theoretical variations are shown in figure 15. Two situations have been simulated, one with a strength of f_{cap} which is ten times larger than the strength of f_c and a second with a strength of f_{cap} hundred times larger than the strength of f_c. The former aims to mimic n_c = 10^3 corresponding to M_w = 27000 and M_w = 110000 while the second aims to simulate the case n_c = 100 corresponding to M_w = 310000.

As shown with these simulations even if our estimation of the number of parallel chains in the neck is below the criterion 11, it does appear that when n_c is an order of magnitude lower, the difference between equation 14 and equation 7 is too small to be unambiguously measured. Not only we do not have the sensitivity to discriminate between the two curves but also tiny accidents, for example very small ruptures, can happen which will make the use of equation 14 doubtless.

For the second case, we clearly see a large difference and the simple arguments given above seem to be sufficient to describe the overall shape observed with M_w = 310000. A fit taking into account the capillarity force gives in many cases exponents between 3/4 and 1 rather than 1/2 (figure 13a). As the description of the neck as an homogeneous cylinder is obviously an oversimplified assumption, it is not very surprising to get exponents different than the one deduced from this assumption. Probably change of the external surface of the neck upon stretching taking into account a possible porosity would be more suitable. We may also note that at a high deformation, the capillarity force becomes negligible as the internal pressure goes to the limit $k_BT/a_u{}^3$, this is also the reason why the tube model applies so well at a very high deformation whatever the size of the polymer neck.

To end this discussion it is interesting to take a look at the way the rupture force f_{rupt} varies as a function of the molecular weight and λ_{max} (figure 16). Here again we note the same type of behavior than the one previously shown with k_s and f_s. For $M_w = 310000$, f_{rupt} varies between 1 and 6 nN without showing a net dependence as a function of λ_{max}, while for the $M_w = 27000$ at a small deformation f_{rupt} can be vary from 5 nN up to 15 nN or more and for $\lambda_{max} > 2$ is mostly below 5 nN, the sample of $M_w = 110000$ exhibiting an intermediate behavior with $f_{rupt} > 5$ nN for λ_{max} around or below 2. At this stage it is worth to recall that our experimental determination of λ is set arbitrarily, but as it was always the same whatever the M_w and the length of the force-displacement curve, the relative evolutions become meaningfull. Therefore leaving aside the quantitative value of λ_{max}, the experimental results show that for the high molecular weight, the force breaking the network does not depend of λ_{max}, or equivalently (figure 14) within the framework of the tube model does not depend of the tube structure a/b. It is much less evident for the two other molecular weights for which the rupture force decreases with a/b. Also it is often suggested that at a high deformation a network of entangled polymer chains may exhibit local crystallisation [20, 27, 28]. If it happens in the neck, this clearly does not influence the elastic behavior before the rupture. We do not know at what location the rupture occurs, within the neck, for example in the middle, or either at the interface polymer-surface or polymer-tip. From the force f_{rupt} and the corresponding diameter ϕ, an estimation of the strength of the adhesive energy gives 0.5 J m^{-2} suggesting that the network breaks at the interface rather than in the network itself. In any case, wherever the rupture happens the elastic responses observed remain identical.

Conclusion.

The multiple ruptures observed on the force-displacement curves have been interpreted as fractures of an entangled network of PDMS chains created between the nanotip and the silica grafted surface. The elastic responses observed at a high deformation of the network always exhibit the same behavior, whatever the molecular weight as long as it is larger than the mass needed to have an entangled network, and whatever the shape, the fragility and size of the network. The use of the tube model taking into account the inextensibility of the chains is able to reproduce most of the elastic responses observed, thus plays a central part in the interpretation of the elastic responses. Also the experimental results bring new information about the respective influence of the force of capillarity with respect to that of the entangled network. From that point of view they might be of some help in understanding

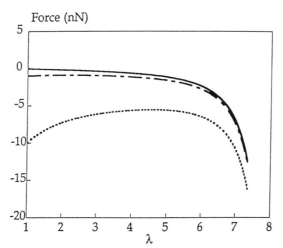

Figure 15 : Theoretical variations of the force versus λ when the Edwards model is used (----) and when a capillarity force f_{cap} is applied. (-- · --) the strength f_{cap} is ten times larger than the strength of the network f_s, (·······) f_{cap} is hundred times larger than the f_s.

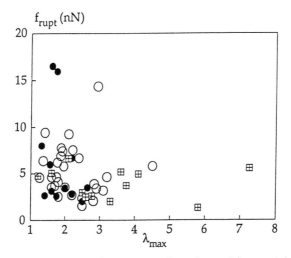

Figure 16 : Measured rupture forces as a function of λ_{max}. (●) M_W=27000, (○) M_W=110000 and (⊞) M_W=310000.

fracture in polymer networks or failure of adhesion when linear polymers are involved [31]. Also the elastic behavior of the PDMS network can provide a useful model for more complex systems such as biological macromolecules since it explicitly shows the entropic contribution of the elastic response.

Acknowledgements:
It is a pleasure to thank Professor M. Fontanille for his collaboration on the general topic "molecules and polymers grafted on silica". We are also very indebted to D. Michel who set the electronic devices and the modifications on the NIII with which the experimental data have been collected.

Literature Cited

[1] Burnham, N.A.; Colton, R. J. *Vac. Sci. Technol. A* **1989,** *4* , 2906.
[2] Salmeron, M. *MRS Bull.* **1993,** *18* , 20,.
[3] Aimé, J. P.; Elkaakour, Z.; Odin, C.; Bouhacina, T.; Michel, D.; Curély, J.; Dautant, A. *J. Appl. Phys.* **1994,** *263* , 1720.
[4] Overney, R. M.; Meyer, E. *M.R.S. Bulletin* **1993,** *18* , 26.
[5] Overney, R. M.; Meyer, E.; Frommer, J.; Brodbeck, D.; Lüthi, R.; Howald, L.; Güntherodt, H. J.; Fujihira, M.; Takano, H.; Gotoh, Y. *Nature* **1992,** *359* , 133.
[6] Radmacher, M.; Tillman, R. W.; Fritz, M.; Gaub, H. E. *Science* **1992,** *257* , 1900.
[7] Singer, L. J. *Vac. Sci. Technol. A* **1994,** *12* , 2605.
[8] Frommer, J. E. *Thin Solid Films* **1996,** *273* , 112.
[9] Johnson, K. L.; Kendall, K.; Roberts, A. D. *Proc. R. Soc. London* **1971,** *A324* , 301.
[10] Maugis D. J. *J. Colloid Interface Sci.* **1992,** *150* , 243.
[11] Carpick, R. W.; Agraït, N.; Ogletree, D. F.; Salmeron, M. *J. Vac. Sci. Technol. B* **1996,** *14* , 1289.
[12] Aimé, J. P.; Elkaakour, Z.; Gauthier, S.,; Michel, D.; Bouhacina, T.; Curély, J. *Surf. Science* **1995,** *329* , 149.
[13] Tsukruk, V. V.; Bliznyuk, V. N.; Hazel, J.; Visser, D.; Everson, M. P. *Langmuir* **1996,** *12* , 4840.
[14] Frisbie, C. D.; Rozsnyai, F. F.; Noy, A.; Wrighton, M.S.; Lieber, C. M. *Science* **1994,** *265* , 2071.
[15] Wilbur, J. L.; Biebuck, H. A.; MacDonald, J. C.; Whitesides, G. M. *Langmuir* **1995,** *11* , 825.
[16] Gauthier, S.; Aimé, J. P.; Bouhacina, T.; Attias, A. J.; Desbat, B. *Langmuir* **1996,** *12,* 5126.
[17] Florin, E.; Moy, V.; Gaub, H. *Science* **1994,** *264* , 415.
[18] Lee, G.; Chrisley, L.; Colton, R.; and al *Science* **1994,** *266* , 771.
[19] Edwards, S. F.; Vilgis, T. A. *Polymer* **1986,** *27* , 483.
[20] Edwards, S. F.; Vilgis, T. A. *Rep. Prog. Phys.* **1988,** *51* , 243.
[21] Bouhacina, T.; Michel, D.; Aimé, J. P.; Gauthier, S. *J. Appl. Phys.* submitted.
[22] Digital Instrument (Santa Barbara).

[23] For instance, the AFM cantilevers (Nanoprobes).

[24] Clevand, J. P.; Manne, S.; Bocek, D.; Hansma, P. K. *Sci. Instrum.* **1993**, *64* , 403.

[25] Israelachvili, J. N. *Intermolecular and Surface Forces* ; Acedemic press limited, London, 1985.

[26] Pincus, P. *Macromolecules* **1976**, *9* , 386.

[27] Treloar, L. R. G. *The Physics of Rubber Elasticity* ; 2° Ed., Oxford University Press, 1976.

[28] Ferry, J. D. *Viscoelastic Properties of Polymers* ; Wiley; New York, 1970.

[29] De Gennes, P. G. *Scaling Concepts in Polymer Physics* ; Cornell University press; Ithaca and London, 1979.

[30] Mazan, J.; Leclerc, B.; Galandrin, N.; Couarraze, G. *Eur. Polym. J.* **1995**, *31* , 803. Macosko, C. (private communication).

[31] Raphaël, E.; De Gennes, P. G. *Journal of Phys. Chemistry*, **1992**, *96,* 4002. Ji, H.; De Gennes, P. G. *Macromolecules* **1993**, *26* , 520.

Chapter 17

Correlating Polymer Viscoelastic Properties with Friction Measures by Scanning Probe Microscopy

Jon A. Hammerschmidt[1], Greg Haugstad[2], Bahram Moasser[1], Richard R. Jones[3], and Wayne L. Gladfelter[1]

[1]Department of Chemistry, University of Minnesota, Minneapolis, MN 55455
[2]Center for Interfacial Engineering, University of Minnesota, 187 Shepherd Laboratories, Minneapolis, MN 55455
[3]Sterling Diagnostic Imaging, Staton Road, Brevard, NC 28712

Scanning probe microscopy (SPM) was used to image as well as quantitatively characterize the viscoelastic character of polymer films through the use of frictional force measurements. Measurements on several polymer systems, which vary in morphology, structure, and hydrophilicity, displayed an increase in friction at specific scan velocities. The velocity dependence on friction is attributed to molecular relaxations occurring within the polymer surface caused by energy imparted, over the contact area, by the probe tip. Determination of the frequency of imparted energy suggested a correlation between bulk viscoelastic relaxations, measured by traditional dynamic mechanical and dielectric techniques, and friction measured by SPM. The ability to further image this frictional dissipation, on the tens of nanometers scale, is also demonstrated.

Scanning probe microscopies have been developed to examine topography and a variety of physical phenomena with resolutions down to the molecular scale.(1-9) In one of the variations, friction force microscopy (FFM), the lateral force on the probe caused by sliding friction is measured.(10) Interpretation of these data for polymers demands that we understand the cause of frictional energy dissipation on polymer surfaces. Friction on polymers is known to have a dominant contribution from internal viscoelastic dissipation which ultimately derives from molecular relaxation. Traditionally, characterization of viscoelastic properties has involved dynamic measurements which measure the loss tangent ($\tan(\delta)$) as a function of temperature, where the δ is the complex phase of viscoelastic moduli.(11, 12) Dynamic mechanical data generally displays several broad peaks corresponding to different classes of molecular relaxation, each class encompassing a sizable range of relaxation times.(11, 12) Typically three broad relaxational peaks are observed in viscoelastic loss tangent measurements of polymers as a function of increasing temperature or time, prior to the melting transition.(12) Secondary relaxations, which can be chain rotations, torsions, or local mode relaxations, are conventionally labeled γ and β, whereas the glass transition (T_g), reflecting rubbery behavior, is commonly assigned as the α relaxation.

288

For polymers it is known from conventional (macroscopic) characterization that the velocity-dependence of frictional force is similar to the rate-dependence of tan(δ).(*13, 14*) The observation of wear, however, especially under sliding conditions complicated the interpretation, as did the presence of a multiasperity surface.(*15*) FFM data acquired on polyethylene terephthalate (PET), polyvinyl alcohol (PVOH), polyvinyl acetate (PVAc), and gelatin surfaces display a correlation between the velocity dependence of frictional force and the viscoelastic loss tangent. In this chapter we will describe the methodology we have developed to collect and interpret FFM data allowing assignment of viscoelastic relaxations. Furthermore we find that for select scanning conditions the velocity dependence of frictional force can be affected by the measurement process, because of energy imparted to the tip-sample contact region. These measurements are designated as "perturbative". The unique strength of FFM to image these effects will also be discussed.

Experimental Measurements using Scanning Probe Microscopy

The Digital Instruments Nanoscope III and the Molecular Imaging PicoSPM scanning force microscopes with Si_3N_4 tips on triangular cantilevers (spring constant = 0.58 N/m) were used to characterize the frictional behavior of the polymer films. The PicoSPM employs an enclosed sample chamber allowing variation of relative humidity (RH) from 5%-95%. Probe tips were characterized by scanning electron microscopy (SEM) in order to determine the radius of curvature (r_c), which ranged from 10 to 300 nm, and by force versus distance (Force Curves) measurements which allow determination of the adhesive, or pull off, force. Stiffness images were recorded using the Nanoscope III force modulation mode.

Quantitative frictional data were collected in the "y-disabled" mode in which the tip scanned along the fast (x) direction. The frictional force was measured by taking the difference (in volts) between lateral forces scanning left-to-right and right-to-left. Absolute frictional force can be obtained with various calibration methods but such techniques were not necessary in this study.(*16, 17*) Scan velocity, calculated from 2 × (scan length) × (scan frequency), was varied with the independent variables of scan length (0.005 - 150 µm), and scan frequency (0.1 - 55 Hz). All data were collected at ambient conditions of ≈ 20°C and relative humidity of ≈ 40%, except for the PVOH and gelatin studies where the relative humidity was varied. PVOH films were prepared by heating a 1 wt % mixture of PVOH (Aldrich, 99% hydrolyzed, M_w = 85,000 - 146,000) and distilled/deionized water (DW) to 90°C. Freshly cleaved muscovite mica (New York Mica Co.) substrates were rinsed in DW, immersed in PVOH solution, immediately removed as to leave a residual puddle of PVOH solution, and dried slowly overnight at 20°C in moderate humidity (30% < RH < 60%). Gelatin (Kind and Knox) films were prepared identically as PVOH except solutions were heated to only 40°C. PVAc (Aldrich, M_w = 113,000) films were prepared similarly using toluene to form a 1 wt % solution. PET (Aldrich) films were prepared by spin coating a 0.08 wt % solution of PET in 2-chlorophenol onto a SiO_2 coated silicon substrate and dried at 60°C for 2 hours.

Correlation of Friction to Viscoelastic Properties

No detectable alteration of the surface morphology occurred during quantitative frictional measurements; this type of scanning is designated as nonperturbative. The velocity dependence of frictional force exhibited by crystalline PET is shown in Figure 1. A maximum was observed at a scanning velocity of ≈ 2000 µm/sec. To convert this velocity into a relaxation rate (frequency) we must divide by the length over which the tip force was applied (Figure 2). This length scale accounts for how often the probe tip accesses a point on the polymer surface. This distance can be approximated by the JKR equation which utilizes the radius of curvature of the tip (r_c), the applied load ($F_{applied}$), and the work of adhesion between the polymer and tip (which is related to the adhesive force).(*18*)

JKR equation:
$$a^3 = \frac{r_c}{K}\left[F + 3\pi r_c W_{12} + \sqrt{6\pi r_c W_{12} F + \left(3\pi r_c W_{12}\right)^2} \right]$$

where a is the contact radius, r_c is the radius of curvature, K is the elastic modulus, R is the applied load, and W_{12} is the work of adhesion. W_{12} was calculated from

$$F_{adhesion} = -\frac{3}{2} W_{12} \pi r_c$$

using the measured adhesive force from a force curve. Future refinements of this method will attempt to account more accurately for the distance over which the force of the tip is felt. Further consideration will have to be given to the contact force, the tip shape, and the possible differences between the surface and bulk mechanical properties of the polymer.

Using the measured r_c (12 nm), $F_{applied}$ (3.5 nN), and $F_{adhesive}$ (26.1 nN) a contact diameter of 15.4 nm, and a corresponding relaxation frequency of $\approx 10^{5.1}$ Hz can be calculated for PET. The dielectric β process in PET has been studied extensively, and was assigned to main chain motions of the polymer involving the COO groups.(*19-22*) Studies of varying degrees of crystallinity have shown that the frequency-temperature location and shape of the β process were independent of crystallinity.(*21, 22*) Takayanagi measured a relaxation frequency of dry PET at -40°C to be ≈ 138 Hz.(*23*) Using the time-temperature conversion for the β process of PET, 30°C per each order of magnitude in frequency,(*11*) the viscoelastic data of Takayanagi and our FFM data can be compared, and are overlaid with the friction versus velocity data (Figure 1). Takayanagi's measured relaxation corresponds to $\approx 10^{4.14}$ Hz at 20°C (Figure 1).(*11*) The loss tangent was scaled arbitrarily to the height of the measured frictional force. The temperature scale corresponds to the literature data only, whereas the relaxation frequency axis is common to the measured FFM data and Takayanagi's result. A qualitative comparison between our frictional data and that reported by Takayanagi also reveals a similar peak shape for the β relaxation.

Effect of Solvent Plasticizer on Friction

PVAc films prepared from toluene solutions do not exhibit a maximum in the friction vs. velocity measurements after > 1 day of drying. The α relaxation for 100% PVAc samples occurs at frequencies less than 0.001 Hz at room temperature,(*11*) and the β mechanism, thought to arise from motions of the -OCOCH$_3$ side chains,(*24*) occurs at $\approx 10^{6.5}$ Hz.(*11*) Therefore, neither relaxation was accessible at room temperature with our velocity range. Figure 3 shows the effect of drying on a PVAc sample prepared

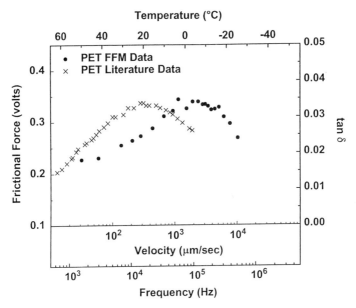

Figure 1. Frictional force versus scanning velocity recorded at a 140 μm scan length for a film of PET, along with an overlay of Takayanagi's viscoelastic data on PET.

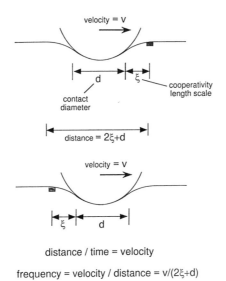

Figure 2. Probe tip scanning across a surface, where the contact diameter along with a cooperatively length scale describes the overall conversion length scale.

from 1 wt % PVAc in toluene. In order to study plasticizing effects a thicker film was used allowing for slower loss of solvent. The results measured 12 hours after preparation show a frictional maximum at $\approx$ 10 μm/sec, as the film ages the peak shifts to lower velocities yielding no distinguishable maximum after 23 days. Similar frequency shifts due to solvent plasticizing were reported on PVAc in diphenylmethane.(25) In this dielectric study 11.3 mole % of diphenylmethane shifted $T_g \approx$ 2 decades, and 24.1 mole % led to T_g shifting $\approx$ 5 decades of rate.

Humidity Dependence on Friction Measurements

Friction force versus scanning velocity at two different humidities for a film of PVOH is shown in Figure 4. No velocity dependence was observed at a relative humidity of 5%, and a gradual increase in friction was observed at velocities < 100 μm/sec. for RH=75%. Takayanagi measured the loss compliance ($E^{''}$) as a function of temperature and assigned the α_a relaxation in PVOH, where the subscript 'a' stands for amorphous.(26) The α_a relaxation corresponded to the glass transition and when measured at a frequency of 138 Hz, occurred at $\approx$ 20°C for 8% water content.(25) Water acts as a plasticizer for PVOH lowering the temperature at which a particular relaxation occurs,(27) or correspondingly decreasing the relaxation time. Under our conditions, 20°C, RH=5% and 75%, the water content in PVOH should approximately be 1% and 12.5% respectively, which corresponds to a T_g of 65°C and 18°C.(27) Therefore at 5% RH the glass transition occurs below our detection limits, whereas at 75% RH the onset of the glass transition is observed. The glass transition for PVOH has a FWHM of $\approx$ 40°C, and a time temperature conversion of $\approx$ 2.5°C per order of magnitude in frequency,(11) therefore we are unable to measure the full relaxation without temperature variation.

   In an earlier FFM study(7) of gelatin films a peak near 1 μm/sec in the velocity dependence of friction was assigned to the glass transition. In Figure 5 we show the dependence of this peak position on gelatin water content, via the latter's known humidity dependence. The data at RH=58% ($\approx$20% water content(28)) are similar to those previously reported by us at similar measurement humidities,(7) i.e. exhibit a broad peak centered near 1 μm/sec. At RH=24% ($\approx$13% water content) this peak apparently has shifted by approximately 4 decades to lower velocity, residing only partially inside our measurement window. At RH=5% ($\approx$6% water content) no evidence of this peak is visible, while a secondary peak apparently has moved into the window at the high velocity end. Interpolation from published gelatin glass transition temperatures yields $T_g \approx$20, 55 and 120°C at the water contents of 20%, 13% and 6%, respectively. Published time-temperature conversion data do not exist for gelatin to our knowledge; our data suggests a peak shift of (55°C-20°C)/4 $\approx$ 9°C per decade of rate, similar in magnitude to tabulated conversions for α relaxations in polyamides like the nylons.(11) The β peak position in amide-linkage polymers typically shifts much less than the α with changing water content.(11)

Perturbative Measurements on Gelatin Films

All of the preceding measurements were performed nonperturbatively, such that the frictional response function (velocity dependence) of the material was not itself modified by frictional energy dissipation (heating). In general this may not be the case. We have identified perturbative effects at high loading forces on gelatin films by

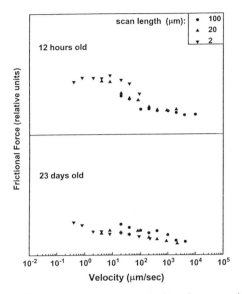

Figure 3. Frictional force versus scanning velocity taken at various scan lengths for a film of PVAc recorded at 12 hours (top), and 23 days (bottom) from preparation.

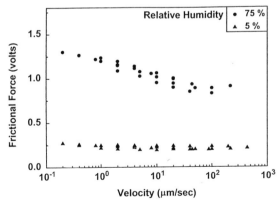

Figure 4. Frictional force versus scanning velocity taken at 5% and 75% relative humidity for a film of PVOH.

comparing friction-loop traces collected successively over a single scan line: the frictional force increases substantially during the first several scan loops, then asymptotically approaches a value reflecting modified frictional response under dynamic equilibrium conditions.(7) The nature of the modified response function is explored in Figure 6a, a comparison of the velocity dependence of friction (under dynamic equilibrium) at five contact forces ranging from 60-180 nN (RH=53%). At the lowest contact force (circles) no evidence of the glass-rubber transition is present; at higher contact forces it is increasingly evident as a rise in friction at decreasing velocity below 1000 μm/sec. As the contact force, and consequently the frictional heating, is increased, the onset of rubbery behavior shifts to higher velocity, in accordance with time-temperature correspondence.

Most remarkable is the observation that the effects of frictional heating, quantified in Figure 6a, do not immediately disappear upon termination of perturbative scanning. Instead, residual elevated friction and reduced stiffness can be imaged in the affected region (for as long as several hours). This is demonstrated in Figure 6b, 5x5-μm topography/friction (top left/right) and topography/stiffness (bottom left/right) images of a 2x2-μm region previously raster scanned perturbatively. The frictional response of the 2x2-μm region just prior to termination of perturbative scanning was approximately that of the inverted triangle data point denoted by an arrow in Figure 6a. The response of the surrounding unperturbed film during imaging was approximately that of the denoted circle data point. The modified region exhibits rubbery characteristics, i.e. elevated friction and reduced stiffness, even under nonperturbative imaging conditions. The fading signature (relaxation) of this modified region has been tracked in images collected at variable elapsed times following termination of perturbative scanning.(7)

Conclusions

The known correlation between friction and viscoelastic dissipation in polymers allowed for the assignments of increased frictional force in velocity dependence measurements, using FFM, to specific viscoelastic relaxations within the polymer. For our allowed range of velocity / frequency the glass transition (α-relaxation) for PVOH, PVAc, and gelatin, and the β relaxation for PET were probed. Effects of solvent plasticizer and humidity on friction-velocity studies were consistent with our interpretation of assigned relaxations. Increased solvent will shorten relaxation times shifting peaks to higher frequencies, as observed. Increasing the contact force, consequently increases frictional heating, which shifts the onset of rubbery behavior in gelatin to a higher velocity, consistent with the time-temperature correspondence. Furthermore we have demonstrated the ability of FFM to image residual effects on frictional characteristics of a polymer film.

Acknowledgments

This work was supported by grants from the Center for Interfacial Engineering, Sterling Diagnostic Imaging, and the donors of The Petroleum Research Fund, administered by the ACS.

Literature Cited

(1) Hues, S. M.; Colton, R. J.; Meyer, E.; Guntherodt, H. J. *MRS Bulletin* **1993**, *18*, pp. 41-49.
(2) Overney, R.; Meyer, E. *MRS Bulletin* **1993**, *18*, pp. 26-34.

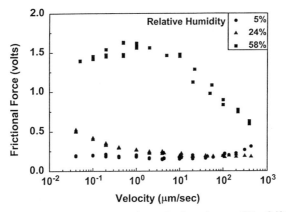

Figure 5. Frictional force versus scanning velocity taken at 5% , 24%, and 58% relative humidity for a gelatin film.

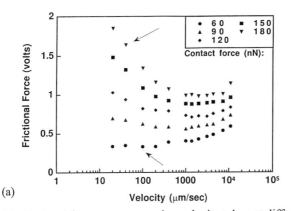

(a)

Figure 6. (a) Frictional force versus scanning velocity taken at different contact forces for a gelatin film. (b) Images 5x5-μm topography/friction (top left/right) and topography/stiffness (bottom left/right) images of a 2x2-μm region previously raster scanned perturbatively. *Continued on next page.*

Topography   Friction

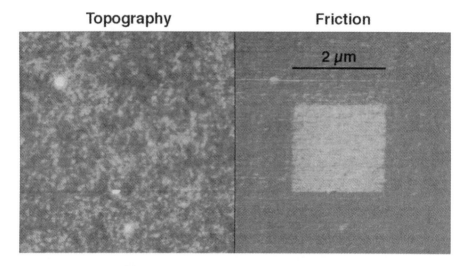

Topography   Stiffness

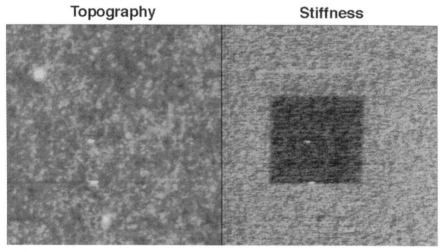

(b)   Figure 6. *Continued.*

(3)   Hartmann, U.; Goddenhenrich, T.; Heiden, C. *J. Magn. Magn. Mater.* **1991**, *101*, pp. 263-270.
(4)   Burnham, N. A.; Dominguez, D. D.; Mowery, R. L.; Colton, R. J. *Phys. Rev. Lett.* **1990**, *64*, pp. 1931-1934.
(5)   Weisenhorn, A. L.; Maivold, P.; Butt, H.J.; Hansma, P. K. *Phys. Rev. B* **1992**, *45*, pp. 11226-11232.
(6)   Blackman, G. S.; Mate, C. M.; Philpott, M. R. *Phys. Rev. Lett.* **1990**, *65*, pp. 2270-2273.
(7)   Haugstad, G.; Gladfelter, W. L.; Weberg, E. B.; Weberg, R. T.; Jones, R. R. *Langmuir* **1995**, *11*, pp. 3473-3482.
(8)   Frisbie, C. D.; Rozsnyai, L. F.; Noy, A.; Wrighton, M. S.; Lieber, C. M. *Science* **1994**, *265*, pp. 2071-2074.
(9)   Florin, E.L.; Moy, V. T.; Gaub, H. E. *Science* **1994**, *264*, pp. 415-417.
(10) Mate, C. M.; McClelland, G. M.; Erlandsson, R.; Chiang, S. *Phys. Rev. Lett.* **1987**, *59*, p. 1942.
(11) McCrum, N. G.; Read, B. E.; Williams, G. *Anelastic and Dielectric Effects in Polymeric Solids*; Wiley: London, 1967.
(12) Aklonis, J.J.; MacKnight W.J. *Introduction to Polymer Viscoelasticity*; 2nd ed.; Wiley: New York, 1983.
(13) *Advances in Polymer Friction and Wear*; Lee, L.; Ed.; Plenum Press: New York 1974; Vol. 5.
(14) Moore, D.F. *The Friction and Lubrication of Elastomers.*; Oxford: Pergamon Press, 1972.
(15) Ludema, K. C.; Tabor, D. *Wear* **1966**, *9*, p. 329.
(16) Haugstad, G.; Gladfelter, W. L.; Weberg, E. B.; Weberg, R. T.; Weatherill, T. D. *Langmuir* **1994**, *10*, pp. 4295-4306.
(17) Noy, A.; Frisbie, C. D.; Rozsnyai, L. F.; Wrighton, M. S.; Lieber, C. M. *J. Am. Chem. Soc.* **1995**, *117*, pp. 7943-7951.
(18) Johnson, K. L.; Kendall, K.; Roberts, A. D. *Proc. R. Soc. London A* **1971**, *324*, pp. 301-313.
(19) Ward, I. M. *Trans. Faraday Soc.* **1960**, *56*, p. 648.
(20) Illers, K. H.; Breuer, H. *J. Colloid. Sci.* **1963**, *18*, p. 1.
(21) Saito, S. *Res. Electrotech. Lab.* **1964** (648).
(22) Ishida, Y.; Yamafuji, K.; Ito, H.; Takayanagi, M. *Kolloid Z* **1962**, *184*, p. 97.
(23) Takayanagi, M. *Mem. Fac. Eng. Kyushu Univ.* **1963**, *23*, p. 1.
(24) Ishida, Y.; Matsuo, M.; Yamafuji, K. *Kolloid Z.* **1962**, *180*, p. 108.
(25) Broens, O.; Muller, F. H. *Kolloid Z* **1955**, *140*, p. 121.
(26) Takayanagi, M.; Nagai, A. *Rept. Progr. Polymer Phys. (Japan)* **1964**, *7*, pp. 249-252.
(27) Finch, C.A., *Polyvinyl Alcohol; Properties and Applications.* **1973**, London: Wiley. p. 346.
(28) Curme, H.G.*Gelatin*, in *The Theory of the Photographic Process*, C.E.K. Mees and T.H. James, Eds.; MacMillan Publishing Co.: New York, 1966; pp. 45-71.

SPM Techniques:
Current Trends

Chapter 18

Resolution Limits of Force Microscopy

R. Lüthi, Meyer, M. Bammerlin, A. Baratoff, J. Lü, M. Guggisberg,
and H.-J. Güntherodt

Institute of Physics, University of Basel, Klingelbergstrasse 82, CH-4056
Basel, Switzerland

Applications of contact force microscopy in ultrahigh vacuum are
presented where the resolution is limited to approximately 1 nm under
optimum conditions. It will be shown that even small van der Waals
forces of 0.1nN are sufficient to explain the finite contact area. One way
to circumvent this problem is to measure in liquid environment where
van der Waals forces can become repulsive. However, this environment
is not compatible with the controlled surface preparation in ultrahigh
vacuum. Recently, progress has been made in non-contact force
microscopy where the cantilever is oscillated at its resonance frequency.
Under appropriate conditions true atomic resolution can be achieved.
Vacancies, adsorbate atoms and step sites are being imaged showing
that individual atoms are resolved by force microscopy.

Resolution limits of contact force microsopy

Force microscopy has been introduced in 1986 by Binnig, Quate and Gerber *(1,2)*.
Contact-mode AFM is accompanied by the jump-in of the soft cantilever. When the
condition

$$\frac{\partial F}{\partial z} < k$$

is met, an instability occurs, where $\partial F/\partial z$ is the attractive force gradient and k the
cantilever spring constant. The probing tip jumps toward the sample. Long-range
attractive forces, $F_{l.r.}$, such as van der Waals forces, capillary forces or electrostatic

forces, have to be compensated by short-range repulsive forces on the foremost tip apex $F_{s.r.}$. After the instability, the operator will try to minimize the forces on the tip apex by compensating the long-range attractive force with the bending of the cantilever $F_L = k\, z_t$

The equilibrium is given by:

$$F_{s.r.} = F_{l.r.} - k\, z_t$$

which determines the force on the tip apex $F_{s.r.}$ (cf. Figure. 1). The minimum $F_{s.r.}(min)$ is achieved close to the jump-out of the contact. Experimentally, it is found that $F_{s.r.}(min)$ is a significant fraction of $F_{l.r.}$ (typically 10% - 50%). Thus, the compensation by the cantilever bending is not complete. This incomplete compensation is explained by local variations of the attractive force. These forces can change quite drastically above hillocks compared to valleys. During scanning, the tip will either jump out of contact above areas with low attractive forces or will experience high forces on the areas with high attractive forces. Thus, the minimum, experimentally achievable force on the tip apex depends also on the roughness and scan area.

The consequences for the resolution of contact force microscopy become evident. Depending on the environment, the long-range forces $F_{l.r.}$ will vary significantly. In ambient pressure, capillary forces, originating from the formation of a liquid meniscus between probing tip and sample, will be dominant. With a tip radius R=100nm, the maximum attractive, capillary force is

$$F_{cap} = 4\, \pi\, \gamma \cos(\theta) \approx 90 \text{ nN}$$

where $\gamma = 0.07$ N/m^2 is the surface tension of water and θ is the contact angle. After the jump to contact this large force acts uncompensated on the tip apex and can deform the tip or sample. Even with optimum bending of the cantilever, the force will be still in the nN-regime.

Capillary forces can be eliminated by measuring in liquids or in vacuum conditions. In liquids, attractive forces can become very small. In best case, van der Waals forces become repulsive by choosing a suitable liquid (3). Such a situation is met, when the refractive index of the liquid, n_l, is between the refractive index of the probing tip, n_t , and the refractive index of the sample, n_s : $n_s < n_l < n_t$. . Ohnesorge and Binnig have shown that it is possible to achieve true molecular resolution on calcite in water, thus demonstrating that the contact diameter is of atomic dimensions (4).

In ultra-high vacuum conditions, van der Waals forces will always be present. Goodman and Garcia (5) have shown that typical van der Waals forces are between 1-10 nN, where the exact values depend on the materials. A collection of their results is shown in table 1. A tip radius of R=100 nm at a distance of z=1 nm has been assumed.

Comparison of Van der Waals Forces

TABLE 1. Van der Waals Forces

Materials	Forces
Graphite-Graphite	8 nN
Diamond-Diamond	17 nN
Metal-Graphite	10 nN
SiO_2-Graphite	1.2 nN

In conclusion, the attractive forces in ambient conditions are the largest: 1-100 nN. In ultra-high vacuum, van der Waals forces are always present and give values between 0.1-10 nN. In liquids, van der Waals forces can become repulsive suitable liquids and forces can become below 100 pN. The Hertz model, gives us an estimate of the contact diameter:

$$a_{min} = 2(D \cdot R \cdot F)^{1/3} \qquad D = \frac{1-v_1^2}{E_1} + \frac{1-v_2^2}{E_2}$$

and v_i and E_i are the Poisson ratii and Youngs modulii of probing tip and sample. With typical parameters ($E_1=E_2=1.7 \cdot 10^{11}$ N/m^2, $v_i= 0.3$ and R = 90 nm), the contact diameter in ambient pressure is 2-10 nm (1 nN<F <100 nN). In ultrahigh vacuum, we estimate contact diameters to be 1-4nm (0.1 nN< F <10 nN). More sophisticated elasticity models for the calculation of the contact area are discussed elsewhere *(6-9)*. In liquids we expect to achieve true atomic resolution under best conditions (a_{min}<1nm). These rough estimates are confirmed experimentally: In ambient pressure, the resolution is typically in the range of 5-10nm. In ultrahigh vacuum, lateral resolutions of about 1nm could be achieved. E.g. Howald et al. could resolve a step edge on NaF(001) with 1nm width *(10)*. Figure. 2 and Figure 3 show contact force microscopy images of C_{60}-islands on NaCl(001). The C_{60}-molecules could be resolved at the step edge both in normal mode and lateral force mode. Considering, that the distance between the C_{60}-molecules is 1nm, it becomes evident that the contact diameter is below 1nm.

For many applications of AFM where a lateral resolution of tens of nanometers is sufficient, the above mentioned resolution limits are not so important. They can be reduced by retracting the cantilever into the attractive range and/or performing measurements in a liquid with a high dielectric constant like water. With perseverance Ohnesorge and Binnig could demonstrate true atomic resolution on the cleavage face of calcite using this approach.

True-atomic resolution with force microscopy in non-contact mode on semiconductors and insulators

In spite of the success of contact force microscopy, it is desirable to study a wide variety of surfaces under clean conditions in the absence of contaminants. A different

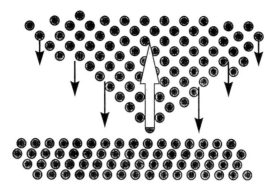

Figure 1
Schematical diagram for contact force microscopy: Long-range forces have
to be balanced by repulsive forces, acting on the tip apex.

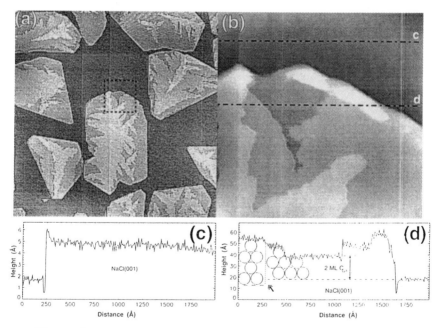

Figure 2
(a) Constant force image of C_{60}-islands grown on NaCl(001). The islands
are 3-6 molecular layers high. (b) A zoom-in image of (a). The profiles (c)
and (d) are indicated in (b).

304

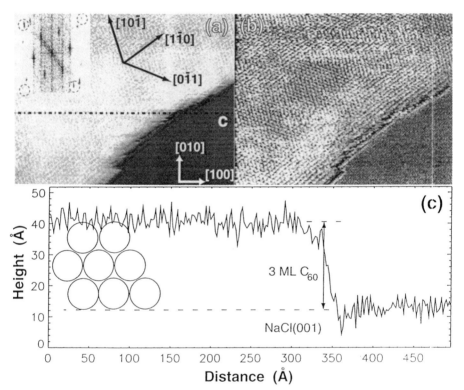

Figure 3
High-resolution constant force image of C_{60} on NaCl(001). The inset shows the FFT-image, showing the spots from both the C_{60}-periodicity and the NaCl(001)-lattice. (b) Corresponding friction force map. Both the molecular structure on the C_{60}-terrace and the NaCl-lattice are visible. The observation of molecular structure at the step edge confirms that the resolution is about 1 nm, which corresponds to the distance between C_{60}-molecules. (c) Profile as indicated in (a).

approach, introduced since the advent of AFM, is to avoid contact altogether and to measure changes in vibrational properties of the cantilever induced by tip-sample interactions *(11-12)* . The majority of such dynamic non-contact investigations have not been aimed at subnanometer resolution and could not reach this limit owing to the long-range of the primarily attractive forces involved. However, quite recently, following the initial work by Giessibl *(13)* , several authors have reported true atomic resolution even on reactive semiconductor surfaces using large-amplitude dynamic AFM in UHV, e.g. on Si(111)7×7 *(13-16)* , Si(100)2×1 *(17)* and InP(110) *(18)* . An important step towards this success of dynamic force microscopy was its adaptation to ultrahigh vacuum (UHV), where high quality factors Q are obtained which increase the detection sensitivity. By identifying the range for non-contact (or rather 'near-contact') operation *(16)* with atomic resolution on Si(111)7×7 (cf. Figure 4), we could understand how to preselect control parameters so as to achieve stable images at constant resonance frequency shift with a quality similar to that known from scanning tunneling microscopy (STM). It was also found that tip preparation is important: Prior to the AFM-experiments, the cantilevers were sputtered in situ with Ar ions to remove the oxide layer from the tip. Cleaning the tip was found to be of importance to achieve stable atomic resolution on semiconductors *(15, 16)* , presumably because Si dangling bonds (DB) are generated near the tip apex. Being laterally localized but polarizable, such DBs might also enhance atomic-scale contrast a few Å from the surface of an ionic crystal even if they show no tendency to form covalent bonds.

The high Q-factor of silicon cantilevers in UHV (Q-factor: The quality factor $Q=f/\Delta f$ is determined from the thermal noise power spectrum, whereby $\Delta f = \gamma_0 / 2\pi$ is the full width of the resonance peak at half maximum and γ_0 the power damping rate of the free lever. Typical Q-factors of Si-cantilevers range from 15,000 to 20,000.) gives high sensitivity, but also requires special operation modes. FM-detection has been suggested by Albrecht et al. and is nowadays used most commonly *(19)* . As emphasized by Albrecht et al. , the response of the FM detector is determined by its bandwidth (<1 kHz) rather than by the damping rate of the cantilever. The measured Δf thus represents an average over about hundred cycles. The vibration amplitude A_0 of the cantilever is kept constant (typically 1-20 nm) by a feedback circuit independent of the slower feedback or external adjustment of the tip-sample separation. Keeping A_0 constant prevents the generation of disturbing amplitude modulation sidebands upon scanning, and also ensures that the closest tip-sample separation z is a linear function of the sample displacement. With a sufficiently large A_0, the static deflection of the cantilever is negligible. Furthermore, judging from the essentially sinusoidal output waveform of the photodiode, the vibrating lever can to a good approximation be described as a driven damped harmonic oscillator weakly perturbed by interactions whenever the tip comes close to the sample *(13,16)* . Besides causing a small frequency shift, interactions can also lead to losses, e.g. owing to partial dissipation of excited deformation pulses into acoustic phonons. Independent of its origin, the additional damping rate, given by $\Delta\gamma=f\ \Delta E/E$, where $E=0.5\ (k\ (A_0)^2)$ is the initial energy of the oscillator and ΔE is the extra energy per cycle required to keep A_0 constant, can be determined from the amplitude feedback signal.

306

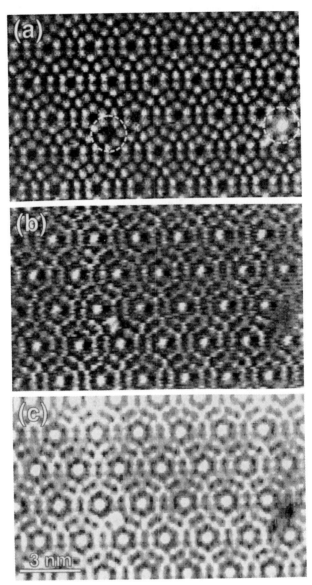

Figure 4
(a) Topography on Si(111)7x7 (constant frequency shift) (b) Damping
image ($\Delta\gamma$) (c) Variations of tunneling current between cantilever tip and
sample.

The application of AFM to insulators, such as ionic crystals, goes back to the first days of AFM. However, it took a long time until true atomic resolution could be achieved on these surfaces. Figure 5 shows approach traces of Δf and $\Delta \gamma$; while the frequency decreases due to attractive forces, the damping signal $\Delta \gamma$ increases monotonically. As discussed in more detail elsewhere (16) three regimes can be distinguished in Δf. Region I dominated by long-range van der Waals plus electrostatic attraction, region II, where true atomic resolution can be reproducibly achieved, and region III where stable constant-A_0 operation becomes impossible, presumably because irreversible material transfer and/or deformations of the tip and sample occur.

For stable scanning operation in region II, the setpoint is chosen on the steeply decreasing part of the $\Delta f(z)$ characteristic somewhat above its negative minimum. Figure 6a shows a grayscale image of the z-feedback signal at a constant $\Delta f = -80$ Hz. The optimum setpoint depends on the tip geometry and can vary about ± 30 Hz. Figure 6b shows a grayscale image of the simultaneously acquired damping signal. Upon correcting the slight distortion due to drift, the protrusions (higher attraction) are consistent with the expected periodic arrangement of the NaCl(001) surface lattice. From Fourier analysis a period of $5.6 \pm 0.6 \text{Å}$ can be determined (cf. Figure 6a) which corresponds to the spacing of equivalent ions. In addition, two seemingly identical atomic-scale defects (marked by arrows) can be identified. Their nature being so far unknown, the apparant asymmetry could be genuine or due to convolution with an irregular tip shape. Note, however, that low-symmetry surface defects, e.g. impurity-vacancy pairs are expected to be more stable than surface vacancies (20). Assuming that both observed defects are identical, their lifetime exceeds the time lag between the scan lines crossing each defect (≈ 50 s). In any case, these data demonstrate that true atomic resolution can be achieved over wide areas and over long times on the surface of a good insulator by non-contact AFM in UHV. Recent simulations of AFM on ionic crystals indicate that in the near-contact range force contrast is dominated by short-range electrostatic interactions between surface ions and charges at the apex of different model-tips (20-21). Since dangling bonds on protruding Si atoms tend to be occupied (negatively charged) (22), we might expect maximum attraction above cations. The same qualitative result was found in a simulation of AFM with a diamond tip on NaCl(001) (23). A definitive assignment of the observed maxima will require a realistic simulation of the system under study or imaging and identifying stable surface defects of known registry with respect to the substrate lattice.

Non-contact dynamic AFM allows one to investigate the dynamics of point defects on time scales in principle as short as the inverse bandwidth of the FM detector (> 1 ms). In Figure 7, selected images show the evolution of a pair of point defects. They correspond to No. 1, 2, 8 and 12 in a sequence of 23 images which were recorded at time lapses of 10 minutes. The first two demonstrate the excellent reproducibility of dynamic force microscope. Both defects seem quite stable; their structure, including distortions in their vicinity appears unchanged. Subsequent images (not shown), revealed no significant changes, until those evident in Figure 7d; in particular the distance between the defects is reduced by one interionic spacing. In the following images the defects were found again to be stable and no further changes occurred.

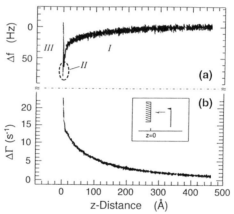

Figure 5

Simultaneously recorded traces of (a) the frequency shift Δf and (b) the additioal damping $\Delta\gamma$ vs. sample displacement. With zero taken at the sharp onset of regime III, z is essentially the distance to the tip at its closest approach. Stable imaging conditions at constant tip vibration amplitude were found in regimes I and II, but only the latter one is suited for true atomic resolution.

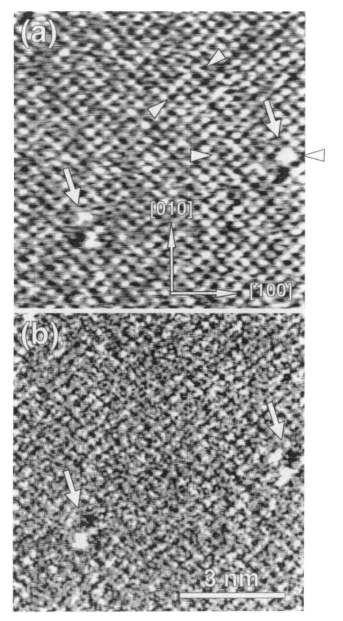

Figure 6
(a) Topography (constant frequency shift of -80Hz) and (b) Damping image
($\Delta\gamma$) on NaCl(001). Atomic scale defects are indicated by arrows.

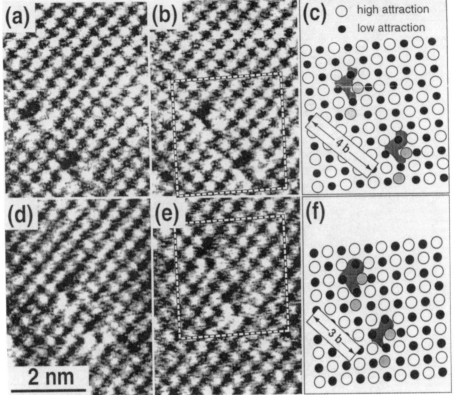

Figure 7

Series of images (damping signal $\Delta\gamma$ at constant frequency shift) of a pair of point defects; (a) and (b) were acquired consecutively (10 min time lapses); (c) is a schematic drawing of the observed structure; (d) and (e) were acquired after a sudden change of the distance between the defects by the amount b=5.6 Å is the lattice constant of NaCl] (70 and 110 min, respectively, after (a); (f) is the corresponding schematic drawing illustrating the reduced defect separation.

Since the structure of both defects is unchanged over long time periods, and only discrete changes in their separation have been observed, we conclude that their apparent evolution is not primarily influenced by the presence of the tip, but most likely corresponds to the initial and final states of an intrinsic thermally activated atomic jump process.

Conclusions

In conclusion, the contact diameters in typical contact force microscopy are between 1-10nm. Generally, atomic-scale images have to interpreted with care. Even, when atomic-scale features are observed in normal or lateral force, the contrast originates from a multiple-atom contact. Exceptions from this rule are contact mode imaging in liquids with ultra-low forces (<100pN) and non-contact imaging in ultrahigh vacuum, where true atomic-resolution can be achieved. We have demonstrated how a dynamic force microscope can be operated so as to achieve true atomic resolution on the surface both insulators and semiconductors. Both the structure and the slow dynamics of point defects can be observed. The assignment of maxima in the topography and damping signals, as well as the nature and migration mechanisms of the defects, require further theoretical and experimental efforts.In the future, dynamic force microscopy with the capability to measure insulating surfaces with a resolution comparable to that of STM on conducting samples gives the opportunity to study fundamental questions such as unknown surface reconstructions of insulators, identification of surface defects, e.g. color centers or catalytic reaction sites on ceramics.

Literature Cited

(1) Binnig, G., Quate, C. F. and Gerber, Ch. *Phys. Rev. Lett.* **1986**, vol. 56, pp. 930-933.
(2) For an overview in force microscopy see: *Forces in Scanning Probe Methods*, Eds. Güntherodt, H.-J., Anselmetti, D. and Meyer, E., NATO ASI Series E: Applied Sciences Vol. 286, Kluwer Academic publishers, 1995.
(3) Israelachvili, J.N.*Intermolecular and Surface Forces*, Academic Press, London, 1985.
(4) Ohnesorge, F. and Binnig, G., *Science* **1993**, vol. 260, pp.1451.
(5) Goodman, F.O. and Garcia, N., *Phys. Rev. B* **1991**, vol. 43, pp. 4728.
(6) Pethica, J.B. and Sutton, A.P, p. 353 in (2).
(7) Meyer E. in *Fundamentals of friction*, Persson, B.N.J. and E. Tosatti, E., Eds., NATO ASI Series E: Applied Sciences, Kluwer Academic Publishers, 1996, vol. 311.
(8) Johnson, K.L., *Contacts Mechanics*, Cambridge University Press, Cambridge, 1994.
(9) Germann, G.J., Cohen, S.R., Neubauer, G., McClelland, G.M., and Seki, H., *J. Appl. Phys.* **1993**, 73, 163-167.
(10) Howald, L., Haefke H. , Lüthi R., Meyer, E. , Gerth, G., Rudin, H. and Güntherodt, H.-J.*Phys. Rev. B* **1993**, vol. 49, pp. 5651-5656.
(11) Mc Clelland G. M., Erlandsson R. and Chiang S., In *Review of the Progress of Qualitative Nondestructive Evaluation*, (Thompson P. O. and Chimenti D. E., Ed., Plenum, New York , 1987, vol. 6b, pp. 1307.
(12) Martin Y., Williams C. C. and Wickramasinghe H. K. *J. Appl. Phys.* **1987**, vol. 61, pp. 4723-4729.
(13) Giessibl F. J., *Science* **1995**, Vol. 267, pp. 68-71.
(14) Kitamura S. and Iwatsuki M., *Jpn. J. Appl. Phys.* **1995**, vol. 34, pp. L145-L148.
(15) Güthner P., *J. of Vac. Sci. Technol. B* **1996** vol. 14, pp.2428-2431.
(16) Lüthi R. et al., *Z. Phys. B* **1996**, vol. 100}, pp. 165-167.
(17) Kitamura S. and Iwatsuki M. *Jpn. J. Appl. Phys.* **1996**, vol. 35, pp. L668-L641.
(18) Sugawara, Y., Ohta M., Ueyama, H., and Morita, S.,*Science* **1995**, vol. 270, pp. 1646.
(19) Albrecht et al., *J. App. Phys.* **1991**, vol. 69, pp. 668.
(20) Shluger A. L., Rohl A. L., Gay D. H., and Williams R. T., *J.Phys.: Condens. Matter*, **1994**, vol. 6, pp. 1825-1846.
(21) Shluger A. L., Wilson R. M.and Williams R. T., *Phys. Rev. B* **1994**, vol. 49, pp.4915-4930.
(22) Chadi D. J., *Phys. Rev. Lett.* **1979**, vol. 43, pp. 43-47.
(23) Tang H., Joachim C., Devillers and Girard C., *Europhys. Lett.*, **1994**, vol. 2, pp. 383-388.

Chapter 19

Chemical Force Microscopy: Probing and Imaging Interactions Between Functional Groups

Dmitri V. Vezenov, Aleksandr Noy, and Charles M. Lieber[1]

Department of Chemistry and Chemical Biology and Division of Engineering and Applied Sciences, Harvard University, Cambridge, MA 02138

Adhesion and friction forces between surfaces with pre-determined terminal groups were measured by chemically modified scanning probe microscopy tips. Surfaces composed of various terminal groups (CH_3, NH_2, and SO_3H) were obtained by direct chemisorption of silane based compounds on silicon/silicon nitrides. Work of adhesion and friction coefficients were obtained for different types of modified tips and surfaces in aqueous solutions with variations in pH. Absolute values of work of adhesion between various surfaces, W_{ad}, are in the range of 0.5 to 8 mJ/m^2. Friction properties vary with pH in a register with adhesive forces showing a broad maximum at intermediate pH values. This behavior can be understood by considering changes in surface charge state which is determined by the zwitterionic nature of silicon nitride surfaces with multiple isoelectric points.

Recently, "chemical force microscopy" (CFM) has been introduced as a new scanning probe microscopy (SPM) mode *(1-8)*. This technique allows discrimination of local surface forces related to intermolecular interactions of different chemical groups with nanometer resolution. CFM involves a chemical modification of the SPM tip by fabricating a robust molecular layer firmly tethered to the tip's surface. The most widely implemented approach in this field uses self-assembly monolayers (SAMs) from thiol molecules on gold *(9, 10)*. Presently, several examples of modified tips with surface CH_3, COOH, CH_2OH, CO_2CH_3, CH_2Br, and NH_2 groups have been demonstrated. "Nanotitration" data were obtained by CFM for different functional surfaces *(5, 10)*. The acid-base behavior of the modified SPM tips is controlled by the nature of the surface terminal groups and can show dramatic changes in the vicinity of isoelectric points. The exhibited

[1]Corresponding author.

friction behavior closely follows variations in adhesive properties. It has been demonstrated that this characteristic can be exploited to identify different microphases on multicomponent surfaces with nanoscale resolution *(6-10)*.

Presently, we discuss the fabrication of SPM nanoprobes applied directly via silane-based modification of the silicon nitride and silicon materials. We will present our results on adhesive interactions and friction forces between chemically modified surfaces (SAMs with different terminal groups) in various aqueous environments.

Experimental

Chemisorption of silane-based molecules with various functional terminal groups is used to modify surface properties of silicon and silicon nitride SPM tips as described previously *(12)*. All titration curves (adhesive forces versus pH) are obtained by variation of pH from 2 to 10. The results are averaged over a total set of 50 - 100 curves. Pull-off forces are determined from the cantilever deflection in a retraction mode. To study frictional properties of the film surfaces, a cross-section of surface topography and variation of torsional deflection (a friction loop) are detected simultaneously according to the well established protocol *(13, 14)*. Calibration plots *(15)* are used to determine normal spring constants, $k_n = 0.235$ N/m, and torsional (lateral) constant, $k_t = 115$ N/m for short, narrow leg V-shaped DI cantilevers used in this study.

Results and discussion

Adhesive forces in aqueous solution. An example of SPM force-distance curves that demonstrates significant variation of adhesive forces (designated as ΔF) for different tip-surface pairs is presented in Figure 1a for modified silicon nitride and silicon surfaces. Strong adhesion is observed for both unmodified tip-surface pairs and NH_2 terminated SAMs. Both SO_3H and CH_3 terminated pairs of tip - substrate possess much smaller adhesive forces. Adhesive forces, ΔF, are in the range of 0.1 to 8 nN for different pairs of modified tips and substrate surfaces. For further comparison, work of adhesion for separating the surfaces and the SPM tip, W_{ad}, is evaluated using the JKR relationship: $W_{ad} = \Delta F/1.5\pi R_c$ *(16)*. Absolute values of W_{ad} are in the range of 0.5 mJ/m^2 to 8 mJ/m^2 at pH = 6 (Figure 2).

Liquid-solid surface energies, γ_{ls}, can be estimated from work of adhesion for symmetrical surfaces according to the known relationship: $W_{ad} = 2\gamma_{ls}$ *(17)*. Absolute values of surface energies for different surfaces in water are substantially (several times) lower than the solid-vapor surface energies, γ_{sv}, for similar surfaces. The γ_{sv} values are in the range of 20 mJ/m^2 (CH_3 SAMs) to 50 mJ/m^2 (NH_2 SAMs) *(17)*. Using the Hamaker constants, A, for different surfaces separated by various media calculated in Ref. 18 and the relationship $\gamma \sim 0.45 \ 10^{21}$ x A *(17)*, the reduction of surface energies at solid-liquid interface by 2 - 8 times can be justified.

Adhesive forces at different pHs. Examples of force-distance curves for the same interacting surfaces (symmetrical NH_2 - NH_2 terminated SAMs) at different pHs are shown in Figure 1b. Strong adhesion is observed at intermediate pH (see examples for pH = 5, 7). At very acidic and basic pH conditions, adhesion becomes very

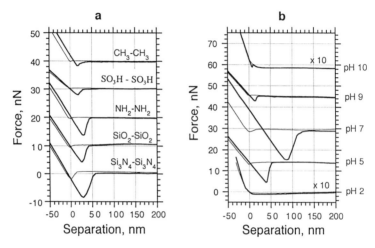

Figure 1. Force-distance curves for five different symmetrical tip-surface pairs in aqueous solution at neutral conditions, pH = 6 (a) and for NH_2 terminated surfaces at different pHs (b). The curves offset along the vertical axis to avoid overlapping. Thin solid lines represents the approaching mode and thick solid line is the retracing mode.

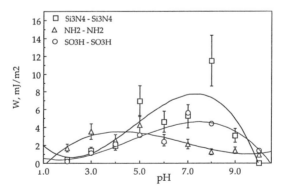

Figure 2. Variations of work of adhesion, W_{ad}, for $Si_3N_4 - Si_3N_4$, $NH_2 - NH_2$, and $SO_3H - SO_3H$ surfaces versus pH.

minute (see data for pH = 2 and 10 in Figure 1b, note that the force scale is magnified by 10 times).

Adhesion behavior of bare silicon nitride, NH , and SO_3H modified surfaces studied at different pH conditions is shown for symmetrical surface pairs (Figures 2). For silicon nitride tip - silicon nitride substrate, W_{ad} (pH) is a non-monotonic function with minima at 2 and 10 (the lowest level of adhesion forces detected here is close to 0.03 nN). A broad maximum of 7 - 8 mJ/m^2 is observed in the pH range from 7 to 8. Adhesion energy between SO_3H terminated surfaces reduces to 4 - 5 mJ/m^2. Amine terminated modification also results in lowering of adhesion energy. Maximum adhesion is shifted to the acid range (pH = 4) for amine terminated tip - substrate pair and a broad plateau of very low energy (~ 1 mJ/m^2) is observed in the pH ranging from 7 to 10 (Figure 2).

The observed pH behavior of the silicon nitride - silicon nitride pair correlates with electrochemical surface properties of these materials.[18-20] Non-monotonic variation of adhesive forces can be qualitatively explained by interplay of electrostatic and van der Waals forces between the surfaces by considering the presence of multiple isoelectric points with different pKs on interacting surfaces (17, 21). Expected behavior for two interacting surfaces with two different expected pKs within the framework of a double-layer theory is presented in Figure 3a. The total balance of electrostatic forces acting between surfaces can be understood by taking into account complex chemical composition of silicon nitride surfaces.

Silicon nitride tip surfaces in water are, in fact, composed of relatively thin silicon oxide layers (SiO_2 bonds, pK = 2 - 3), silanol groups (SiOH, pK ~ 6), and silylamine groups (SiNH, $SiNH_2$, pK = 10 - 11) (18, 22). In addition, a partial dissolution of the substrate and tribochemical reactions can take place during shearing contact at high pH with the formation of silicic acid ($Si(OH)_4$, pK = 9.8) (23). Studies of various silicon nitrides showed a wide range of isoelectric points from 3 to 9 depending upon composition. Such zwitterionic behavior is caused by variation of acid-base equilibrium for such multicomponent surfaces.

An illustration of combined electrostatic and van der Waals interactions at different pHs is presented in Figure 4. Tip-surface interactions at various pHs can be rationalized as follows: at very low pH, the silicon nitride surfaces bear strong *positive charges* due to ionization of both silanol and silylamine groups that compensates for strong van der Waals forces within the contact area and results in repulsive behavior at small separation (Figure 3). Gradual decrease of surface group ionization at higher pH results in increase the role of van der Waals attraction due to diminishing electrostatic contribution. Within the pH range of 6 - 8, surfaces are essentially *neutral* which should result in maximum adhesion caused by non-compensated van der Waals forces. Further pH increase causes gradual decrease of adhesive forces because of additional repulsive contribution between *negatively charged* silicon nitride surfaces. At pH higher 8.5, increased repulsion between highly negatively charged surfaces results in decreasing net adhesion.

For silicon nitride surfaces, we observe a very broad maximum in pH ranging from 7 to 8 that indicates a minor shift of the acid-base balance towards larger contribution of the basic groups. This difference apparently is caused by excess amine groups in surface composition of the SPM tip. A minute percentage increase in basic amine groups on a surface can shift pK of silicon nitrides by several pH units. Using the results obtained through theoretical calculations in Ref. 18, an estimate of the silicon nitride surface composition of up to 35 - 40% of silylamine groups (SiNH and $SiNH_2$) and 60 - 65% of silanol groups (SiOH) can be made.

Coverage of silicon nitride surfaces with a thin molecular layer bearing acid terminal groups leads to a shift of W_{ad} (pH) maximum to higher pH (Figure 2). Replacement of Si-OH groups with SO_3H terminal groups (expected pK is in the range 1 to 2) does not significantly change the overall acid-base balance of

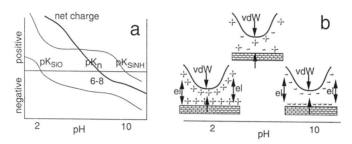

Figure 3. Surface potential variation for silicon nitride tip-surface pair studied in this work (a), and a scheme of tip-surface pairs with different surface charge distribution and force balance at different pH (b).

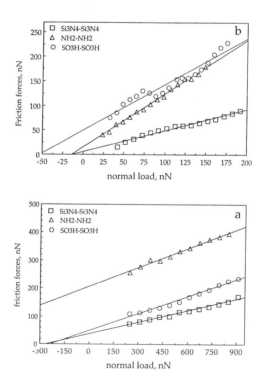

Figure 4. Loading curves (friction forces versus normal spring load) and their linear fits for different mating pairs in air (a) and in aqueous solution at normal conditions (b).

intermolecular interactions for multicomponent surfaces of silicon nitrides. Amine termination significantly modifies the force behavior of silicon nitride surfaces. The maximum adhesive interactions between NH_2 - NH_2 terminated SAMs is observed at pH = 4. This change cannot be explained by variation of ionized state of the modified surfaces alone. In fact, if we accept usual value pK = 7 - 9 for amine groups *(24)*, we can expect the maximum force range to be shifted to higher pH. A reasonable explanation of the observed adhesive maximum at low pH can be a significant shift of pK for amine groups confined by SAM surface. Actual pK values for amine terminated SAMs can be 3 - 5 units pH lower, due to the difficulty of forming charged state at constrained SAM surfaces *(24)*. Substantial decrease of adhesive forces between modified surfaces at high pH can be related to ongoing tribochemical reaction into the contact area that results in the formation of silicic acid $Si(OH)_4$ surface groups.

Friction behavior of modified surfaces. Loading curves F_f (F_n) (friction versus normal load) for silicon nitride surfaces and two modified surfaces at neutral conditions in aqueous solution and in air are shown in Figure 4 along with linear regression analysis. General frictional behavior of the surfaces studied here is very similar to earlier reports for modified surfaces *(5-10)* revealing the generalized Amontos law:

$$F_f = f_o + \mu F_n$$

where μ is a friction coefficient defined as $\mu = \partial F_F / \partial F_n$ and F_o is "residual force" (i. e., normal force not accounted in a spring normal load) *(25-27)*. "Residual" force, F_o , is related to adhesion between mating surfaces and can be thought as amount of "negative" spring forces to be applied to overcome this attraction. The value of F_o usually correlates but is not equal to adhesion forces obtained from approaching-retracing cycles. Values of residual forces, F_o , determined from intersection of corresponding linear fits with the horizontal axes (Figure 4) are much higher in air (200 - 600 nN) than in aqueous solution (10 - 50 nN) which is related to the overwhelming contribution of capillary forces in humid air.

Friction coefficients for identical surfaces in aqueous solutions vary in a wide range from 0.3 for silicon nitrides in the basic range to 2.3 for SO_3H terminated SAMs in the acidic range (Figure 5). Friction coefficient is in the range 1.0 - 1.6 for amine surfaces if measured in the acidic range. These values are within the parameters reported for thiol SAMs with similar terminal groups *(10)*. Absolute values of friction coefficients for SAMs are very high when compared to standard values reported for organic surfaces from macroscopic measurements *(28)*. Friction coefficients for all surfaces studied in humid air (estimated humidity was about 30%) are significantly lower with typical values being 0.14 for silicon nitrides and 0.23 for NH_2 terminated surfaces *(26)*. On the other hand, friction coefficient for any of NH_2 - NH_2 and SO_3H - SO_3H surfaces is substantially higher than for silicon nitride - SAMs pairs and other SAMs studied previously *(27)*. This is due to strong intermolecular interactions (e. g., hydrogen bonding) between these chemical groups and high shear strength of short-chain monolayers.

pH variation of aqueous solution results in significant changes of both F_o and μ for all mating surfaces. Friction coefficients possess maxima at intermediate values of pH for all mating surfaces studied (Figure 6). For bare silicon nitride surfaces, friction coefficient of 0.2 - 0.4 is the lowest at pH = 2 and 10 and a maximum of 0.5 is reached within 6 - 7 range (Figure 5).

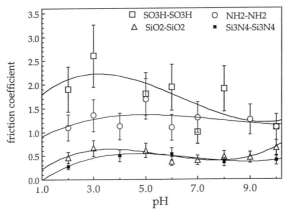

Figure 5. Variation of friction coefficient versus pH for silicon-silicon, Si_3N_4 - Si_3N_4, NH_2 - NH_2, and SO_3H - SO_3H tip-substrate pairs.

Conclusions

Functional alkylsilane SAMs were used to modify surfaces of silicon nitride and silicon SPM tips. Adhesive and friction forces between CH_3, NH_2, and SO_3H terminated surfaces were investigated at different environments by SPM technique. We measured work of adhesion, effective residual forces, and friction coefficients for different chemically modified SPM tips and surfaces in air, neutral water, and aqueous solutions with pH ranging from 2 to 10. Absolute values of work of adhesion between these surfaces, W_{ad}, are in the range 0.5 to 8 mJ/m^2 at normal conditions. Maximum adhesive forces revealed at intermediate pHs are due to a predominant role of van der Waals interaction between essentially neutral surfaces with balanced charges of acidic and basic terminal groups on the zwitterionic silicon nitride surfaces. The silicon nitride surfaces possess complex composition with approximately 2 : 1 ratio of silanol (SiOH) and silylamine (SiNH and SiNH$_2$) groups.

Absolute values of the friction coefficient in aqueous solution vary in a wide range from 0.2 for silicon nitride-silicon nitride mating pair at low pH to 2.3 for SO_3H surfaces at intermediate pH. The friction properties vary in a register with adhesive behavior showing a broad maximum at intermediate pH values for silicon nitride - silicon nitride pair. This maximum is shifted from the intermediate position for NH_2 and SO_3H terminated SAMs

Acknowledgments

This work is supported by The Surface Engineering and Tribology Program, The National Science Foundation, CMS-94-09431 Grant and by US Air Force Office for Scientific Research, Contract F49620-93-C-0063. The authors thank T. Nguyen for technical assistance.

References

1. Frisbie, C. D.; Rozsnyai, L. W.; Noy, A.; Wrington, M. S.; Lieber, C. M. *Science* **1994**, *265*, 2071.
2. Noy, A.; Frisbie, C. D.; Rozsnyai, L. F.; Wrighton, M. S.; Lieber, C. M. *J. Am. Chem. Soc.* **1995**, *117*, 7943.
3. Akari, S., Horn, D., Keller, H., Schrepp, W. *Adv. Materials,* **1995**, *7*, 549.
4. Nakagawa, T.; Ogawa, K; Kurumizawa, T.; Ozaki, S. *Jpn. J. Appl. Phys.* **1993**, *32*, L294; Alley, R. L.; Komvopoulos, K.; Howe, R. T. *J. Appl. Phys.,* **1994**, *76*, 5731.
5. Vezenov, D. V.; Noy, A.; Frisbie, C. D.; Rozsnyai, L. F.; Lieber, C. M. *J. Am. Chem. Soc.* **1997**, *119*, 2006.
6. Green, J.-B.; McDermott, M. T.; Porter, M. C.; Siperko, L. M. *J. Phys. Chem.* **1995**, *99*, 10960.
7. Berger, C. E.; van der Werf, K. O.; Kooyman, R. P.; de Grooth, B. G.; Greeve, J. *Langmuir* **1995**, *11*, 4188.
8. Sinniah, S. K.; Steel, A. B.; Miller, C. J.; Reutt-Robey, J. E. *J. Amer. Chem. Soc.* **1996**, *118*, 8925.
9. Frommer, J. E. *Thin Solid Films* **1996**, *273*, 112.
10. Noy, A.; Vezenov, D. V.; Lieber, C. M. *Annu. Rev. Mater. Sci.* **1997**, *27*, 381.
11. Bain, C. D.; Evall, J.; Whitesides, G. M. *J. Am. Chem. Soc.* **1989**, *111*, 7155; Bain, C. D.; Whitesides, G. M. *J. Am. Chem. Soc.* **1989**, *111*, 7164; Bain, C. D.; Troughton, E. B.; Tao, Yu.; Evall, J.; Whitesides, G. M.; Nuzzo, R. G. *J. Am. Chem. Soc.* **1989**, *111*, 321.
12. Tsukruk, V. V.; Bliznyuk, V. N. *Langmuir,* **1997**, submitted.
13. Ruan, J.; Bhushan, B. *J. Tribology,* **1994**, *116*, 378.

14. Liu, Y.; Wu, T.; Evans, D. E. *Langmuir,* **1994,** *10,* 2241.
15. Hazel, J.; Bliznyuk, V. N.; Tsukruk, V. V. *J. Tribology,* **1997,** accepted.
16. Johnson, K. L.; Kendal, K.; Roberts, A. D. *Proc. R. Soc., London,* **1971,** *A324,* 301.
17. Israelachvili, J. *Intermolecular and Surface Forces,* Academic Press, San Diego, 1992.
18. Senden, T. J.; Drummond, C. J. *Colloids Surf.,* **1995,** *94A,* 29.
19. Marti, A.; Hahner, G.; Spencer, N. D. *Langmuir,* **1995,** *11,* 4632.
20. Murray, B. S.; Godfrey, J. S.; Grieser, F.; Healy, T. W.; Lovelock, B.; Scales, P. J. *Langmuir,* **1991,** *7,* 3057; Williams, J. M.; Han, T., Beebe, T. P. *Langmuir,* **1996,** *12,* 1291.
21. Lin, X-Y.; Creuzet, F.; Arribart, H. *J. Phys. Chem.,* **1993,** *97,* 7272.
22. Bergstrom, L.; Bostedt, E. *Colloids Surf.,* **1990,** *49,* 183.
23. Fisher, T. E. *Ann. Rev. Mater. Sci.,* **1988,** *18,* 303.
24. Lee, T. R.; Carey, R. I.; Biebuyck, H. A.; Whitesides, G. M. *Langmuir,* **1994,** *10,* 741.
25. Berman, A.; Israelachvili, J. in: *Micro/Nanotribology and Its Applications,* Bhushan, B., Ed., NATO ASI Series, Kluwer Acad. Publ., Dordrecht, 1997, p. 317.
26. Tsukruk, V. V.; Bliznyuk, V. N.; Hazel, J.; Visser, D.; Everson, M. P. *Langmuir* **1996,** *12,* 4840.
27. Tsukruk, V. V.; Everson, M. P.; Lander, L. M.; Brittain, W. J. *Langmuir* **1996,** *12,* 3905.
28. Bowden, F. P.; Tabor, D. *The Friction and Lubrication of Solids,* Claredon Press, Oxford, 1950.

Chapter 20

pH Variations of Surface Forces as Probed by Chemically Modified Tips

Vladimir V. Tsukruk, Valery N. Bliznyuk[1], and John Wu

College of Engineering and Applied Sciences, Western Michigan University, Kalamazoo. MI 49008

Chemical information about systems studied by atomic force microscopy can be obtained with probe tips covalently modified with self-assembled monolayers (SAMs) that terminate in well-defined functional groups. This new chemical force microscopy technique (CFM) has been used to probe adhesion and friction forces between distinct chemical groups (COOH, OH, NH_2 and CH_3) in organic and aqueous solvents. These forces varied with the chemical identity of the tip and sample surface, solvent composition and changes in the ionization state of the terminal functionalities and followed trends expected from the strengths of the intermolecular interactions. Contact mechanics provides a framework to model adhesion between molecular groups. Adhesion and friction measurements made as a function of pH (force titrations) on ionizable surfaces were used to assess the pK_a of surface groups. The basic knowledge of adhesive and frictional forces provides a basis for rationally-interpretable mapping of a wide range of chemical functionality.

Understanding of a wide spectrum of phenomena, extending from capillarity and lubrication (1) at macroscopic length scales to molecular recognition and protein folding at the nanoscale (2) requires detailed knowledge of the magnitude and range of intermolecular forces (3). Atomic force microscopy (AFM) (4) has evolved into a powerful tool for probing intermolecular interactions with piconewton sensitivity and a spatial resolution of nanometers (5).

Although force microscopy can provide nanoscale information about friction, adhesion and elasticity, it has not been possible to probe directly the interactions between specific chemical groups that ultimately determine these phenomena at the molecular level. While the absolute force resolution in AFM (6) suggests that it should be possible to measure individual molecular interactions, the chemical groups on a tip interacting with a surface are typically ill-defined. To overcome this inherent

---

[1]Also: Institute of Semiconductor Physics, Ukrainian Academy of Science, Kiev 252650, Ukraine.

limitation of the AFM, we introduced the concept of chemical modification of probes to make them sensitive to specific types of molecular interactions (7). By utilizing chemically functionalized tips, force microscope can be used to 1) probe forces between different groups, 2) measure surface energetics on a nanoscale, 3) determine local pK values of the surface acid/base groups and 4) map the spatial distribution of specific functional groups and their ionization state. This ability to discriminate between chemically distinct groups has led us to name this variation of AFM carried out with specifically functionalized tips "chemical force microscopy" (CFM) (7).

Chemically Sensitive Adhesion Forces

Measuring Interactions Between Functional Groups. Probe modification with functional groups can be accomplished in a rational manner by using well-defined molecular layers of amphiphilic molecules adsorbed on the surface of a tip. Different types of interactions can in principle be studied by varying the head group of the amphiphile. A successful method we have reported for covalently modifying AFM tips (7-9) involves monolayer self-assembly of organic thiols onto the surfaces of Au-coated Si_3N_4 tips. Stable, robust and crystalline monolayers of alkyl thiols or disulfides containing a variety of terminal groups can be readily prepared (10), and thus it is possible to carry out systematic studies of the interactions between basic chemical groups on the probe tip and similarly modified Au substrates. Covalent modification of AFM probes with thiols and reactive silanes has also been reported by other groups (11-14).

The medium, in which surface groups interact can play a crucial role in determining measured forces. To probe bare interactions determined solely by solid surface free energies, adhesion forces must be measured in ultrahigh vacuum. Force measurements carried out in ambient air are more difficult to interpret, because capillary forces (3) are usually 1-2 orders of magnitude higher than specific chemical interactions and thus will obscure relatively small differences in molecular forces. Caution should be used in interpreting measurements performed in dry inert gas atmosphere, since it is difficult to exclude or account for the presence of adsorbed vapor on high energy surfaces. When experiments are conducted in liquid rather than in air, the capillary effect is eliminated (15). Adhesion force measurements with both surfaces immersed in liquid will then reflect the interplay between surface free energies of solvated functional groups.

The magnitude of intermolecular interactions can be assessed directly in an adhesion measurement. Adhesion measurements were made with a Digital Instruments (Santa Barbara, CA) Nanoscope III Multi-Mode scanning force microscope equipped with a fluid cell. The adhesive interaction between different functional groups is determined from force versus sample displacement (F-D) curves. In these measurements, the deflection of the cantilever is recorded during the sample approach-withdrawal cycle (16). The observed cantilever deflection is converted into a force using the cantilever spring constant. The pull-off force determined from the sample retracting trace corresponds to the adhesion between functional groups on the tip and sample surface. To obtain absolute values of forces, the sensitivity of the detector and cantilever spring constants were calibrated (8). The tip radii affect the number of molecular contacts and were assessed by inspection of electron microscope images.

Measurements in Organic Solvents. Our CFM experiments carried out in organic solvents have focused on probing van der Waals and hydrogen bonding interactions, while those performed in electrolyte solution were used to assess hydrophobic and electrostatic forces. Representative F-D curves obtained in ethanol using Au-coated tips and samples that were functionalized with SAMs terminating in either CH_3 or COOH groups readily reveal the difference between the individual interactions (Fig. 1A). To quantify the differences and uncertainties in the adhesive interactions between different functional groups, however, it is necessary to record multiple F-D curves for each type of intermolecular interaction. Histograms of the adhesive force versus the number of times that this force is observed, typically exhibit Gaussian distributions (Fig. 1B) and yield mean adhesion forces (± its experimental uncertainty) of 2.3±0.8, 1.0±0.4, and 0.3±0.2 nN for interactions between COOH/COOH, CH_3/CH_3, and CH_3/COOH groups, respectively. Since the mean values do not overlap, it is possible to differentiate between these chemically distinct functional groups by measuring the adhesion forces with a tip that terminates in a defined functionality. The observed trend in the magnitudes of the adhesive interactions between tip/sample functional groups, i.e. COOH/COOH > CH_3/CH_3 > CH_3/COOH, agrees with the qualitative expectation that interactions between hydrogen bonding groups (COOH) will be greater than between non-hydrogen bonding groups (CH_3). The forces observed in dry, inert gas between these and similar functional group combinations parallel our results in ethanol (*12,13*).

Continuum and Microscopic Models.

JKR Model. The adhesion data can be used to assess the energetics of the different intermolecular interactions and to estimate the absolute number of functional groups contributing to experimentally observed forces. This requires consideration of the tip-surface contact deformations. The theory of adhesion mechanics (*17-19*) can be used to assess quantitatively the above two points. These models predict that the pull-off force, F_{ad}, required to separate a tip of radius R from a planar surface will be given by:

$$F_{ad} = n\pi R W_{SLT} \qquad (3/2 < n < 2) \qquad [1]$$

where

$$W_{SLT} = \gamma_{SL} + \gamma_{TL} - \gamma_{ST} \qquad [2]$$

is the thermodynamic work of adhesion for separating the sample and tip, γ_{SL} and γ_{TL} are the surface free energies of the sample (S) and tip (T), respectively, in contact with the liquid (L) and γ_{ST} is the tip-sample interfacial free energy of the two contacting solids. In the case when the sample and tip bear identical functional groups (e.g. CH_3/CH_3 interactions), $\gamma_{ST} = 0$ and $\gamma_{SL} = \gamma_{TL}$, and Eq. 2 simplifies to $W_{SLT} = 2\gamma$, where γ corresponds to the free energy of the surface in equilibrium with solvent. Therefore, it is the solid-liquid surface free energy that should determine the adhesive force between tip and sample pairs modified with the same molecular groups. The value of numerical coefficient n in Eq. 1 (3/2 vs. 2, i.e. JKR (17) vs. DMT (18) limit) depends on the relative magnitude of surface forces and surface deformations. It is beyond the experimental uncertainty (width of adhesion distribution plus uncertainties in k and R) in CFM measurements and both approaches can and have been used to derive surface free energies from adhesion data. However, application of the JKR model to correlate adhesion force and surface free energy allows for estimates of the number of interacting groups to be made as well and is thus used here.

To justify this approach, we compared the expected value of F_{ad} for CH_3 terminated surfaces and tips with the experiment. The value of F_{ad}=1.2 nN calculated using γ=2.5 mJ/m^2 (20) and the experimentally determined tip radius is in a good agreement with the measured value of 1.0±0.4 nN. Thus, this continuum approach provides a reasonable interpretation of microscopic CFM measurements.

Another verification comes from experiments in several different solvents. Solid-liquid free energies can be readily varied with our studies by using a series of solvents with a wide range of surface tensions, at the same time geometrical parameters of the system are preserved. For example, we have used a CH_3-terminated tip-sample pair in methanol-water mixtures with methanol at the low surface energy end and water at the high energy end. At relatively small adhesion forces in this system (<30 nN), we always observe a direct proportionality between adhesion and the surface free energies estimated from contact angles of the same composition liquids on methyl surfaces (Fig. 2A). In this case, γ_{SL} and, hence, F_{ad} are directly proportional to $\gamma_{LV}\cos(CA)$. Plots extrapolated to zero adhesion force will yield γ_{SV}. We have used this approach with the data for CH_3 terminated SAMs (Fig. 2B), to obtain a value of γ_{SV}, 18.2±1.0 mJ/m^2, in close agreement with the literature (20) value of 19.3 mJ/m^2.

Alternatively, this analysis can yield important information about the surface free energies, in cases where conventional contact angle measurements fail to yield γ. In particular, contact angle measurements cannot be used to probe high free energy surfaces that are wet by most liquids. Adhesion measurements are free of this limitation, and in the case of COOH-terminated SAMs, CFM data yield a value of γ=4.5 mJ/m^2 in ethanol. Another unique application of the CFM is the determination of interfacial free energies γ_{ST} between two dissimilar solids. By first determining γ values for homogeneous interactions, γ_{ST} can be determined using Eq. 2 and relevant experimental F_{ad} values. For example, a small adhesion force in the case of COOH/CH_3 pair in ethanol is consistent with a high value of γ_{ST}=5.8 mJ/m^2, as expected for unfavorable termination of a hydrogen-binding interface (COOH) with a nonpolar (CH_3) one.

The JKR model can also provide an estimate of the number of molecular interactions contributing to the measured adhesive forces. The contact radius at pull-off, a, for surfaces terminating in the same functional groups is $a=(3\pi\gamma R^2/K)^{1/3}$, where $K=(2/3)(E/(1-v^2))$ is the effective elastic modulus of the tip and sample, E is the Young modulus and v is the Poisson ratio. With approximation of K by the bulk value for Au, 64 GPa (21), we obtain a=1.0 nm and a contact area of 3.1 nm^2 for the CH_3/CH_3 interaction in ethanol. Since the area occupied by a single functional group of close-packed thiolate SAMs on Au is 0.2 nm^2 (22), this corresponds to an

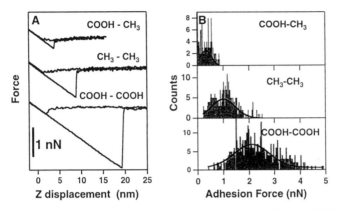

Figure 1. (A) Representative F-D curves in ethanol recorded for COOH/COOH, CH_3/CH_3 and $CH_3/COOH$ tip/sample functionalization (tip R~60nm). (B) A histogram showing the number of times that a given adhesion force was observed in repetitive measurements using COOH terminated samples and tips. The histogram represents ~ 400 tip-sample contacts for one functionalized tip in ethanol.

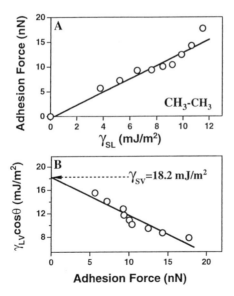

Figure 2. (A) Measured adhesion forces for CH_3-terminated tip and sample versus solid-liquid surface free energies determined for a series of solvents (methanol-water mixtures). (B) Extracting solid-vapor surface free energy from combined adhesion force and contact angle measurements.

interaction between only 15 molecules on the tip and sample. A similar analysis of the COOH/COOH data shows that the adhesion is determined by 25 molecular contacts.

Role of the Solvent. The role that solvation plays in these adhesion measurements can be further analyzed by using the Fowkes-van Oss-Chaudhury-Good (FOCG) surface tension component model *(23-25)*. In this model, the total surface tension of a polar system is separated into dispersion, γ^{LW}, (Lifshitz - van der Waals) and Lewis acid, γ^+, and base, γ^-, components: $\gamma_{total} = \gamma^{LW} + \gamma^{AB}$ where $\gamma^{AB} = 2\sqrt{\gamma^+\gamma^-}$. The work of adhesion between identical surfaces, S, immersed in liquid, L, then becomes:

$$W_{SLS} = 2\gamma_{SL} = 2\left[\left(\sqrt{\gamma_L^{LW}} - \sqrt{\gamma_S^{LW}}\right)^2 + 2\left(\sqrt{\gamma_L^+\gamma_L^-} + \sqrt{\gamma_S^+\gamma_S^-} - \sqrt{\gamma_L^+\gamma_S^-} - \sqrt{\gamma_S^+\gamma_L^-}\right)\right] \quad [3]$$

With apolar surfaces, such as the CH_3-terminated SAMs, the last three terms in the Eq. 3 are zero. The dispersion component of CH_3-terminated SAMs determined from CA measurements with liquid hydrocarbons $\gamma^{LW}=19.3$ mJ/m^2 *(20)* is close to the $\gamma^{LW} = 20.1$ mJ/m^2 for ethanol *(23)*. Therefore, the first term in Eq. 3 is also negligible and the value of adhesion force between two methyl surfaces in ethanol is essentially a measure of the strength of hydrogen bonding interaction between ethanol molecules, $W_{CH_3/EtOH/CH_3} = 2\gamma^{AB}_{EtOH}$. Given the total surface tension of ethanol of 22.75 mJ/m^2, we conclude that the CH_3 SAM/EtOH interfacial tension, $\gamma_{CH_3/EtOH}$, derived from force measurements should be $\gamma_{CH_3/EtOH} \approx \gamma^{AB}_{EtOH} = 2.65$ mJ/m^2. This value compares favorably with the 2.1±0.8 mJ/m^2 determined from adhesion experiments.

Additional corroboration of this interpretation comes from comparisons of the adhesion forces in ethanol and methanol, since these solvents have virtually identical surface tensions (22.75 and 22.61 mJ/m^2, respectively *(26)*) but different hydrogen bonding ($\gamma^{AB}=2.65$ and 4.11 mJ/m^2, respectively *(23)*). By exchanging solvents with the same tip and sample we found that forces between CH_3 groups in MeOH are consistently 1.5-2 times greater than those in EtOH, in agreement with the ratio of 1.6 between corresponding γ^{AB} values. Comparison between forces in ethyleneglycol (EG) and dimethylsulfoxide (DMSO) provide another example ($\gamma^{AB}=19$ and 4 mJ/m^2, $\gamma^{LW}=29$ and 36 mJ/m^2, respectively *(23)*). Adhesion between methyl surfaces in EG is predicted (Eq. 3) to be greater than in DMSO by a factor of 3.0, while we observed a factor of 2.8 difference in the experiment. These data independently confirm the validity of the FOCG approach for treating the interactions in solvents. Analogous analysis suggests that the small magnitude of the adhesion force between CH_3/COOH terminated SAMs in ethanol is due to the similarity in the hydrogen bonding energetics for the COOH/EtOH system.

Adhesion in Aqueous Solutions

Water is an important medium for CFM experiments. In aqueous solutions the surface free energy depends on the ionization state of functional groups and reflects their degree of ionization. The change in free energy can be monitored by measuring CA versus pH; that is, there is a large change in CA of the buffered droplet at a pH $\approx$ pK of the surface group *(27)*. Chemically-modified AFM tips and samples can be used to probe directly the changes in γ_{SL} with pH. The surface charge induced by the dissociation of acidic/basic groups can be detected by monitoring the adhesive force with an AFM probe sensitive to electrostatic interactions, with variations in the sign and magnitude of the force indicating changes in the surface charge.

Adhesion Force and Contact Angle Titrations. Adhesion force values obtained at different solution pH values for tips and samples functionalized with aminopropyl-triethoxysilane (APTES) SAMs terminating with amine groups show a sharp drop to zero (indicating a repulsive interaction) below a pH of 4 (Fig. 3A). Contact angle values measured using buffered solution droplets on this same surface (Fig. 3B) also show a sharp transition (an increase in wetability) at pH<4.5. The decrease and elimination of an attractive force between the tip and sample and an increase in wetability are consistent with protonation of the amine groups under these conditions. Hence, local force microscopy measurements using a modified probe tip and macroscopic wetting studies provide very similar values for the pK of the surface amine group in the APTES derived SAMs. The AFM approach to determining local pKs has been termed *force titrations* (*9*).

The apparent pK obtained from force microscopy, 3.9, and contact angle wetting, 4.3, for the surface amine group is 6-7 pK units lower than bulk solution values (*26*). Large shifts in dissociation constants observed for mixed acid-methyl monolayers have been attributed to unfavorable solvation of the carboxylate anion at the monolayer interface (*28*). Lower pK value relative to solution has also been observed in studies of amino groups grafted onto the surface of a hydrophobic polymer (*29*). The relatively high CAs and large adhesion forces at pH>4 observed in AFM experiments indicate that the APTES derived monolayers are relatively hydrophobic, possibly due to a partially disordered structure that exposes CH_2 groups at the surface. Hence, it is reasonable to attribute the large observed pK shift to a hydrophobic environment surrounding the amine groups. The ability to detect such pK changes locally by AFM should be of significant utility to studies of biological and polymeric systems.

Force Titrations on Hydrophilic Surfaces. The contact angle technique is limited to surfaces sufficiently hydrophobic so that they are not wet completely in both non-ionized and ionized states. In the case of high free energy surfaces, such as COOH SAM, it is necessary to dilute the hydrophilic groups with a hydrophobic surface component to perform a CA experiment. However, the incorporation of a hydrophobic component produces large pK shifts of surface COOH relative to bulk solution (*28*). Hence, it has not been possible to determine the pK of a homogeneous COOH-terminated surface by the CA approach.

Force titrations provide a direct measure of the γ_{SL} and thus bypass the above limitations. A force titration curve obtained for COOH-terminated sample and tip is shown in the Fig. 4A. The sharp transition from positive adhesion forces at low pH to zero (indicating repulsion) at high pH is a prominent feature in this plot. The observed repulsion at pH>6 can be attributed to electrostatic repulsion between negatively charged carboxylate groups, while the adhesive interaction at low pH values originates from hydrogen bonding between uncharged COOH groups. The F-

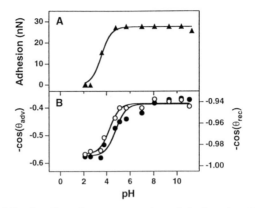

Figure 3. (A) Adhesion force between sample and tip functionalized with amino groups versus pH. (B) Negative cosines of the advancing (O) and receding (●) contact angles of phosphate buffer drops on a sample modified with APTES as a function of pH.

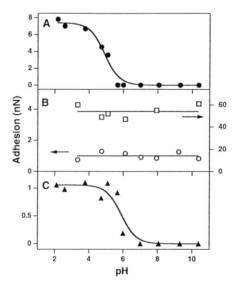

Figure 4. Adhesion force titration curves recorded in buffered solutions. Adhesion force versus pH for (A) COOH/COOH; (B) CH_3/CH_3 (□) and OH/OH (O); and (C) COOH/OH contacts.

D curves become fully reversible and practically identical at all pH values higher than 7. This indicates that the surface charge density is saturated under these conditions. Based on these data, we estimate that the pK_a of the surface-confined carboxylic acid is 5.5 ± 0.5, which is within 0.75 pK units of the solution value (26). The similarity of surface confined and solution pK_a's strongly indicates that solvation effects do not play a significant role in determining the ionization behavior of pure COOH terminated SAMs.

Control experiments with both hydrophilic (OH/OH) and hydrophobic (CH_3/CH_3) groups that do not dissociate in aqueous solutions do not display pH-dependent transitions and show an approximately constant, finite adhesive interaction throughout the whole pH range studied (Fig. 4B). It is worth noting that the measured adhesion forces between CH_3 terminated SAMs are typically 1-2 orders of magnitude larger than the forces observed for hydrophilic groups (COOH and OH). This large difference shows clearly the importance of hydrophobic forces in aqueous media and supports our conclusion that solvation effects are responsible for the anomalously high adhesion force (and large shift in pK) between the APTES monolayers.

Lastly, to probe pKs on unknown surfaces we suggest the use of functionalized tips with SAMs that 1) do not exhibit a pH-dependent change in ionization and 2) are hydrophilic. The hydroxyl-terminated SAM meets these requirements, and has been used to determine the pK_a of carboxyl terminated SAM as shown in Fig. 4C. The dissociation constant is within 0.5 pK units of that determined from the data in Fig. 4A using COOH terminated SAMs on both sample and tip.

Modeling the Charged Interface and Probing Double Layer Forces. The JKR theory of contact mechanics can serve as a reasonable basis for understanding adhesion data in aqueous medium. To interpret pH dependent adhesion data in electrolyte solutions, long-range electrostatic forces between the tip and sample surfaces have to be considered. Since the JKR theory is based on energy balance, one expects no adhesion when the free energy per unit area of a double layer w_{DL} balances the attractive surface free energy component γ_{SL}. The corresponding surface potential $\psi=[\lambda\gamma_{SL}/(\varepsilon_0\varepsilon)]^{1/2}$ is independent of the tip radius for $\lambda<<R$ (λ is the Debye length, ε is the dielectric constant of water) (9). Therefore, the change from adhesive to repulsive behavior is characteristic of the ionization state of the interacting surfaces and can be used to estimate the surface potential.

From this model, the surface free energies and surface potentials of hydrophilic SAMs can be calculated. The values of γ_{SL} determined from adhesion data for OH and COOH (fully protonated) terminated surfaces were 8 ± 3 mJ/m^2 and 16 ± 4 mJ/m^2, in a good agreement with the values determined from interfacial tension measurements using two-phase systems consisting of water and melts of either long-chained alcohols (7-8 mJ/m^2) (30) or carboxylic acids (10-11 mJ/m^2) (31). The calculated surface potential of the carboxylate SAM at pH>6 and ionic strength IS=0.01 M, 140 ± 20 mV, is in reasonable agreement with an independent analysis of long-chain, fatty-acid monolayers at the air-water interface under the same condition (32) and is further substantiated by analysis of F-D curves.

The electrostatic origin of the pH-dependent repulsive forces can be verified by changing the Debye screening length λ; that is the solution ionic strength. Fig. 5A

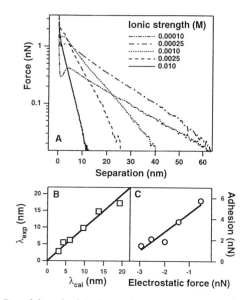

Figure 5. (A) Repulsive double layer force versus separation recorded upon approach between COOH modified tips and samples at different solution ionic strengths at pH=7.2. (B) Debye length obtained from experimental data in (A) and calculated from solution ionic strength. (C) Observed relationship between the adhesion force and repulsive electrostatic force, P_{DL}, for COOH terminated tips and samples.

shows that the repulsive interaction becomes progressively longer-ranged as the solution IS decreases. A detailed analysis of the electrostatic force entails taking into account the surface charge-potential regulation imposed by the potential dependent binding of H^+ and Na^+ ions at the interface (33). For COOH surfaces, the experimental data were well-fit down to separations of 1 nm (IS=0.01 M) using linearized charge-potential regulation condition (34). The condition of constant charge (vs. constant potential) was approached in these experiments independent of IS. The values of the screening length, λ_{exp}, extracted from experimental F-D curves agree well with the values calculated from the solution ionic strengths, λ_{cal} (Fig. 5B). Significantly, the surface potential calculated using this analysis for the carboxy terminated surface at pH=7.2 and IS=0.01 M, 120±5 mV, is close to the value estimated from adhesion measurements. These results suggest that it should be possible to determine double layer parameters of systems being imaged by simultaneously recording F-D curves, using a tip bearing a functional group with a predetermined ionization behavior.

The magnitude of the electrostatic force at contact decreases for IS<0.01M, whereas adhesion can occur between surfaces despite the repulsive interaction; a plot of the measured adhesion force as a function of the electrostatic force is shown in Fig. 5C. This phenomenon is fully reversible: by varying the ionic strength one can go from pure repulsion at high IS to repulsion on approach and adhesion on separation at low IS. A decrease in ionic strength (larger λ) diminishes the double layer repulsion (Fig. 5A), resulting in a larger adhesion force as found in Fig. 5C.

Chemical Effects in Friction

Friction in Organic Solvents. Recently, we (7-9) and others (12) observed that microscopic friction and adhesive forces correlate with each other for organic monolayer systems. The friction force between functional groups on a sample surface and tip is usually determined by recording the lateral deflection of the cantilever as the sample is scanned in a forwards/backwards cycle along the direction perpendicular to the cantilever axis to produce a friction loop. The externally applied load is controlled independently through the cantilever normal deflection. The friction coefficients are obtained from the slopes of the corresponding friction versus load (F-L) curves.

Friction forces between tips and samples modified with different functional groups have been measured in ethanol (8) and water (9) as a function of an applied load. In all of these measurements, the friction forces were found to increase linearly with the applied load. For a fixed external load the absolute friction force decreased as COOH/COOH > CH_3/CH_3 > COOH/CH_3 (8). The trend in the magnitude of the friction forces and friction coefficients (2.5; 0.8 and 0.4 for the above sequence) is the same as that observed for the adhesion forces: COOH/COOH terminated tips and samples yield large friction and adhesion forces, while the COOH/CH_3 combination displayed the lowest friction and adhesion. Thus, there is a direct correlation between the friction and adhesion forces measured between well-defined SAM surfaces. SFA studies of structurally similar layers also show that the friction force correlates with the force of adhesion (35).

Friction in Aqueous Solutions. Frictional forces in aqueous solutions at different pHs were also found to be dependent on the ionization state of functional groups (Fig. 6). F-L curves for COOH-terminated tips and samples are linear, but fall into two distinct categories: larger friction forces and friction coefficients are found in solutions at pHs<6 compared to pHs>6. This cross-over in behavior occurs at the same region of pH where the adhesion forces exhibit a transition from attraction to repulsion. There is also a finite load (~ 4 nN) necessary to achieve non-zero friction at high pH, which can be interpreted as the load required to overcome the double layer repulsion (between charged surfaces) and bring the tip into physical contact with the sample surface.

The frictional behavior of tips and samples functionalized with ionizable and non-ionizable SAMs can be summarized in plots of the friction coefficient versus pH (Fig. 7). The friction coefficients determined for OH- and CH_3-terminated SAMs (not shown) are independent of pH as expected for neutral, nonionizable functional groups. In contrast, the friction coefficients determined for cases, in which one or both SAM surfaces terminate in COOH groups show significant decreases at pH above the pK_a of the surface COOH group. Since friction coefficients exhibit similar pH dependencies to those observed in adhesion measurements, these results suggest that the drop in friction coefficient with changes in pH can be attributed to ionization of surface groups.

We also found that the magnitudes of the friction force and friction coefficient for hydroxyl-terminated surfaces were the lowest of those investigated in aqueous solution. Analysis of F-L curves for methyl terminated tip/sample combinations also yielded low friction coefficients (~ 0.3), however, for comparable tip radii the magnitude of the friction force was an order of magnitude greater (~ 60 nN) than for either carboxyl or hydroxyl functionalized surfaces. The large magnitude of the friction force between methyl surfaces in aqueous media originates from the large contact area between hydrophobic surfaces.

It is important to note that the JKR model used to interpret the adhesion data predicts a nonlinear dependence for friction versus load when a single spherical tip contacts a planar surface. The nonlinear behavior should be detectable in systems that exhibit a large adhesion, such as hydrophobic SAMs in water, at small or tensile loads. Notably, F-L data acquired for CH_3-terminated SAMs in water (Fig. 8) show a nonlinear dependence on applied load, and are well-fit to the contact area versus load dependence predicted by the JKR model (*17*). These results demonstrate that CFM probes friction arising from interactions within a single asperity contact.

Chemically Sensitive Imaging.

Friction Force Imaging. To achieve chemical sensitivity in scanning probe microscopy imaging several key issues must be addressed. The contrast in AFM images is dependent upon many factors such as surface chemistry, morphology, mechanical properties and the nature of the surrounding medium. To have chemical contrast dominate observed images it is necessary to identify and enhance forces that are chemically specific and reduce or eliminate other forces. One can envision fine-

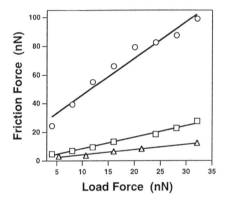

Figure 6. Summary of the friction force versus applied load data recorded for functionalized samples and tips terminating in COOH/COOH (O), CH_3/CH_3 (□) and COOH/CH_3 (△) in ethanol.

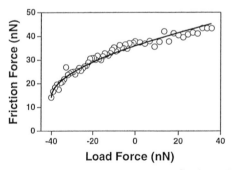

Figure 7. Friction force versus applied load curve (O) for a CH_3 terminated tip and sample in water. The concave shape of the curve is consistent with the non-linear dependence of the contact area on external load predicted by the JKR model (—).

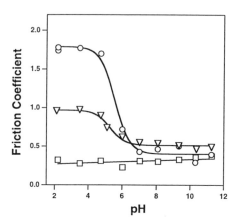

Figure 8. Friction coefficient versus pH for COOH/COOH (O), COOH/OH (□) and OH/OH (Δ) contacts.

tuning the probe-surface interactions as a way of enhancing the imaging sensitivity and specificity, if the origin of the chemical contrast is understood.

The results presented above show that chemical modification of probe tips is often sufficient to observe chemical sensitivity in normal and lateral forces. The observed differences in friction can be exploited to produce lateral force images of heterogeneous surfaces with predictable contrast (7-9,12). Patterned SAMs of alkane thiols on Au surfaces present a convenient model system for such studies, because they are readily prepared, can incorporate a variety of different terminal functionalities and have similar mechanical properties regardless of the terminal functional group. In addition, the flatness of patterned SAM surfaces eliminates unwanted topographic contributions to lateral force images.

We imaged a photochemically patterned SAM (36) having 10 μm x 10 μm square regions that terminate with COOH groups; the regions of the SAM surrounding these squares terminate with CH_3 groups. Topographical images failed to reveal this pattern since such surfaces exhibit almost flat topography across the CH_3 and COOH terminated regions of the sample. Friction maps of these samples taken with different tip functionality readily show chemical information about the surfaces (Fig. 9A,B). Friction maps recorded with COOH tips display high friction on the area of the sample that terminates in COOH groups, and low friction on the CH_3 terminated regions. Images recorded with CH_3 tips exhibit a reversal in the friction contrast: low friction is found in the area of the sample that contains the COOH terminated SAM, and higher friction is observed in the surrounding CH_3 regions.

As expected, this reversal in friction contrast is consistent with the F-L curves obtained on homogenous SAM samples and occurs only with changes in the probe tip functionality. The image resolution is not that of a single functional group but rather an ensemble of groups defined by the tip contact area. Moreover, in the images shown in Fig. 9, the resolution is limited by the photopatterning method itself (~ 200 nm). Employing a microcontact printing technique for pattern generation improves the resolution by a factor of 4-5 (37). In addition, a resolution on the order of 10-20 nm was demonstrated in a monolayer-bilayer COOH/CH_3 system with functionalized tips in dry Ar atmosphere (12).

Changing the solvent characteristics such as pH can present a general approach for further enhancing chemical sensitivity of imaging (9). The validity of this approach was demonstrated by mapping changes in functional group ionization states with varying solution pH values. Images of COOH/OH patterned surfaces obtained with a COOH-terminated tip in different pH solutions show that the friction contrast between COOH/OH regions can be reversibly inverted, with the change in contrast occurring near the pK_a of the surface carboxyl (Fig. 9C,D).

The reversals in friction contrast presented above occur only with changes in the probe tip functionality or a change in the probe tip ionization state in both ethanol and aqueous solvents. These results demonstrate that the imaging is chemically sensitive. For this reason we have termed this approach to imaging (i.e. with specifically functionalized tips) chemical force microscopy (CFM). The approach is reproducible and may serve as a method for mapping more complex and chemically heterogeneous surfaces.

The conditions under which the imaging is performed are important, since dominant interactions depend on the media even for the same tip-surface system. Fig. 9E and 9F shows lateral force images of a COOH/CH$_3$ pattern acquired in water and air using a CH$_3$-functionalized tip. The friction contrast between COOH and CH$_3$ functional groups is relatively large in both images. In water, this pattern shows high friction on the methyl terminated regions and low friction over the carboxyl terminated regions of the sample. This result is readily understood on the basis of our friction studies discussed above. The contrast reflects the dominant effect of hydrophobic forces that mask other chemical interactions. Thus, imaging under water with hydrophobic CH$_3$-terminated tips constitutes an approach to construct hydrophobicity maps of sample surfaces such as biomedical polymers. Images acquired in air exhibit friction contrast opposite to that obtained in aqueous solution. These images do not reflect chemical interactions directly, but rather are due to large capillary forces between the sample and tip over the hydrophilic COOH terminated areas of the surface that are wet more readily than CH$_3$ regions. This high friction on the hydrophilic patches of the surface is always observed in air regardless of whether the tip is bare Si$_3$N$_4$, coated with Au, or derivatized with COOH or CH$_3$ functional groups (8).

Chemical Sensitivity in Tapping Mode Imaging. One can envision that bringing our CFM approach to tapping mode imaging (38) would provide a possibility for nondestructive chemical mapping of delicate samples. Advantages of tapping mode over conventional contact mode such as complete elimination of lateral forces and overall reduction of the tip-sample force combined with the ability to image in physiologically relevant media (39), make it ideally suitable for the imaging soft polymer surfaces and biological samples. Recently, it was also shown that phase-sensitive tapping mode imaging can enhance resolution and is sensitive to surface properties (40). A tapping mode phase image obtained on a patterned SAM sample bearing a uniform box-within-a-box pattern of alternating CH$_3$ and COOH groups with a COOH-terminated tip is shown in Fig. 9G. The regions of different chemical composition are readily identified on the image: phase imaging is clearly sensitive to differences in tip-sample interactions. We propose that the phase contrast observed with soft cantilevers in liquids can be attributed in certain cases to differences in adhesion forces.

When the tip functionality was switched to methyl, the phase image contrast inverted (Fig. 9H). Moreover, when the media was changed from ethanol to water the magnitude of the phase lag increased several fold, while no contrast reversal took place upon varying tip functionality. These latter results are consistent with the adhesion-friction data discussed above. This clearly shows that phase lag imaging is sensitive to the nature of the chemical interactions between the sample surface and the probe: regions that interact more strongly produce greater phase lag (darker areas), while regions with weaker interactions produce smaller phase lag (lighter areas). Therefore, mapping functional group distributions in a predictable way can also be achieved with tapping mode.

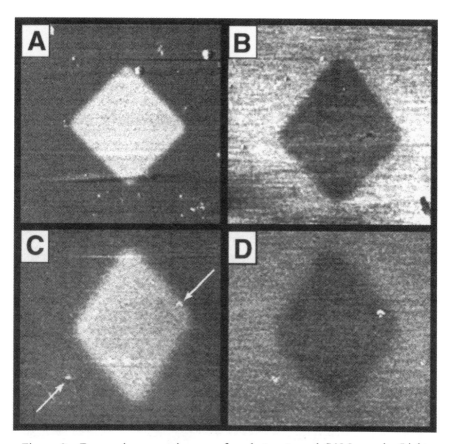

Figure 9. Force microscopy images of a photopatterned SAM sample. Light regions in (A) - (F) indicate high friction; dark regions indicate low friction. Dark regions in (G) and (H) indicate greater phase lag. The friction force images are of a 10μm x 10μm square region terminating in COOH and the surrounding region terminating in CH_3 acquired using a tip modified with a COOH-terminated SAM (A) and using a tip modified with a CH_3-terminated SAM (B). Lateral force images of a 10μm x 10μm COOH region of a patterned SAM sample surrounded with OH groups background recorded with pH values of 2.2 (C) and 10.3 (D). Lateral force image of a sample patterned with regions terminating in CH_3 and COOH functional groups recorded in DI water (E) and air (F) with a CH_3-functionalized tip. The images are 100μm x 100μm. Tapping mode images recorded in ethanol with a tip terminating in a CH_3 SAM (G) and with a tip terminating in COOH groups (H). Image sizes are 20μm x 20μm.

Continued on next page.

338

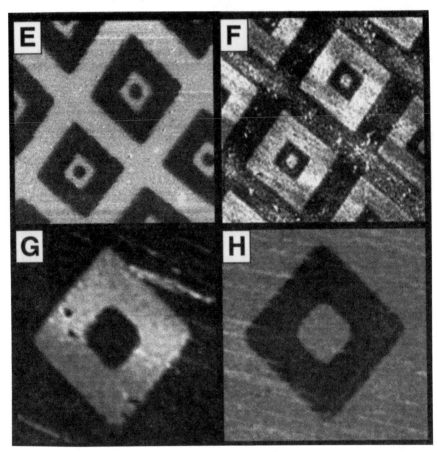

Figure 9. *Continued.*

Conclusions

We have shown that the chemically-modified tips produced by covalently linking molecules to force microscope probe tips may be used to measure and quantify adhesion and friction forces between the functional groups on a tip and sample. Adhesion studies between SAMs that terminate with hydrogen-bonding and hydrophobic functional groups have shown that the interaction between these groups follows chemical intuition and can be rationally interpreted using a surface tension component model. These adhesion forces between tips and samples agree well with forces predicted by the JKR theory of adhesive contact and correspond to an interaction between only 15-25 molecular pairs. Surface free energies for organic surface/liquid as well as solid/solid interfacial free energies that are not readily obtainable by CA measurements were determined from analyses of the available CFM data. A new method has also been developed to determine local pKs of surface ionizable functional groups through force titrations. The interactions observed between modified tip and sample surfaces in aqueous solutions were also shown to agree well with the predictions of double layer and JKR models. These models have been used to extract surface free energies and double layer parameters that are essential to understanding interactions in aqueous media.

In addition, the friction forces between modified tips and samples were found to be chemically specific. The magnitudes of the friction forces follow the same trend as adhesion forces and depend on the ionization states of interacting functionalities (for ionizable groups). The predictable dependence of friction forces on the tip and sample functionality is the basis for chemical force microscopy where lateral force images are interpreted in terms of the strength of adhesive and shear interactions between functional groups. Thus, in conjunction with adhesion data, CFM can unambiguously distinguish different functional group domains in organic and aqueous solvents. When present, hydrophobic effects dominate both adhesion and friction forces and, hence, lateral force images taken with hydrophobic tips in aqueous solutions can map hydrophobic regions on a sample. On hydrophilic surfaces, observed pH-dependent changes in friction forces of ionizable groups can be exploited to map spatially surfaces terminating in hydrophilic functional groups and to define the ionization state as a function of pH.

We believe that many opportunities for basic and applied research are opened up by these studies. A wide range of intermolecular interactions can be studied by the CFM technique and basic thermodynamic information relevant to chemists and biologists can be obtained by the analysis of these data. CFM imaging of systems such as polymers, biomolecules and other materials could lead to new insights into the spatial distribution of functional groups, hydrophobic versus hydrophilic domains and/or improved resolution. Our approach of using force titrations to determine the local pK of acidic and basic groups can be used to probe the local electrostatic properties of protein surfaces in their native environments and the ionization of colloidal particles at the nanoscale. Lastly, new insights into the molecular mechanisms of dissipative processes relevant to tribology can be gained from studies utilizing chemically-modified tips.

340

Literature Cited

1. Bhushan, B. *Handbook of Micro/Nano Tribology*; 1 ed.; Bhushan, B., Ed.; CRC Press: Boca Raton, 1995.
2. Creighton, T. E. *Proteins: Structure and Molecular Properties*; 2 ed.; W. H. Freeman: New York, 1993.
3. Israelachvili, J. *Intermolecular and Surface Forces*; Academic Press: New York, 1992.
4. Binnig, G.; Quate, C. F.; Gerber, C. *Phys. Rev. Lett.* **1986**, *56*, 930.
5. Colton, R. J. *Procedures in Scanning Probe Microscopy*; Colton, R. J., Ed.; John Wiley and Sons: NY, 1996.
6. Smith, D. P. E. *Rev. Sci. Instr.* **1995**, *66*, 3191.
7. Frisbie, C. D.; Rozsnyai, L. F.; Noy, A.; Wrighton, M. S.; Lieber, C. M. *Science* **1994**, *265*, 2071.
8. Noy, A.; Frisbie, C. D.; Rosznyai, L. F.; Wrighton, M. S.; Lieber, C. M. *J. Am. Chem. Soc.* **1995**, *117*, 7943.
9. Vezenov, D. V.; Noy, A.; Rosznyai, L. F.; Lieber, C. M. *J. Am. Chem. Soc.* **1997**, *119*, 2006.
10. Nuzzo, R. G.; Allara, D. L. *J. Am. Chem. Soc.* **1983**, *105*, 4481.
11. Nakagawa, T.; Ogawa, K.; Kurumizawa, T. *J. Vac. Sci. Technol. B* **1994**, *12*, 2215.
12. Green, J.-B. D.; McDermott, M. T.; Porter, M. D.; Siperko, L. M. *J. Phys. Chem.* **1995**, *99*, 10960.
13. Thomas, R. C.; Tangyunyong, P.; Houston, J. E.; Michalske, T. A.; Crooks, R. M. *J. Am. Chem. Soc.* **1995**, *98*, 4493.
14. Sinniah, S. K.; Steel, A. B.; Miller, C. J.; Reutt-Robey, J. E. *J. Am. Chem. Soc.* **1996**, *118*, 8925.
15. Weisenhorn, A. L.; Hansma, P. K.; Albrecht, T. R.; Quate, C. F. *Appl. Phys. Lett.* **1989**, *54*, 2651.
16. Burnham, N. A.; Colton, R. J.; Pollock, H. M. *J. Vac. Sci. Technol. A* **1991**, *9*, 2548.
17. Johnson, K. L.; Kendall, K.; Roberts, A. D. *Proc. Royal Soc. London A* **1971**, *324*, 301.
18. Derjaguin, B. V.; Muller, V. M.; Toporov, Y. P. *J. Colloid Interface Sci.* **1975**, *53*, 314.
19. Muller, V. M.; Yushchenko, V. S.; Derjaguin, B. V. *J. Colloid Interface Sci.* **1980**, *77*, 91.
20. Bain, C. D.; Troughton, E. B.; Tao, Y.-T.; Evall, J.; Whitesides, G. M.; Nuzzo, R. G. *J. Am. Chem. Soc.* **1989**, *111*, 321.
21. Lynch, C. T. *CRC Handbook of Materials Science*; Lynch, C. T., Ed.; CRC Press: Boca Raton, FL, 1975; Vol. II.
22. Ulman, A. *Introduction to Ultrathin Organic Films*; Academic Press: New York, 1991.
23. van Oss, C. J.; Chaudhury, M. K.; Good, R. J. *Chem. Rev.* **1988**, *88*, 927.
24. van Oss, C. J.; Good, R. J.; Chaudhury, M. K. *Langmuir* **1988**, *4*, 884.
25. Fowkes, F. M. *J. Phys. Chem.* **1962**, *66*, 682.
26. Lide, D. R. *CRC Handbook of Chemistry and Physics*; 72 ed. ed.; Lide, D. R., Ed.; CRC Press: Boca Raton, FL, 1991.
27. Holmes-Farley, S. R.; Reamey, R. H.; McCarthy, T. J.; Deutch, J.; Whitesides, G. M. *Langmuir* **1988**, *4*, 921.
28. Creager, S. E.; Clark, J. *Langmuir* **1994**, *10*, 3675.
29. Chatelier, R.; Drummond, C.; Chan, D.; Vasic, Z.; Gengenbach, T.; Griesser, H. *Langmuir* **1995**, *11*, 4122.
30. Glinski, J.; Chavepeyer, G.; Platten, J. K.; De Saedeleer, C. *J. Colloid Interface Sci.* **1993**, *158*, 382.

31. Chavepeyer, G.; De Saedeleer, C.; Platten, J. *J. Colloid Interface Sci.* **1994**, *167*, 464.
32. Yazdanian, M.; Yu, H.; Zografi, G. *Langmuir* **1990**, *6*, 1093.
33. Pashley, R. M. *J. Colloid Interface Sci.* **1981**, *83*, 531.
34. Reiner, E. S.; Radke, C. J. *Adv. Colloid Interface Sci.* **1993**, *58*, 87.
35. Yoshizawa, H.; Israelachvili, J. *Thin Solid Films* **1994**, *246*, 1-2.
36. Wollman, E. W.; Kang, D.; Frisbie, C. D.; Lorkovic, I. M.; Wrighton, M. S. *J. Am. Chem. Soc.* **1994**, *116*, 4395.
37. Wilbur, J.; Biebuyck, H. A.; MacDonald, J. C.; Whitesides, J. M. *Langmuir* **1995**, *11*, 825.
38. Zhong, Q.; Inniss, D.; Kjoller, K.; Elings, V. B. *Surf. Sci.* **1993**, *290*, 1.
39. Hansma, P. K.; Cleveland, J. P.; Radmacher, M.; Walters, D. A.; Hillner, P. E.; Bezanilla, M.; Fritz, M.; Vie, D.; Hansma, H. G.; Prater, C. B.; Massie, J.; Fukunaga, L.; Gurley, J.; Elings, V. *Appl. Phys. Lett.* **1994**, *64*, 1738.
40. Digital Instruments, Inc., Santa Barbara, CA; 1995.

Chapter 21

Recognition and Nanolithography with the Atomic Force Microscope

T. Boland, E. E. Johnston, A. Huber, and Buddy D. Ratner

Departments of Chemical Engineering and Bioengineering, University of Washington, Seattle, WA 98195

The abilities of the atomic force microscope (AFM) to apply forces in precisely defined surface regions and to measure forces in the piconewton range have been applied to spatially distribute and specifically recognize biomolecules on surfaces. First, an RF-plasma deposited 70Å thick triglyme film on a gold substrate was AFM-patterned and the pattern filled with a self-assembled adenine layer. Thymine modified AFM tips were used to probe the pattern. Second, a plasma deposited film was patterned with a short chain, carboxylic acid terminated thiol. The usefulness of this pattern as a template for protein adsorption was investigated.

Living biological systems routinely make use of recognition events that are based upon biomolecules in ordered (oriented) assemblies with spacings in the range from 0.05µm to 10 µm. Technologies that can create materials and surfaces with precise incorporation of spatial information at this scale represent a powerful approach to the engineering control of biological systems. Control of chemistry at surfaces in this mezzo-scale regime is technically and intellectually challenging and has led, in recent years, to a veritable flood of literature exploring molecular self-assembly and surface patterning at the nanometer and micron scales.

Self-assembled monolayers of alkane thiols, where the n-alkane chain is 11 or more carbons in length, have been widely explored as a readily implementable means to precision ordered surfaces [1]. Films prepared by spontaneous assembly of mixed alkyl thiols segregate into islands [2]. However, there was no direct control of the size and shape of the islands other than by varying the relative solution concentration. Some methods have recently been developed to pattern self-assembled films including soft X-ray lithography [3-6], near-UV irradiation through a Cr-glass mask [7], or using elastomeric stamps [8]. Usually, the masking techniques will produce features with dimensions in the range of 1-10 microns that are well defined chemically and spatially.

Higher resolution in patterning can be achieved with scanning probe instruments. Many studies using the scanning tunneling microscope for direct spatial modification of surfaces have been published [e.g., 9-11]. Most studies address semiconductor or conductor surface modification. Atomic force microscopy (AFM) has also been used for spatially controlled surface modification [12], but its application to the nanopatterning of organic films is novel. By slightly increasing the scanning force, AFM has been used to manipulate adsorbed organic layers [13]. Increasing the AFM loading force even further has been shown to damage overlayer films and expose bare substrate [14].

It is possible by AFM to rupture amorphous, weakly crosslinked films such as those resulting from glow discharge plasma deposition of organic precursors. Films of this type have been extensively studied and characterized by ESCA, static SIMS, contact angle, and AFM measurements [13-16]. By using patterns created by film removal as templates for self-assembled monolayers, this couples the chemical precision available through self-assembly with the durable film properties of plasma deposits. The precursor molecule to create the plasma deposited films in the studies reported here was triethylene glycol dimethyl ether (triglyme). Layers formed from this precursor have been shown to exhibit unusually low protein adsorption and cell adhesion [13,16].

Two examples that illustrate how surface patterning of plasma deposited films and self-assembled monolayers may be used are presented here [17]. First, a plasma deposited triglyme film on a gold substrate was AFM-patterned and the pattern filled with self-assembling adenine. Chemically sensitive, thymine modified AFM tips were used to probe the pattern. Second, a plasma deposited film was patterned with a short chain, carboxylic acid terminated thiol. The usefulness of this pattern as a template for protein adsorption was investigated.

Experimental

A Nanoscope II AFM instrument (Digital Instruments, Santa Barbara, CA) equipped with a 10 μm x 10 μm scanner and fluid cell was used for all experiments. The set-up included cantilevers possessing spring constants ranging from 0.6 to 0.06 N/m with integrated tips (r ~ 50 nm) (commercially available - Digital Instruments, Santa Barbara, CA). Spring constants were calibrated using the room temperature oscillation method [17, 18]. The NIH Image software package (version 1.49) was employed in image analysis and particle quantification.

In the first experiment, gold surfaces were prepared by gold coating freshly cleaved mica using an argon plasma sputtering technique (dc, 200 mtorr, 40 W, 90 sec.) followed by annealing at 300°C for 7 minutes on a hot plate. Triglyme was deposited as a 8-10 nm thick film on these gold samples using the radio-frequency plasma-deposition technique previously described [19, 20]. The AFM tip, mounted in the instrument, was then used to rupture the film, and expose micron sized areas of the underlying substrate, with loading forces exceeding 100 nN [14]. The removal of plasma deposited material was visualized by imaging larger areas, at loading below 10 nN.

The AFM-spatially etched gold surfaces were subsequently used as substrates for spontaneous self-assembly of purines, pyrimidines and dithiodipropionic acid (Aldrich Chemical) dissolved in ethanol at a concentration of 1 mM [17, 21]. Functionalized AFM tips were then used to scan the surface. The functionalized tips were prepared by gold coating with the technique described above, followed by immersion in dilute solutions of adenine and thymine (Sigma Chemical) in absolute ethanol for several days to deposit a self-assembled nucleotide monolayer [22].

The second example of nanopatterning involved the adsorption of the protein lysozyme from aqueous solution. A triglyme plasma coated gold surface was prepared as in the previous experiment. Using an AFM probe tip, two square regions, 1 μm on each side, were scraped into the surface. A lysozyme solution, with a final concentration of 100 μg/ml, was prepared by dissolving a known amount of triply crystallized, dialyzed and lyophilized egg white lysozyme (Sigma, Lot 21F-8081) in 140 mM citrate PBS buffer at pH 7.4. This solution was adsorbed to the scraped surface for 30 minutes before rinsing with buffer, followed by repeated scans under buffer. The deposited lysozyme was then removed from the 1 μm square regions by scraping, re-exposing the underlying gold surface. A dithiodipropionic acid solution was introduced into the fluid cell, and allowed to react for 30 minutes, forming a thiol-gold bonded acidic film on the surface. The specimen was then rinsed with buffer. This dicarboxylic acid was chosen since lysozyme has a positive net charge at physiological pH, and thus would be expected to be strongly adsorptive to lysozyme. The surface was imaged once again by AFM. A final adsorption of lysozyme solution was performed followed by rinsing and AFM imaging.

Results and Discussion

An example demonstrating the efficacy of the AFM scraping technique can be seen in Figure 1. The image shows the result of removing a portion of the triglyme film with the AFM tip. Two 600 Å parallel lines were "written" less than 1 μm apart. The width of the lines is directly related to the probe tip diameter. This shows the feasibility of using triglyme plasma deposited films on gold as substrates for patterning. Typically, triglyme films resist the adsorption of biomolecules [13]. Following are the results of two experiments demonstrating how these scratched tryglyme surfaces can be used to form templates for protein or nucleotide adsorption.

In the first experiment, a 1 mM adenine solution in ethanol was adsorbed onto the scraped surface. The surface was then imaged with an unmodified tip and a thymine functionalized tip, as shown in Figure 2. When imaged with the unmodified silicon nitride tip (image A), a negative contrast (darker color) indicates the scraped region. Image B shows that there is an attraction between the thymine tip and the adenine surface because the scraped/adsorbed region appears light. This confirmed the presence of an adenine film in the scraped regions, and its absence on the plasma triglyme portion of the surface. Thus, AFM is useful for both topographic imaging and "recognition" imaging.

The second experiment was used to show that the patterned samples could be used as templates for protein adsorption. The series of AFM images displayed in Figure 3 shows the various steps of the procedure. Image A displays the triglyme plasma surface with square regions of the gold substrate exposed by the scraping technique. The plasma film remaining in the unscraped areas is approximately 8 nm thick. This surface was adsorbed with lysozyme, and Figure 3B suggests that little protein is present on the plasma surface, but the exposed gold regions are covered with a thin protein film of approximately 6 nm (*i.e.*, the scraped areas from Figure 3A are filled in to approximately the thickness of the triglyme film). The adsorbed proteins are then removed by the scraping technique as imaged in Figure 3C. A

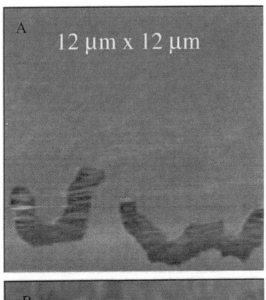

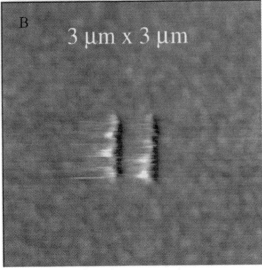

Figure 1. Patterning of a tryglyme plasma deposited thin film with AFM tip forces greater than 100 nN. After scraping, the images were acquired with a tip force of 10 nN. (A) The initials "UW." (B) Two 600Å wide lines engraved less than 1μm apart. The etched features are approximately 70Å deep.

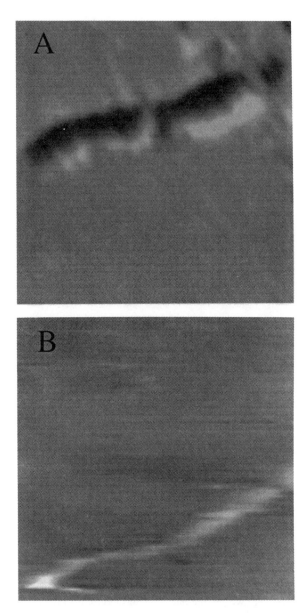

Figure 2. AFM images (2.8 µm x 2.8 µm) of a triglyme film deposited onto gold. The images were obtained using (a) an unmodified tip (silicon nitride) and (b) a thymine-coated tip. In both cases, the triglyme overlayer film was selectively removed with the AFM and the bare gold thus exposed was interacted with adenine.

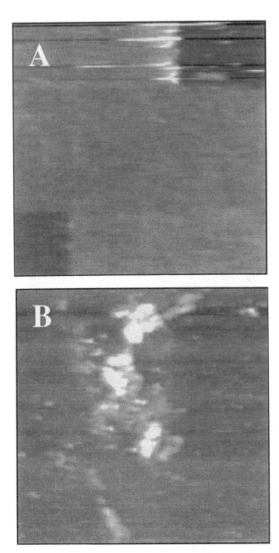

Figure 3. AFM images (12 μm x 12 μm) of a triglyme film deposited onto gold. (A) Two square regions of film have been scraped off. (B) Lysozyme has been adsorbed into the scraped features. (C) Dithiodipropionic is adsorbed to the gold exposed by scaping the film. (D) High levels of adsorbed lysozyme are noted where the diacid has adsorbed.

Continued on next page.

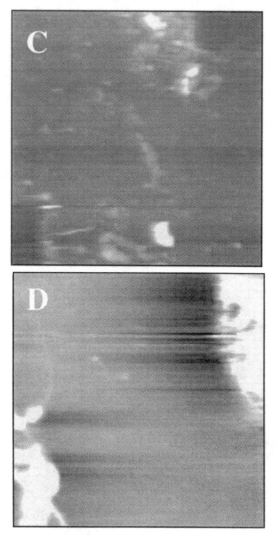

Figure 3. *Continued.*

dithiodipropionic acid film is adsorbed to the surface followed by a repeat of the lysozyme adsorption. Figure 3D displays the preferred adsorption of the positively charged lysozyme to the carboxylic acid-derivatized areas over the plasma film. It has been observed previously that lysozyme adsorbs in amounts greater than monolayers onto acidic surfaces at physiological pH[23- 25].

Conclusions

These experiments offer insights useful in the development of nanopatterned of surfaces. Because the triglyme plasma film is known to resist the adsorption of organic molecules, selective removal of the triglyme enabled controlled patterning on the surface. The addition of a functionalized tip in the first experiment allowed for biorecognition information at the molecular level. In the second experiment, precision spatial control of protein adsorption was shown. Biology normally uses biorecognition and spatially localized receptor events to control processes in living systems. Thus, this nanotechnology may ultimately be useful for precision control of biological processes at interfaces with applications for biomaterials, cell culture, affinity separations and biosensors.

Acknowledgement

Generous support from the NIH (NESAC/BIO grant RR01296), the National Science Foundation (UWEB ERC grant EEC-9529161) and the Center for Process Analytical Chemistry, University of Washington is acknowledged.

References

1.   Bain, C.D.; Whitesides, G.M. *Science*, **1988**, *240*, pp. 62-63.

2.   Lopez, G.P.; Biebuyck, H.A.; Whitesides, G.M. *Langmuir*, **1993**, *9(6)*, pp. 1513-1516.

3.   Dressick, W.J.; Calvert, J.M. *Jpn. J. of Appl. Physics*, **1993**, *32(12B)*, pp. 5829-5839.

4.   Calvert, J.M.; et al. *SPIE*, **1993**, pp. 30-41.

5.   Calvert, J.M.; et al. *Optical Engineering*, **1993**. *32(10)*, pp. 2437-2445.

6.   Dulcey, C.S., et al. SPIE, **1993**. *1925*(657-666).

7.   Frisbie, C.D., et al. *J. Vac. Sci. Technol. A*, **1993**. *11(4)*, pp. 2368-2372.

8.   Kumar, A.; Whitesides, G.M. *Appl. Phys. Lett.*, **1993**. *63(14)*, pp. 2002-2004.

9.   Zeppenfeld, P.; Lutz, C.P.; Eigler, D.M. *Ultramicroscopy*, **1992**. *42-44(pt.A)*, pp. 128-133.

10.   Kobayashi, A.; et al. *Science*, **1993**, *259(5102)*, pp. 128-133.

11. Yang, R.; Evans, D.F.; Hedrickson, W.A. *Langmuir*, **1995**, *11(1)*, pp. 211-213.

12. Kim, Y.; Lieber, C.M. *Science,* **1992**. *256(5068)*, pp. 375-377.

13. Lopez, G.P.; Ratner, B.D.; Tidwell, C.D.; Haycox, C.L.; Rapoza, R.J. Horbett,T.A. *J. Biomed. Mater. Res.,* **1992**, *26*, pp. 415-439.

14. Favia, P.; Perez-Luna, V.H.; Boland, T.; Castner, D.G.; Ratner, B.D. *Plasmas and Polymers,* **1996**, *1(4)*, pp. 299-326.

15. Clark, D.T.; Abu, S.M.M. *J. Polym. Sci. A*, **1984**, *22(1)*, pp. 1-16.

16. Lopez, G.P.; Rapoza, R.J.; Ratner, B.D.; Horbett, T.A., *Preparation of non-fouling biomaterial surfaces by plasma deposition of poly(ethylene glycol) oligomers and precursors.* In *Extended Abstracts of the 2nd Topical Conference on Emerging Technologies in Materials*, AIChE: San Francisco, CA, **1989**.

17. Boland, T., *Analysis and Engineering of Two-Dimensional Assemblies of Purines and Pyrimidines on Gold,* Ph.D. Thesis in Chemical Engineering, University of Washington: Seattle, Washington, **1995**.

18. Hutter, J.L.; Bechhofer, J. *Rev. Sci. Instrum.,* **1993**. *64*, pp. 1868.

19. Lopez, G.P., *Effect of precursor adsorption on film chemistry and plasma deposition of organics.,* Ph.D. Thesis in Chemical Engineering, University of Washington: Seattle, Washington, **1991**.

20. Ratner, B.D.; Chilkoti, A.; Lopez, G.P, *Plasma deposition and treatment for biomaterial applications.,* In *Plasma-Materials Interactions*; Auciello, O.; Flamm, D.L.., Eds.; Academic Press: San Diego, CA, **1990**, pp. 463-516.

21. Boland, T.; Ratner, B.D. *Langmuir*, **1994**, 10, pp. 3845-3852.

22. Boland, T.; Ratner, B.D. *Proc. Matl. Acad. Sci.* (USA), **1995**, *92(12)*, pp. 5297-5301.

23. Horsley, D., et al., *Human and hen lysozyme adsorption: a comparative study using total internal reflection fluorescence spectroscopy and molecular graphics,* In *Proteins at Interfaces: physiochemical and biochemical studies.,* Brash, J.L.; Horbett, T.A., Eds.; American Chemical Society: Washington, D.C. **1987**, pp. 290-305.

24. Schmidt, C.F.; Zimmerman, R.M.; Gaub, H.E. *Biophys. J.,* **1990**, *57(3)*, pp. 577-588.

25. Bohnert, J.L.; Horbett, T.A.; Ratner, B.D.; Royce, F.H. *Invest. Ophthalom. Vis. Sci.,* **1988**, *29(3)*, pp. 362-373.

INDEXES

Author Index

Subject Index

Scanning Probe Microscopy of Polymers

Edited by Buddy D. Ratner and Vladimir V. Tsukruk

ACS Symposium Series 694

Errata

On page 312, Chapter 19, the authors should be listed as Vladimir V. Tsukruk, Valery N. Bliznyuk, and John Wu. On page 312, Chapter 20, the authors should be listed as Dmitri V. Vezenov, Aleksandr Noy, and Charles M. Lieber. The authors for the two chapters were reversed.

Bestsellers from ACS Books

The ACS Style Guide: A Manual for Authors and Editors (2nd Edition)
Edited by Janet S. Dodd
470 pp; clothbound ISBN 0–8412–3461–2; paperback ISBN 0–8412–3462–0

Writing the Laboratory Notebook
By Howard M. Kanare
145 pp; clothbound ISBN 0–8412–0906–5; paperback ISBN 0–8412–0933–2

Career Transitions for Chemists
By Dorothy P. Rodmann, Donald D. Bly, Frederick H. Owens, and Anne-Claire Anderson
240 pp; clothbound ISBN 0–8412–3052–8; paperback ISBN 0–8412–3038–2

Chemical Activities (student and teacher editions)
By Christie L. Borgford and Lee R. Summerlin
330 pp; spiralbound ISBN 0–8412–1417–4; teacher edition, ISBN 0–8412–1416–6

Chemical Demonstrations: A Sourcebook for Teachers, Volumes 1 and 2, Second Edition
Volume 1 by Lee R. Summerlin and James L. Ealy, Jr.
198 pp; spiralbound ISBN 0–8412–1481–6
Volume 2 by Lee R. Summerlin, Christie L. Borgford, and Julie B. Ealy
234 pp; spiralbound ISBN 0–8412–1535–9

The Internet: A Guide for Chemists
Edited by Steven M. Bachrach
360 pp; clothbound ISBN 0–8412–3223–7; paperback ISBN 0–8412–3224–5

Laboratory Waste Management: A Guidebook
ACS Task Force on Laboratory Waste Management
250 pp; clothbound ISBN 0–8412–2735–7; paperback ISBN 0–8412–2849–3

Reagent Chemicals, Eighth Edition
700 pp; clothbound ISBN 0–8412–2502–8

Good Laboratory Practice Standards: Applications for Field and Laboratory Studies
Edited by Willa Y. Garner, Maureen S. Barge, and James P. Ussary
571 pp; clothbound ISBN 0–8412–2192–8

For further information contact:
Order Department
Oxford University Press
2001 Evans Road
Cary, NC 27513
Phone: 1-800-445-9714 or 919-677-0977
Fax: 919-677-1303

Highlights from ACS Books

Desk Reference of Functional Polymers: Syntheses and Applications
Reza Arshady, Editor
832 pages, clothbound, ISBN 0–8412–3469–8

Chemical Engineering for Chemists
Richard G. Griskey
352 pages, clothbound, ISBN 0–8412–2215–0

Controlled Drug Delivery: Challenges and Strategies
Kinam Park, Editor
720 pages, clothbound, ISBN 0–8412–3470–1

Chemistry Today and Tomorrow: The Central, Useful, and Creative Science
Ronald Breslow
144 pages, paperbound, ISBN 0–8412–3460–4

Eilhard Mitscherlich: Prince of Prussian Chemistry
Hans-Werner Schutt
Co-published with the Chemical Heritage Foundation
256 pages, clothbound, ISBN 0–8412–3345–4

Chiral Separations: Applications and Technology
Satinder Ahuja, Editor
368 pages, clothbound, ISBN 0–8412–3407–8

Molecular Diversity and Combinatorial Chemistry: Libraries and Drug Discovery
Irwin M. Chaiken and Kim D. Janda, Editors
336 pages, clothbound, ISBN 0–8412–3450–7

A Lifetime of Synergy with Theory and Experiment
Andrew Streitwieser, Jr.
320 pages, clothbound, ISBN 0–8412–1836–6

Chemical Research Faculties, An International Directory
1,300 pages, clothbound, ISBN 0–8412–3301–2

For further information contact:
Order Department
Oxford University Press
2001 Evans Road
Cary, NC 27513
Phone: 1-800-445-9714 or 919-677-0977
Fax: 919-677-1303